Lecture Notes in Civil Engineering

Volume 116

Lecture Notes in Civil Engineering (LNCE) publishes the latest developments in Civil Engineering - quickly, informally and in top quality. Though original research reported in proceedings and post-proceedings represents the core of LNCE, edited volumes of exceptionally high quality and interest may also be considered for publication. Volumes published in LNCE embrace all aspects and subfields of, as well as new challenges in, Civil Engineering. Topics in the series include:

- Construction and Structural Mechanics
- Building Materials
- Concrete, Steel and Timber Structures
- Geotechnical Engineering
- Earthquake Engineering
- Coastal Engineering
- Ocean and Offshore Engineering; Ships and Floating Structures
- Hydraulics, Hydrology and Water Resources Engineering
- Environmental Engineering and Sustainability
- Structural Health and Monitoring
- Surveying and Geographical Information Systems
- Indoor Environments
- Transportation and Traffic
- Risk Analysis
- Safety and Security

To submit a proposal or request further information, please contact the appropriate Springer Editor:

- Pierpaolo Riva at pierpaolo.riva@springer.com (Europe and Americas);
- Swati Meherishi at swati.meherishi@springer.com (Asia - except China, and Australia, New Zealand);
- Wayne Hu at wayne.hu@springer.com (China).

All books in the series now indexed by Scopus and EI Compendex database!

More information about this series at http://www.springer.com/series/15087

T. G. Sitharam · Sreevalsa Kolathayar ·
Mukat Lal Sharma
Editors

Seismic Hazards and Risk

Select Proceedings of 7th ICRAGEE 2020

Springer

Editors
T. G. Sitharam
Indian Institute of Technology Guwahati
Guwahati, Assam, India

Sreevalsa Kolathayar
Department of Civil Engineering
National Institute of Technology Karnataka
Mangalore, Karnataka, India

Mukat Lal Sharma
Department of Earthquake Engineering
Indian Institute of Technology Roorkee
Roorkee, Uttarakhand, India

ISSN 2366-2557 ISSN 2366-2565 (electronic)
Lecture Notes in Civil Engineering
ISBN 978-981-15-9975-0 ISBN 978-981-15-9976-7 (eBook)
https://doi.org/10.1007/978-981-15-9976-7

This Springer imprint is published by the registered company Springer Nature Singapore Pte Ltd.
The registered company address is: 152 Beach Road, #21-01/04 Gateway East, Singapore 189721, Singapore

Preface

Vulnerability of structures exposed to earthquake hazards in locations of high seismic activity leads to high disaster risk. In this regard, the estimation of seismic hazard is important to design new earthquake-resistant structures as well as to retrofit existing structures. A reliable estimate of seismic hazard also aids in land-use planning, location of critical infrastructures, liquefaction potential evaluation and seismic landslide susceptibility analysis for a region. This book volume contains latest research papers on engineering seismology, wave propagation, strong ground motions, seismic hazard analysis, seismic risk assessment and vibration isolation, selected from the proceedings of 7th International Conference on Recent Advances in Geotechnical Earthquake Engineering and Soil Dynamics, 2021.

We thank all the staff of Springer for their full support and cooperation at all the stages of the publication of this book. We do hope that this book will be beneficial to students, researchers and professionals working in the field of earthquake hazards. The comments and suggestions from the readers and users of this book are most welcome.

Guwahati, India
Mangalore, India
Roorkee, India

T. G. Sitharam
Sreevalsa Kolathayar
Mukat Lal Sharma

Acknowledgements

We (editors) want to thank all the authors who have contributed to the book. We could bring this book out due to all the authors' timely contribution and cooperation. We thank and acknowledge the service of following reviewers for their valuable time and efforts.

Ajay Chourasia, CSIR-CBRI
Amarnath Hegde, Indian Institute of Technology, Patna
Amit Verma, IIT (BHU)
Anil Cherian, Strainstall
Anitha Kumari S. D., Ramaiah University of Applied Sciences
Arvind Kumar Jha, Indian Institute of Technology (IIT), Patna
Asha Nair, CMR Institute of Technology, Bengaluru
Babloo Chaudhary, NITK Surathkal
Bal Rastogi, Indian Society of Earthquake Science
Chittaranjan Birabar Nayak, Vidya Pratishthan's Kamalnayan Bajaj Institute of Engineering and Technology
Dauji Saha, Bhabha Atomic Research Centre; Homi Bhabha National Institute
Deepankar Choudhury, Indian Institute of Technology Bombay
Dhanaji Chavan, IISc
Gopal Santana Phani Madabhushi, University of Cambridge
Jagdish Sahoo, IIT Kanpur
Kalyan Kumar G, NIT Warangal
Karthik Reddy Konala S. K., IIT Hyderabad
Ketan Bajaj, Risk Management Solutions
Manas Kumar Bhoi, PDPU
Md Mizanur Rahman, University of South Australia
Padmanabhan G, Indira Gandhi Centre for Atomic Research
Pradeep Kumar Singh Chauhan, CSIR-Central Building Research Institute, Roorkee
Premalatha Krishnamurthy, Anna University
Prishati Raychowdhury, IIT Kanpur
Purnanand Savoikar, Goa Engineering College

Rajib Saha, NIT Agartala
Rajib Sarkar, IIT (ISM) Dhanbad
Ramkrishnan R., Amrita Vishwa Vidyapeetham
Rangaswamy K., NIT Calicut
Ravi K., IIT Guwahati
Renjitha Varghese, National Institute of Technology, Calicut, Kerala, India
Sanjay Verma, Indian Geotechnical Society, Jabalpur Chapter
Sarat Kumar Das, Indian Institute of Technology (ISM) Dhanbad
Shreyasvi C., National Institute of Technology Karnataka
Snehal Kaushik, Girijananda Chowdhury Institute of Management and Technology, Guwahati
Supriya Mohanty, Indian Institute of Technology (BHU), Varanasi
Surya Muthukumar, Amrita School of Engineering, Amrita Vishwa Vidyapeetham
Vinay Srivastava, Retd. IIT (ISM) Dhanbad
Vipin K. S., Swiss Re
Visuvasam Joseph Antony, Vellore Institute of Technology

Contents

About the Editors

Prof. T. G. Sitharam is currently the Director of Indian Institute of Technology Guwahati, India. He is a KSIIDC Chair Professor in the area of Energy and Mechanical Sciences and Senior Professor at the Department of Civil Engineering, Indian Institute of Science, Bengaluru (IISc). He was the founder Chairman of the Center for Infrastructure, Sustainable Transport and Urban Planning (CiSTUP) at IISc, and is presently the Chairman of the AICTE South Western Zonal Committee, Regional office at Bengaluru and Vice President of the Indian Society for Earthquake Technology (ISET). Prof Sitharam is the founder President of the International Association for Coastal Reservoir Research (IACRR). He has been a Visiting Professor at Yamaguchi University, Japan; University of Waterloo, Canada; University of Dolhousie, Halifax, Canada; and ISM Dhanbad, Jharkhand, and was a Research Scientist at the Center for Earth Sciences and Engineering, University of Texas at Austin, Texas, USA until 1994.

Dr. Sreevalsa Kolathayar pursued M.Tech. from Indian Institute of Technology (IIT) Kanpur, Ph.D. from Indian Institute of Science (IISc) and served as International Research Staff at UPC BarcelonaTech Spain. He is presently Assistant Professor in the Department of Civil Engineering, National Institute of Technology, Karnataka. Dr. Kolathayar has authored three books and over 65 research papers. His broad research areas are geotechnical earthquake engineering, geosynthetics & geonaturals, and water geotechnics. He is currently the Secretary of the Indian chapter of International Association for Coastal Reservoir Research (IACRR), and Executive Committee Member of Indian Society of Earthquake Technology. In 2017, The New Indian Express honored Dr. Kolathayar with 40 under 40—South India's Most Inspiring Young Teachers Award. He is the recipient of ISET DK Paul Research Award from Indian Society of Earthquake Technology, IIT Roorkee. He received "IEI Young Engineers Award" by The Institution of Engineers (India), in recognition of his contributions in the field of Civil Engineering.

Dr. Mukat Lal Sharma is a Professor at the Department of Earthquake Engineering, Indian Institute of Technology (IIT) Roorkee, India with over 35 years of experience in teaching, research and consultancy. Prof. Sharma has rendered expert advice to more than 500 engineering sites regarding seismic hazard and risk assessment. His long association with strong ground motion program of Department of Science and Technology has resulted in the development of Ground Motion Prediction Equations for Himalayas. He initiated the Earthquake Early Warning system approach for disaster mitigation and management in India. Based on his contributions towards disaster mitigation he was conferred with the A. S. Arya-IITR Disaster prevention award in 2012. He was president of Indian Society of Earthquake Technology, India during 2015-2019 and Head, Department of Earthquake Engineering, IIT Roorkee during 2012-2016.

Deterministic Seismic Hazard Assessment of Dhanbad City, India

Rahul Sinha and Rajib Sarkar

Abstract Dhanbad is one of the most important districts in India. It is rich in ores, minerals and specially known for its coal production. The city is experiencing rapid industrialization and growth in population. This paper aims at developing microzonation maps for seismic hazard of the city. The seismic hazard maps will be helpful for mining authorities as well as city planners. In this study, *deterministic seismic hazard analysis* (*DSHA*) is carried out for the city. Epistemic uncertainties in the hazard definition have been handled by considering the logic tree framework comprising of four different prediction equations. It is estimated that the *peak ground acceleration* (*PGA*) at bedrock level in the city can vary in range of 0.13–0.22 g. Furthermore, places in the outskirts like *Baghmara*, *Topchachi*, *Baliapur* and *Nirsa* are found to be most prone to seismic activities. Based on the developed microzonation map, the whole district is segregated into three major seismic zones. Next, site-specific response spectrum for every zone is developed to aid the seismic design of critical and lifeline structures. Finally, these spectra are compared with that of recommended by Indian codes.

Keywords Deterministic seismic hazard analysis (DSHA) · Ground motion prediction equation (GMPE) · Microzonation · Response spectrum · Dhanbad

1 Introduction

Due to huge coal production, Dhanbad city has become hub of the coal business and related industrial establishments. Due to rapid industrialization, the city has also been witnessing an increase in the population. As per census 2011, published by Govt. of India, Dhanbad is the second most populous city in Jharkhand. Hence, proper city planning and urbanization are need of the hour. Besides, the mining activities need to be planned and executed in accordance with the expected seismicity of the city. Indian Standard [14] classifies Dhanbad in seismic *Zone-III* and recommends *peak ground*

R. Sinha · R. Sarkar (✉)
Department of Civil Engineering, IIT(ISM) Dhanbad, Dhanbad, India
e-mail: rajib@iitism.ac.in

T. G. Sitharam et al. (eds.), *Seismic Hazards and Risk*, Lecture Notes in Civil Engineering 116, https://doi.org/10.1007/978-981-15-9976-7_1

acceleration (*PGA*) 0.16 g for the whole city. However, Iyengar [15] suggested that ignoring the spatial variability in seismic hazard can be fatal for important structures or mines around the city. To aid the seismic design, microzonation is highly needed. However, till date, no hazard map is available for the city.

Seismic hazard analysis (*SHA*) is the prime step in microzonation where micro-level variation in seismic hazard is quantified and mapped at bedrock level [1]. This can be done by two frameworks (a) *deterministic seismic hazard analysis* (*DSHA*) (b) *probabilistic seismic hazard analysis* (*PSHA*). Hence, this study aims to develop hazard maps for Dhanbad city showing variations in PGA values at bedrock level. Attempts have been made to develop the worst-case seismic hazard maps at bedrock level following *DSHA*. All types of geological discontinuities present around the city are taken into consideration. Subsequent section describes the methodology adopted in the current study.

2 Methodology

The major steps for the present study are stipulated below.

1. **Collection of the Past Seismic Events**: The past earthquake data are collected from the *United States Geological Survey* (*USGS*). Next, a homogeneous catalogue is prepared by changing the magnitude scale of all events to moment magnitude (M_w). Then, data are checked for dependent events and analysed for completeness.
2. **Preparation of Seismotectonic Map**: Seismotectonic Atlas (SEISAT) [10] published by *Geological Survey of India* (*GSI*) is used for preparation of seismotectonic maps in *ArcGIS* (*Version 10.5*) using the concepts of digitization and georeferencing. Here, *World Geodetic System* (*WGS*)*1984* has been used while georeferencing.
3. **Input for DSHA**: DSHA needs two inputs (a) maximum probable magnitude ($M_{\max}$) of earthquake that the fault or source can produce (see Sect. 5.2) (b) shortest distance of the fault from the site of interest, computed from *ArcGIS*.
4. **Evaluation of Seismic Hazard**: Considering the most suitable *ground motion prediction equations* (*GMPE*) (see Sect. 6), DSHA has been carried out for the whole Dhanbad district after dividing it into numbers of grid points. *MATLAB* (*Version R2019a*) has been used for all the computations required. Finally, using kriging interpolation, hazard maps showing variation in PGA at bedrock level are prepared.

3 Regional Seismicity

Dhanbad is surrounded by many geological discontinuities. Shear zones such as *South Purulia Shear Zone* (*SPSZ*) and *Singhbhum shear zone* (*SSZ*) are present near the southern parts of the city. These may become a major source of seismic threats to opencast coal mining in the outskirts of Dhanbad. *SSZ* is a belt of approximately 200 km in length and is one of the most well-known mineral abundant zones in the country. The deformational history of this shear zone is highly complex as shown by repeated phases of folding [29]. To the north, subsurface faults like *East Patna, Munger–Saharsa* ridge and *West Patna* faults are there. Furthermore, subsurface faults like *Pingla* fault, *Garhmayna–Khandaghosh* fault and *Eocene Hinge* zones are present to the east side of the city.

Recently, Dhanbad has experienced some medium earthquakes of magnitude in the range of 4.0–5.0 (e.g. *Mushabani* 2019, *Bankura* 2019, *Gobindpur* 2019 earthquakes). Furthermore, the seismological station in *Indian Institute of Technology (Indian School of Mines)* (*IITDHN*) recently recorded two earthquakes viz. *Singhbhum* ($M_w = 4.4$) on 20 April 2019 and *Jamtara* ($M_w = 4.0$) on 12 April 2019 at distances of 120 km and 35 km, respectively. During 2007–08, 160 local earthquake events were recorded by the station with moment magnitude confined between $1 \leq M_w \leq 2$. Few events were also recorded having magnitude $M_w \geq 4$ [17].

For carrying SHA, details about various seismic sources are collected from *SEISAT*. The *SEISAT* maps contain details about every geological discontinuity present in the country. High-quality map nos. *11–13*, *22–24* and *30–31* are digitized and georeferenced in ArcGIS. The prepared seismotectonic map containing 394 geological discontinuities overlapped with the past seismic events is shown in Fig. 1. It extends between 81°0′0″ E–90°0′0″ E and 17°0′0″ N–29°0′0″ N.

4 Data Collection and Completeness

In this study, seismic data have been compiled from *USGS* website for the period 1934–Nov 2019. The details of 1112 events have been collected containing information about co-ordinates, month, year, magnitude and hypocentral depth. After that, to overcome the saturation problem of different magnitude scales, homogenous data have been generated by converting all magnitude scales into moment magnitude (M_w). Equations suggested in Das et al. [9], given by Eqs. (1–3), are used.

$$M_w = 0.67(\pm 0.00005)M_s + 2.12(\pm 0.00001) \quad \text{for } 3.0 \leq M_s \leq 6.1 \quad (1)$$

$$M_w = 1.06(\pm 0.0002)M_s - 0.38(\pm 0.006) \quad \text{for } 6.2 \leq M_s \leq 8.4 \quad (2)$$

$$M_b = 0.65(\pm 0.003)M_w + 1.65(\pm 0.02) \quad \text{for } 2.9 \leq M_b \leq 6.5 \quad (3)$$

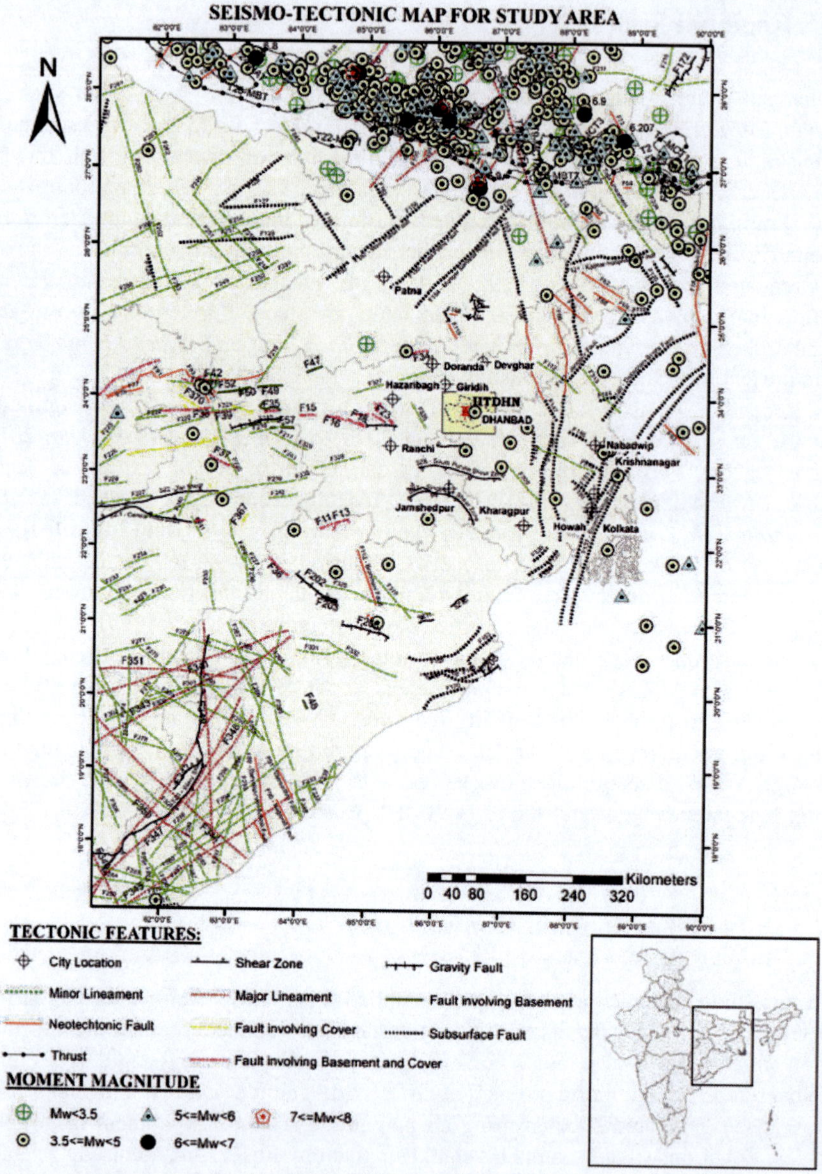

Fig. 1 Seismotectonic map of Dhanbad overlapped with the past seismic events ($M_w > 2.5$)

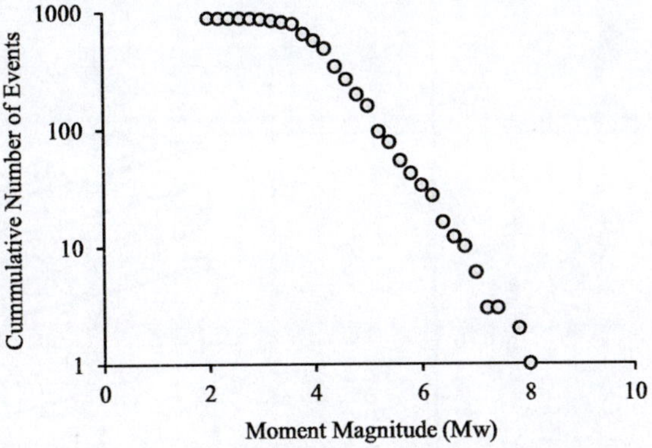

Fig. 2 Cumulative number events against magnitude

where M_w is the moment, M_s is the surface wave and M_b is the body wave magnitudes. Dependent events such as aftershocks and foreshocks are removed by declustering by using open-source software *ZMAP* (Version 7) following Reasenberg [25]. From the analysis, 21% of data (16 clusters) are found dependent and hence removed. Then, the remaining data (878 events) are checked for completeness with respect to both time and magnitude. For analysing the completeness of data with respect to magnitude, *maximum curvature method* (*MAXC*) suggested by Wiemer and Wyss [33] is followed. The detailed procedure of the analysis can also be found in [5]. Figure 2 shows the computed frequency magnitude curve. It can be inferred from the figure that $M_w = 4$ is the required magnitude of completeness (M_c). Hence, data with $M_w \geq 4$ are considered to be complete with regard to magnitude. Next, following Stepp [30], temporal completeness is checked. For this purpose, time range is divided into bins and treated as a single process. Figure 3 shows the standard deviation of the mean annual number of events versus time intervals (T) or bins plotted for every magnitude class. Table 1 summarizes the year of completeness for different magnitude classes. A detailed discussion on the procedure can also be found in [5,8].

5 Focal Depth Analysis and Estimation of M_{max}

5.1 Focal Depth Analysis

Computation of focal depth (h) of earthquake expected in future is supposed to be an important parameter in *GMPE*. Most of the SHA is carried out considering the least possible focal depth [4]. In this study, the prepared catalogue has been analysed for obtaining the most appropriate value of focal depth (similar to Anbazhagan et al. [2]). Focal depths of events with $M_w \geq 4.0$ are analysed with respect to magnitude and epicentral distance from *IITDHN*. Figure 4 shows the variation of hypocentral depths of earthquake data. From Fig. 4, for epicentral distance greater than or equal

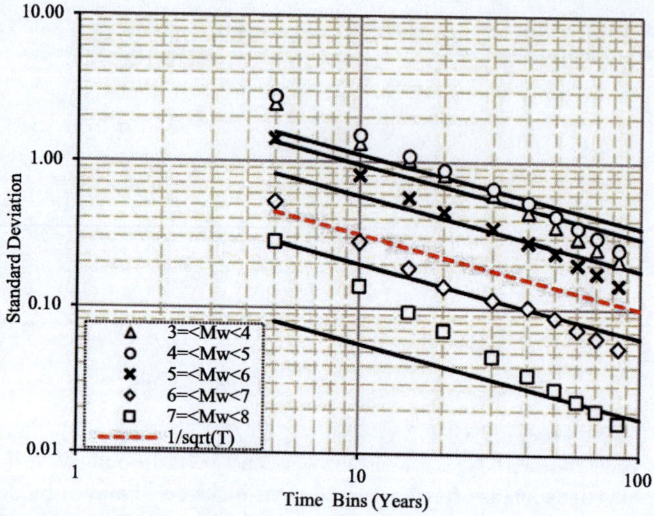

Fig. 3 Standard deviation of magnitude class versus time intervals

Table 1 Completeness of homogeneous catalogue with respect to time

Magnitude class	Approximate year interval of completeness
$3 \le M_w < 4$	20–40
$4 \le M_w < 5$	30–40
$5 \le M_w < 6$	50–60
$6 \le M_w < 7$	50–60
$7 \le M_w < 8$	>70

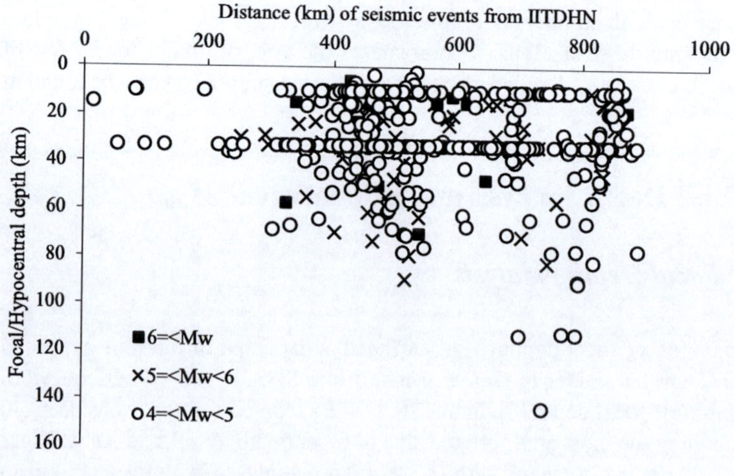

Fig. 4 Focal depth events considered in the study with distance from *IITDHN*

to 250 km with magnitude in the range $4 < M_w \leq 5$, focal depth $h = 5$ km is obtained while $h = 10$ km is found for the events with magnitude $M_w > 5$. Furthermore, for events with epicentral distance less than or equal to 250 km and magnitude in the range $4 < M_w \leq 5$, $h = 10$ km is assigned. A study of seismic events in the stable continental regions across the world [27] shows that the average expected depth of earthquakes varies from 3 to 70 km [22].

5.2 *Estimation of Maximum Probable Earthquake Magnitude (M_{max})*

The maximum probable magnitude ($M_{\max}$) of each linear source has been calculated by the two methods mentioned below.

1. By adding a constant value to maximum observed magnitude $\left(m_{\max}^{\text{obs}}\right)$, as shown below [2,20,24].

$$M_{\max} = \begin{cases} m_{\max}^{\text{obs}} + 0.3, m_{\max}^{\text{obs}} < 5 \\ m_{\max}^{\text{obs}} + 0.5, m_{\max}^{\text{obs}} \geq 5 \end{cases} \tag{4}$$

2. Using empirical relation [32] given by Eq. (5).

$$M_w(\pm 0.28) = 5.08(\pm 0.10) + 1.16(\pm 0.07)\log(L) \tag{5}$$

where $m_{\max}^{\text{obs}}$ is the maximum observed magnitude on each source, L is surface rupture length in km and M_w is the expected moment magnitude. Based on worldwide data, rupture length may be adopted as 1/3 to 1/2 of the total fault length [21,28]. Here, the rupture length is considered to be one-third of the fault length. $M_{\max}$ for a particular source is adopted as the maximum of the two values obtained from both the approaches.

6 Ground Motion Prediction Equation (GMPE)

GMPEs or attenuation equations are the empirical relations that correlate various factors like local soil characteristics, type of sources, magnitude of an earthquake, hypocentral distance, etc., with ground motion parameters at the site of interest [31]. However, no GMPE is available for Dhanbad geological region (Chota Nagpur Plateau). Therefore, the concept of logic tree has been used that allows using alternative prediction equations with suitable weight factors. Anbazhagan et al. [2] carried out SHA for Patna taking an area within 500 km radius around Patna city centre that covered the entire state of Jharkhand including Dhanbad. Furthermore, considering the geographical closeness of Dhanbad and Patna, GMPEs used in the reference

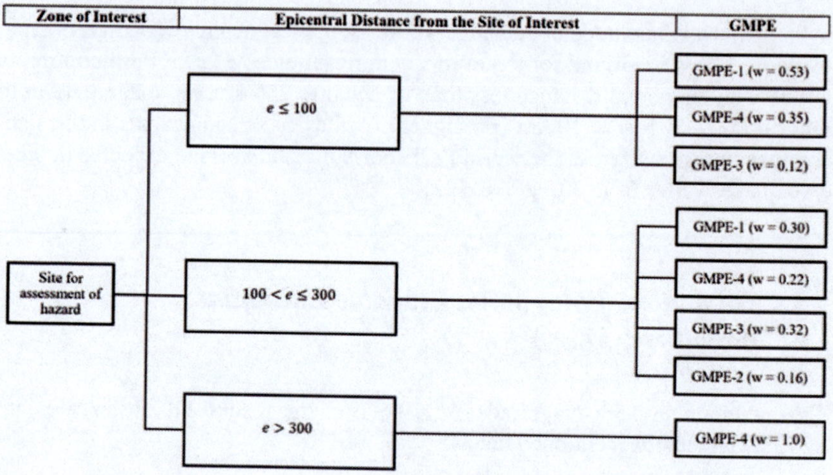

Fig. 5 GMPE sets considered in the study with weight factors (*w*) (*e:* epicentral dist. in km)

have been adopted in the current study. The best-suited GMPEs were selected by Anbazhagan et al. [2] using the criteria proposed by Bommer et al. [6] and efficacy test suggested by Delavaud et al. [11] and Scherbaum et al. [26]. After that, the weight factors for selected GMPEs were computed based on log-likelihood (LLH) concept as proposed by Delavaud et al. [12]. In the present study, GMPEs proposed by Anbazhagan et al. [3] (*GMPE-1*), Boore and Atkinson [7] (*GMPE-2*), Kanno et al. [16] (*GMPE-3*), NDMA [24] (*GMPE-4*) are used as in Anbazhagan et al. [2]. Figure 5 shows the logic tree used with appropriate adopted weight factors from [2].

6.1 Seismic Site Conditions

Seismic hazard analyses for a particular site are generally carried out at bedrock level [15, 18, 23, 24]. Furthermore, Kumar et al. [20] recommended the seismic zonation at bedrock level as the first step to determine the level of ground shaking. Particularly, Nath and Thingbaijam [22] recommended boundary site class BC (standard engineering bedrock conditions with V_{s30} ~ 760 m/s) to be considered for any regional hazard computation. In the present study, GMPEs are adopted following the above recommendation. The attenuation behaviours of GMPEs for $M_w = 7.0$ are shown in Fig. 6.

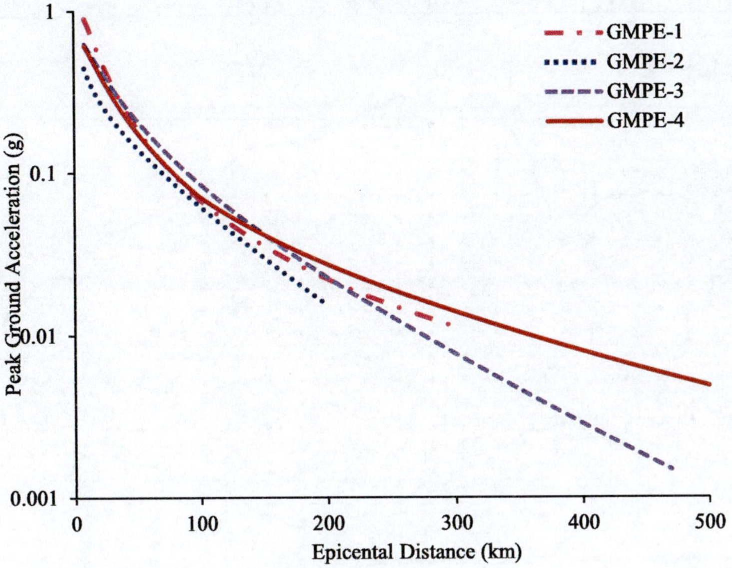

Fig. 6 Attenuation behaviour of GMPE models used for $M_w = 7.0$

7 Hazard Assessment of Dhanbad

For obtaining the political boundary of Dhanbad, Geological Map of Jharkhand published by *Jharkhand Space Application Centre* (*JSAC*) has been digitized and georeferenced. Next, for computation of variation in seismic hazard at bedrock level, the city has been divided using a grid of 0.0313° × 0.0238° (i.e. 3.46 km × 2.63 km) comprising of 296 grid points as shown in Fig. 7. Every point inside the grid is considered as a site of interest. The shortest distance of every geological discontinuity from each grid point is obtained using the '*Near*' tool under '*Proximity Analysis*' in *ArcGIS*. Detailed procedure of carrying out DSHA can be found in Kramer [19]. PGA values are evaluated at all the grid points using the GMPEs considered in the study and discussed in Sects. 5 and 6. *Kriging* interpolation technique has been used (similar to Anbazhagan et al. [2]) for the development of hazard maps.

Figure 8 presents the worst-case possible hazard map of Dhanbad district showing variation in expected PGA at bedrock level. From the map developed, it can be easily inferred that PGA of the district at *bedrock level* is expected to vary in the range 0.13–0.22 g. Kolathayar et al. [18] developed the deterministic macrozonation map at bedrock level for the entire Indian continent and reported PGA in the range 0.15–0.20 g around Dhanbad. NDMA [24] developed probabilistic PGA contours at bedrock level for the entire country. PGA in range 0.12–0.25 g was reported around Dhanbad with 2% probability of exceedance in 50 years (i.e. *maximum considered earthquake* or *MCE*). Moreover, Nath and Thingbaijam [23] after carrying probabilistic SHA found that the PGA at bedrock level around Dhanbad varies in between

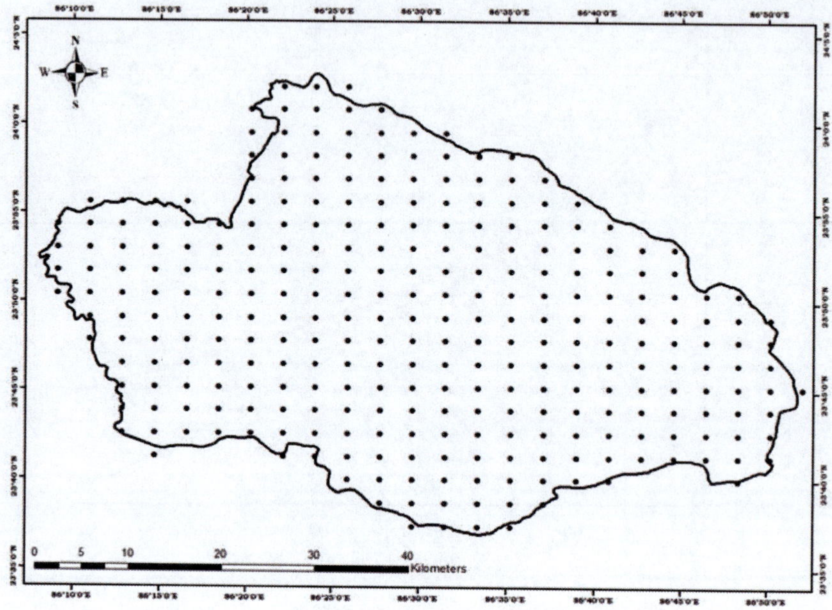

Fig. 7 Grid points considered for carrying out DSHA

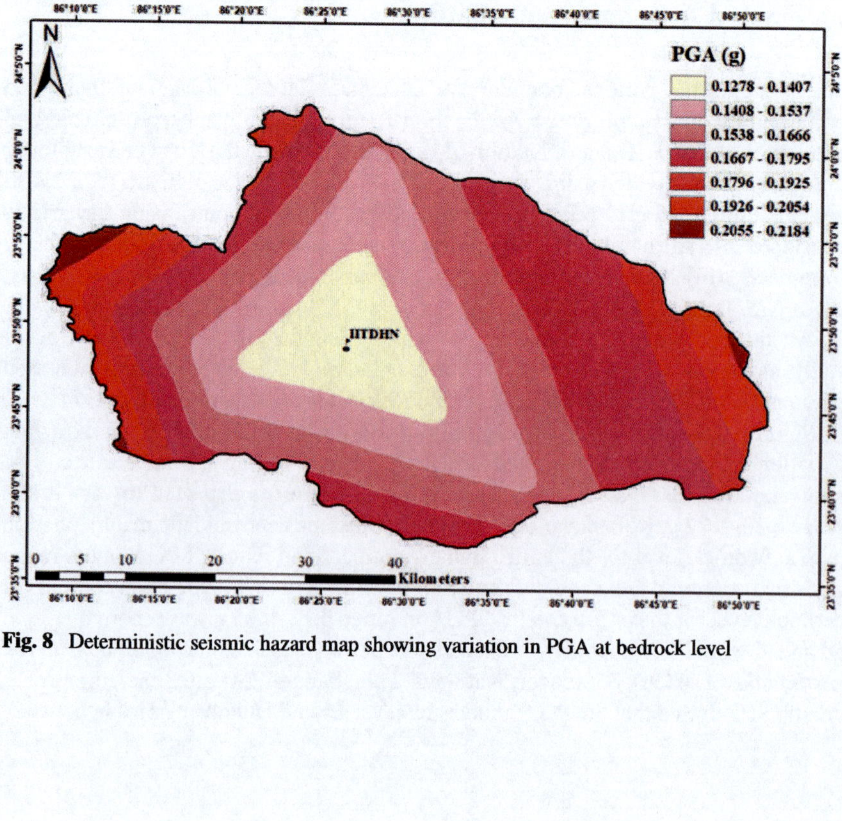

Fig. 8 Deterministic seismic hazard map showing variation in PGA at bedrock level

Table 2 Comparison of the present study with the previous macrozonation

Hazard study	Methodology	Reported PGA around Dhanbad City (g)
Present study	DSHA	0.13–0.22
NDMA [24]	PSHA (MCE)	0.12–0.25
Kolathayar et al. [18]	DSHA	0.15–0.20
Nath and Thingbaijam [23]	PSHA (CLE)	0.12–0.16

0.12 and 0.16 g for CLE conditions. Hence, it can be seen that the PGA range obtained in the present study is very close to that of reported in the previous macro-SHAs. Table 2 summarizes and compares the obtained results with major past hazard studies.

The hazard map developed in the study is also reasonable in respect of the seismotectonic set-up of the city (see Fig. 2). For most of the grid points in the southern region, *SZ-6 South Purulia Shear Zone* is found to be the controlling linear source while for the grid points in east or north-east, *F95—Sainthia Brahmani Fault* (neotectonic fault) and *F146—Pingla Fault* (subsurface fault) are found to be the most hazardous. Similarly, for grid points located in western and north-western parts, minor lineaments *F-327* and *F-326* are found to be contributing most to the overall hazard. It is observed that some areas in Topchanchi, Baghmara and Nirsa are most prone to seismic hazard (0.20 g $\leq PGA \leq$ 0.22 g). Moreover, southern part of Baliapur, Jharia and Sindri is also found relatively more vulnerable (0.17 g $\leq PGA \leq$ 0.19 g) when compared to block like Govindpur.

8 Site-Specific Response Spectrum at Bedrock Level

For the development of site-specific response spectrum, the developed hazard map is segregated into three major seismic zones based on PGA as shown in Fig. 9. Following the recommendations of *Eurocode 8* [13], *Type-1* horizontal design response spectra (as estimated M_{max} of controlling sources are > 5.5) for bedrock motion with 5% damping are developed for all three major seismic zones of the city. Next, the obtained response spectra are compared with spectra recommended by IS code for rocky or hard soil as shown in Fig. 10. From the figure, it can be seen that for Zone A, recommendation by Indian code can be safely used. However, for Zone B and Zone C, the Indian code may underestimate the seismic hazard. It is observed that the developed design acceleration spectra for the city are in between the response spectra suggested by *IS 1893* [14] for *Zones-III* and *IV*, respectively.

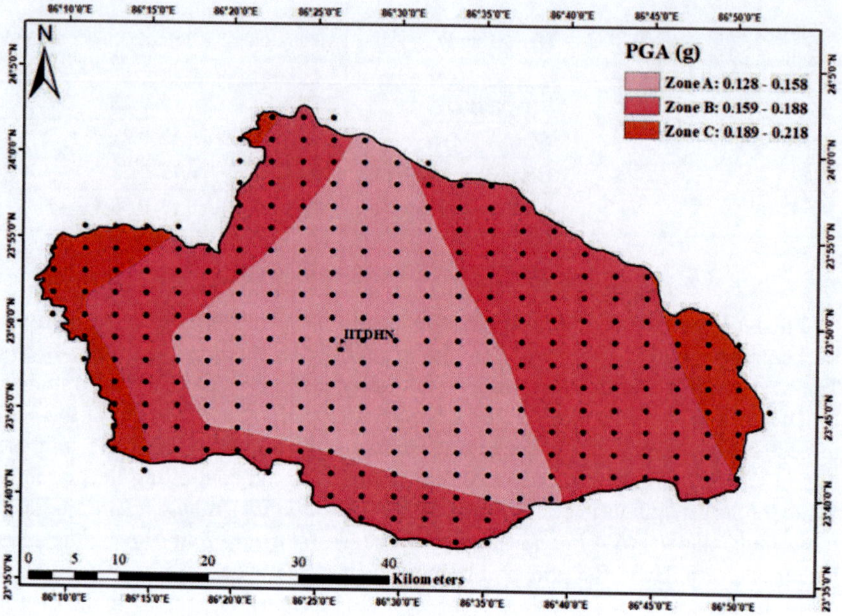

Fig. 9 Hazard map of Dhanbad city for development of site-specific response spectra

Fig. 10 Developed site-specific response spectra for major seismic zones of Dhanbad city

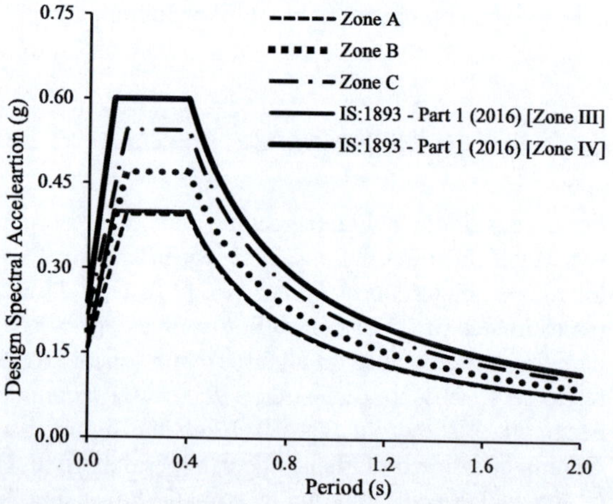

9 Summary and Discussion

In the present study, microzonation map for Dhanbad city is developed based on detailed deterministic SHA at bedrock level. The study may be summarized as below.

- It is observed that expected PGA in the city can vary from 0.13 to 0.22 g.

- It is observed that places like Baghmara, Topchanchi and Nirsa are more vulnerable to seismic hazard relative to other blocks in the city.
- Site-specific response spectra are developed for three major seismic zones of the city showing the exceedance of spectra values than that suggested by Indian standard.

Acknowledgements The first author is highly grateful MHRD, Govt. of India for providing financial support. The second author thankfully acknowledges, Mr. S. Dasgupta, GSI for providing high-quality SEISAT maps.

References

1. Anbazhagan P, Bajaj K, Dutta N, Moustafa SSR, Al-arifi NSN (2017) Region-specific deterministic and probabilistic seismic hazard analysis of Kanpur city. J Earth Syst Sci 126:12. https://doi.org/10.1007/s12040-016-0779-6
2. Anbazhagan P, Bajaj K, Patel S (2015) Seismic hazard maps and spectrum for Patna considering region-specific seismotectonic parameters. Nat Hazards. https://doi.org/10.1007/s11069-015-1764-0
3. Anbazhagan P, Kumar A, Sitharam TG (2013) Ground motion prediction equation considering combined data set of recorded and simulated ground motions. Soil Dyn Earthq Eng 53:92–108
4. Anbazhagan P, Smitha CV, Kumar A, Chandran D (2013) Seismic hazard assessment of NPP site at Kalpakkam, Tamil Nadu. India. Nucl Eng Des 259:41–64
5. Baro O, Kumar A (2017) Seismic source characterisation for the Shillong Plateau in Northeastern India. J Seismol. https://doi.org/10.1007/s10950-017-9664-2
6. Bommer JJ, Douglas J, Scherbaum F, Cotton F, Bungum H, Fah D (2010) On the selection of ground-motion prediction equations for seismic hazard analysis. Seismol Res Lett 81(5):783–793
7. Boore DM, Atkinson GM (2008) Ground-motion prediction equations for the average horizontal component of PGA, PGV and 5% damped PSA at spectral periods between 0.01 and 10.0s. Earthq Spectra 24(1):99–138
8. Borah N, Kumar A (2018) Studying and comparing the declustered EQ catalogue obtained from different methods for Guwahati region NE India. In: Proceedings of Indian geotechnical conference 2018, IISc Bangalore
9. Das R, Wason HR, Sharma ML (2011) Global regression relations for conversion of surface wave and body wave magnitudes to moment magnitude. Nat Hazards 59:801–810
10. Dasgupta S, Pande P, Ganguly D, Iqbal Z, Sanyal K, Venkatraman NV, Dasgupta S, Sural B, Harendranath L, Mazumdar K, Sanyal S, Roy A, Das LK, Misra PS, Gupta H (2000) Seismotectonic atlas of India and its Environs. Geological Survey of India, Special Publication, No. 59, p 87
11. Delavaud E, Scherbaum F, Kuehn N, Riggelsen C (2009) Information-theoretic selection of ground-motion prediction equations for seismic hazard analysis: an applicability study using Californian data. Bull Seismol Soc Am 99:3248–3263
12. Delavaud E, Scherbaum F, Kuehn N, Allen T (2012) Testing the global applicability of ground-motion prediction equations for active shallow crustal regions. Bull Seism Soc Am 102(2):702–721
13. Eurocode 8 (2005) Eurocode 8: BS-EN 1998-1: design of structures for earthquake resistance—part 1: General rules, seismic actions and rules for buildings. European Committee for Standardization, Brussels

14. IS 1893 (2016) Indian standard criteria for earthquake resistant design of structures. Part 1: General provisions and buildings. Bureau of Indian Standards, New Delhi
15. Iyengar RN, Ghosh S (2004) Microzonation of earthquake hazard in greater Delhi region. Curr Sci 87(9):1193–1202
16. Kanno T, Narita A, Morikawa N, Fujiwara H, Fukushima Y (2006) A new attenuation relation for strong ground motion in Japan based on recorded data. Bull Seismol Soc Am 96:879–897
17. Khan PK, Biswas B, Samdarshi P, Prasad R (2011) Seismicity and the coda-Q variation in eastern Indian shield region. Indian J Geosci 65:43–50
18. Kolathayer S, Sitharam TG, Vipin KS (2012) Deterministic seismic hazard macrozonation of India. J Earth Syst Sci 121:1351–1364
19. Kramer SL (1996) Geotechnical earthquake engineering, 1st edn. Prentice Hall, Upper Saddle River
20. Kumar A, Anbazhagan P, Sitharam TG (2013) Seismic hazard analysis of Lucknow considering local and active seismic gaps. Nat Hazards 69:327–350
21. Mark RK (1977) Application of linear statistical model of earthquake magnitude versus fault length in estimating maximum expectable earthquakes. Geology 5:464–466
22. Naik N, Choudhury D (2014) Deterministic seismic hazard analysis considering different seismicity levels for the state of Goa. India. Nat Hazards 75(1):557–580
23. Nath SK, Thingbaijam, KKS.: Peak ground motion predictions in India: an appraisal for rock sites. J Seismolog 15,295–315 (2011)
24. NDMA (2010) Development of probabilistic seismic hazard map of India. Technical report by National Disaster Management Authority, Government of India, New Delhi
25. Reasenberg P (1985) Second order moment of central California seismicity, 1969–1982. J Geophys Res Solid Earth 90(B7):5479–5495
26. Scherbaum F, Delavaud E, Riggelsen C (2009) Model selection in seismic hazard analysis: an information theoretic perspective. Bull Seismol Soc Am 99:3234–3247
27. Schulte SM, Mooney WD (2004) An updated earthquake catalog for stable continental regions, Intraplate earthquakes. United States Geological Survey, pp 495–2002 https://earthquake.usgs.gov/research/data
28. Shukla J, Choudhury D (2012) Estimation of seismic ground motions using deterministic approach for major cities of Gujarat. Nat Hazards Earth Syst Sci 12:2019–2037
29. Singh B, Dowerah J (2010) Geospatial mapping of Singhbhum Shear Zone (SSZ) with respect to mineral prospecting. J Geogr Inf Syst 20:177–184
30. Stepp JC (1972) Analysis of completeness of the earthquake sample in the Puget sound area and its effect on statistical estimates of earthquake hazard. In: Proceeding of the international conference on Microzonation, Seattle, USA, vol 2, pp 897–910
31. Villaverde R (2009) Fundamental concepts of earthquake engineering, 1st edn. CRC Press
32. Wells DL, Coppersmith KJ (1994) New empirical relationships among magnitude, rupture length, rupture width, rupture area and surface displacement. Bull Seismol Soc Am 84(4):974–1002
33. Wiemer S, Wyss M (2000) Minimum magnitude of completeness in earthquake catalogs: examples from Alaska, the Western United States, and Japan. Bull Seismol Soc Am 90(4):859–869. https://doi.org/10.1785/0119990114

Seismic Hazard Assessment of Nuclear Power Plant Site in Jaitapur: Deterministic and Probabilistic Approaches

Sreevalsa Kolathayar and T. G. Sitharam

Abstract This paper presents the seismic hazard analysis of Jaitapur region on the west coast of India, where a nuclear power plant is being planned to locate. A homogenized earthquake catalog in unified moment magnitude scale for the region has been prepared compiling updated data from different agencies. The seismicity parameters were estimated for Jaitapur region, and the hazard evaluation has been performed using different methodologies with different source models and attenuation relations. The deterministic and probabilistic seismic hazard analyses were performed with currently available data and their best possible scientific interpretation by considering alternative models (source models, maximum magnitude, and attenuation relationships). The hazard maps have been produced for horizontal ground motion on the bedrock level using both deterministic and probabilistic methods. The value of Peak Horizontal Acceleration (PHA) for the study region estimated probabilistically for 475 year return period varies from 0.05 to 0.2 g, whereas with deterministic approach, PHA reaches value as high as 0.3 g.

Keywords First keyword · Second keyword · Third keyword

1 Introduction

Nuclear power plants are considered to be the most sensitive structures, and hence, they have to be designed to survive major natural disasters. The safety of nuclear power plants is always put to question against natural hazards, especially against earthquakes. With the greater demand for nuclear power in India, more and more nuclear power plants are being set up at various locations in the country. In a densely populated country like India, any damage to the nuclear power plant during an earthquake will result in a cataclysm. Hence, it is imperative to carry out a detailed

S. Kolathayar (✉)
National Institute of Technology Karnataka, Surathkal, India
e-mail: sreevalsakolathayar@gmail.com

T. G. Sitharam
Indian Institute of Technology Guwahati, Guwahati, India

T. G. Sitharam et al. (eds.), *Seismic Hazards and Risk*, Lecture Notes in Civil Engineering 116, https://doi.org/10.1007/978-981-15-9976-7_2

assessment of the probable hazard in the location identified for a nuclear power plant.

Jaitapur is a small port situated in Rajapur Tehsil of Ratnagiri district, Maharashtra State, India on the Arabian sea coast located at 16.59°N 73.35°E. Jaitapur came into limelight due to the proposed Jaitapur nuclear power project by nuclear power corporation of India. A lot of discussions have been happening in the seismic safety of Jaitapur region because of the nuclear power plant to be set up in the region. We would like to throw more light on the seismicity of Jaitapur and adjoining areas based on scientific hazard analysis. Jaitapur is placed in Zone 3 as per Indian seismic code BIS 1893-2016 (though various reports include Jaitapur in Zone 2). The Indian seismic code BIS-1893 (2016) divides the country into four zones, Viz. Zones II, III, IV, and V. The Jaitapur region is placed in Zone III with a zone factor of 0.16 g. The BIS 1893 was based on past earthquake data and geology and is being updated from time to time after major earthquakes). This map provides the main input regarding the peak ground acceleration values at each region, which is based on the past seismic activity and not based on a systematic evaluation of seismic hazard. These things point toward the need for evaluation of seismic hazard of the Jaitapur region with a proper scientific approach to demarcate the seismic zones. In the present paper, we try to present a better picture of seismic hazard in Jaitapur using both deterministic and probabilistic methods.

2 Seismicity of Jaitapur

Earthquake events within a radius of 300 km were compiled for the period till 2018 from various sources like USGS/NEIC catalog, International Seismological Center (ISC), Gauribidanur Array (GBA), Indian Meteorological Department (IMD), and International Data Center (IDC). The events which were in different magnitude scales were converted into unified magnitude scale MW using the region-specific magnitude conversion relations. Figure 1 shows the distribution of declustered earthquake events within a radius of 300 km from Jaitapur.

3 Delineation of Seismic Source Zones and Source Models

Identification and delineation of regional seismic source zones are important to do a seismic hazard analysis. The zones of different seismicity characteristics should be defined based on space limits, and separate catalogs should be used for each of these zones. Three types of source models are considered in the present study; Linear seismic sources, gridded seismicity model, and areal sources.

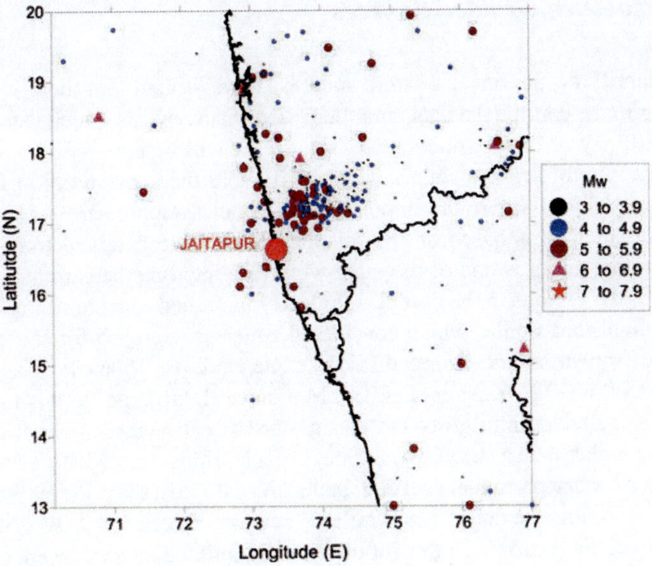

Fig. 1 Distribution of seismic events in and around Jaitapur

3.1 Linear Sources

Geological Survey of India (GSI) has published the Seismotectonic Atlas [1], which is one of the best documents listing the linear seismic sources in India and adjoining areas. This map was prepared after extensive research using remote sensing technique and then cross verification by geological explorations. The details of the faults, lineaments, shear zones, and the geological features in India and adjoining areas are presented in SEISAT [1]. The required sheets of SEISAT [1] were scanned and georeferenced using MapInfo. Then, these individual images were merged to form the complete source map of Jaitapur. After this exercise, the declustered earthquake data with MW 4.0 and above were superimposed in the map of extracted tectonic features. To characterize the seismic sources, the maximum reported magnitude and number of earthquake events associated with each of these sources were noted. In case, if any event falls within 15 km radius of two or more sources, then it is assumed that the event is associated with the source nearest to the event. There are some earthquake events which are not falling along any of the identified faults. In those cases, the linear source model alone may not be able to give the correct picture of seismic hazard levels. To overcome this limitation, two more types of source models were selected in this study—smoothed gridded sources and areal sources. These are sources which are spread over a large area and widely accepted methods for evaluation of seismic hazard for regions, where the sources are not delineated.

3.2 Smoothed Gridded Sources

While identifying the linear seismic sources, it was found that there were lots of scattering in the earthquake data, and there were many events which were not associated with any of the seismic sources. Hence, it will be a better idea to consider a different type of seismic source which will avoid the subjectivity in identifying the sources. This is where the importance of zoneless approach [2–4] comes into focus. This method is based on the seismic activity rate obtained from the earthquake catalog, and it is one of the most widely adopted methods to model seismic sources for the regions in the absence of clearly identified seismic sources. Some of the seismic hazard studies which considered zoneless approach for source identification are Wahlstrom and Grunthal [5], Lapajne et al. [6], Vilanova and Fonseca 7, Jaiswal and Sinha [8], Kalkan et al. [9], Menon et al. [10], etc. In this method, the study area is divided into grids and the number of earthquakes, which are having magnitude higher than a cutoff magnitude, in each grid is counted. This will give the activity rate for that particular grid cell. In the present study, the grid size was adopted as 0.1° × 0.1°, and the cutoff magnitude (Mcut) was selected as MW = 4.0. Based on this value, the recurrence rates for different magnitude intervals were calculated, and these values were smoothed using a Gaussian function to get the final activity rate for each grid cell (Eq. 5). The uncertainty involved in estimating the location of the earthquake event and the size of the seismic source can be accounted for by this smoothing.

$$\underset{i}{\hat{n}} = \frac{\sum_j n_j e^{-\Delta_{ij}^2/c^2}}{\sum_j e^{-\Delta_{ij}^2/c^2}} \tag{1}$$

where n_j—number of earthquakes in the jth grid cell, $\hat{n}_i$—smoothed number of earthquakes in ith cell, c—correlation distance to account for the location uncertainties, and Δ_{ij}—distance between the ith and jth cells.

4 Evaluation of Seismicity Parameters

The seismic activity of a region is given by the Gutenberg and Richter [11] earthquake recurrence law. The recurrence rate given by this to this law is:

$$\text{Log10}\, N = a - bM \tag{2}$$

where N is the total number of earthquakes with magnitude M and above which will occur in a year, and a and b are the seismicity parameters of the region. These values signify the background seismicity and the magnitude size distribution for the region, respectively. The values of seismicity parameters were evaluated using the maximum likelihood estimation technique [12, 13].

We have estimated the seismicity parameters based on the minimum magnitude of completeness [14]. The magnitude of completeness is the lowest magnitude above which the earthquake recording is assumed to be complete [15]. Below this magnitude, a fraction of events is missed by the network because they are either too small to be recorded by enough stations, or because they are below the magnitude of interest, or because they are mixed with the coda of a larger event, and therefore, they passed undetected. There will be a spatial and temporal variation of M_c, and it will decrease with time, mainly because of the increase in the number of seismographs in the region. It is standard practice to consider the minimum magnitude of completeness as the threshold for any analysis that uses a well-constrained earthquake catalog, prohibiting error due to the presence of low-magnitude earthquakes in the region. Various researchers in the past have adopted a similar method to estimate seismicity parameters for different parts of the world. The "b" value can be evaluated using the maximum likelihood estimate as given below.

$$b = \frac{\log_{10}(e)}{\left[m_{\text{mean}} - \left(M_c - \frac{\Delta m_{\text{bin}}}{2}\right)\right]} \tag{3}$$

where m_{mean}—mean magnitude of the sample; M_c—the magnitude of completeness, and Δm_{bin}—magnitude bin size. The bin size (incremental MW values) used in the study is 0.1.

One of the most widely used methods to evaluate M_c is based on a power law fit for the Frequency Magnitude Distribution (FMD), which was suggested by Wiemer and Wyss [14]. In this method, a series of synthetic magnitude distributions are developed for each magnitude interval using maximum likely hood estimate. These "a" and "b" values obtained in the synthetic distribution are compared with the observed distribution, and the goodness of fit is calculated. For calculating the goodness of fit, the absolute difference (R) between the observed and synthetic distribution has to be calculated.

$$R(a, b, M_i) = 100 - \left(\frac{\sum_{m_i}^{m_{\max}} \text{mod}(B_i - S_i)}{\sum_i B_i} * 100 \right) \tag{4}$$

where B_i and S_i are the observed and predicted cumulative number of events in each magnitude bin. The seismicity analysis was done using ZMAP software [16]. The $M_{\max}$ was estimated for all the zones based on the generic equation given by Kijko [17]. For linear sources, two different maximum magnitudes have been considered, the maximum historical magnitude ($M_{\max 1}$) and the previous increase by 0.5 units ($M_{\max 2}$). This is the criterion followed and defined by Gupta [18] for assigning a maximum magnitude to a linear source. Similar methods were adopted by Iyengar and Ghosh [19] and Raghu Kanth and Iyengar [20] for the seismic hazard analysis of

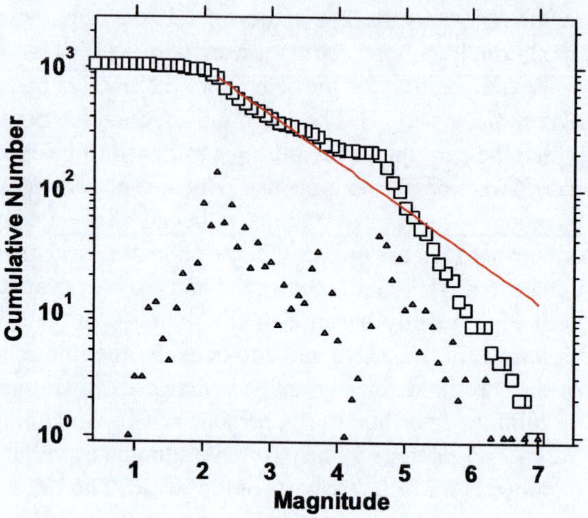

Fig. 2 Frequency magnitude distribution plot for the study region

Delhi and Mumbai, respectively. Bozzoni et al. [21] used the same criteria to assign a maximum magnitude to a regional source zone. The b value for the region is found to be 0.38, and a value is found to be 1.13. Frequency magnitude distribution plot for the study region is presented in Fig. 2.

5 Seismic Hazard Analysis

To define the seismicity of an area, it is more appropriate to predict the earthquake hazard in terms of ground motion parameters. The seismic hazard evaluation of the Jaitapur region based on a state-of-the-art Deterministic Seismic Hazard Analysis (DSHA) as well as Probabilistic Seismic Hazard Analysis (PSHA) has been performed using different source models and attenuation relations. The analysis has been performed with currently available data and their best possible scientific interpretation using an appropriate instrument such as the logic tree to explicitly account for epistemic uncertainty by considering alternative models (source models, the maximum magnitude in hazard computations, and ground-motion attenuation relationships). In the present analysis, three different GMPEs each were used to model the attenuation properties of the region. The relations used are Raghu Kanth and Iyengar [22], Atkinson and Boore [23], and Graizer [24]. The relation by Raghu Kanth and Iyengar [22] was developed for the peninsular Indian shield regions. Attenuation relations by Atkinson and Boore [23] and Graizer [24] were developed for Eastern North America (ENA) and Central and Eastern North America (CENA), respectively. Ground-motion attenuation in ENA and peninsular Indian shield is comparable. Two types of seismic sources: linear and gridded seismicity models were employed in the analysis. The results obtained from different models were combined in a logic tree framework giving equal weightages to all the models.

5.1 *Logic Tree Structure*

There are lots of uncertainties involved in models used for seismic hazard assessment, and this makes the selection of a seismic hazard model difficult. The use of the logic tree approach allows characterization of epistemic uncertainties in various models by including alternative models in the analysis [25–27]. Bommer et al. [26] had studied the need for the use of the logic tree approach in predicting the hazard to ensure that the epistemic uncertainty is fully captured. Logic tree consists of a series of nodes and branches, and these branches denote different models (hypothesis). A subjective weightage, based on engineering judgment, can be given to each of these branches depending on the likelihood of being correct. The weightage for all the branches at a particular node should be equal to unity. The weightage of the terminal branch of the logic tree can be obtained by multiplying the weightage of all the branches leading to it. The present study considers two types of source models and three different attenuation relations. Equal weightages have been assigned to each of the source models as a predefined criterion is not available for assigning the weightages. The weights applied to different ground-motion prediction equations in a logic tree formulation reflect the degree to which each equation is judged to be the best estimate of earthquake ground motions in that particular region [26]. The GMPE by Raghu Kanth and Iyengar [22] was given a higher weightage of 0.4 as this relation was developed for southern peninsula. Other two GMPEs were given equal weightages of 0.3 each. All the values obtained using different models have to be combined using the logic tree framework. The logic tree branch used for DSHA and PSHA with the weightage assigned to each model is shown in Figs. 3 and 4, respectively.

6 Results and Discussion

The seismic hazard analysis of the region was done by dividing the entire study area into grids of size 0.01° × 0.01° (about 1 km × 1 km). For each grid point, the Peak Horizontal Acceleration (PHA) values were evaluated for bedrock level using both deterministic as well probabilistic methods (corresponding to the Probability of Exceedance (PE) of 10% in 50 years which correspond to return periods of 475 years. This was done by deaggregating the hypocentral distance into small intervals of 1 km and the magnitude range (between minimum and maximum magnitude) into small incremental values of 0.1 (a lower value of the hypocentral incremental distance (less than 1 km) and magnitude interval (0.1) was not improving the results significantly, and it was increasing the computational time). For each grid point, all the sources

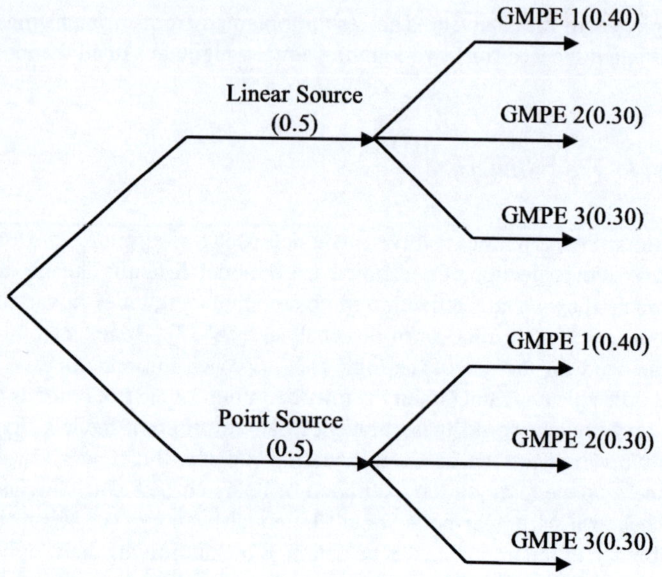

Fig. 3 Parameters and weighting factors adopted in the logic tree for DSHA

within a radius of 300 km were considered for evaluation of PHA values. Evaluation of hazard using linear sources and zoneless approach [2] was done using a program coded in MATLAB. The spatial variation of PHA values obtained using DSHA and PSHA is shown in Figs. 5 and 6, respectively. The PGA value at each grid point was estimated using 6 different methods for DSHA and 12 different methods for PSHA that were later combined to get the final value using the logic tree framework. The PGA contours obtained through independent methods show little deviation as the source models and attenuation relations used are different but are comparable.

The value of PHA for the study region for 475 year return period varies from 0.05 to 0.2 g, whereas for DSHA, PHA reaches value as high as 0.3 g. The Indian seismic code BIS-1893 [28] divides the country into four zones, Viz. Zones II, II, IV, and V. The maximum expected accelerations in each of these zones are 0.1 g, 0.16 g, 0.24 g, and 0.36 g, respectively. The Jaitapur region is placed in Zone III with a zone factor of 0.16 g.

7 Conclusions

An updated catalog of earthquakes has been prepared for Jaitapur and adjoining regions. The catalog was homogenized to a unified magnitude scale, and declustering of the catalog was performed to remove aftershocks and foreshocks. The seismicity parameters were estimated for each of the regions which is a necessary

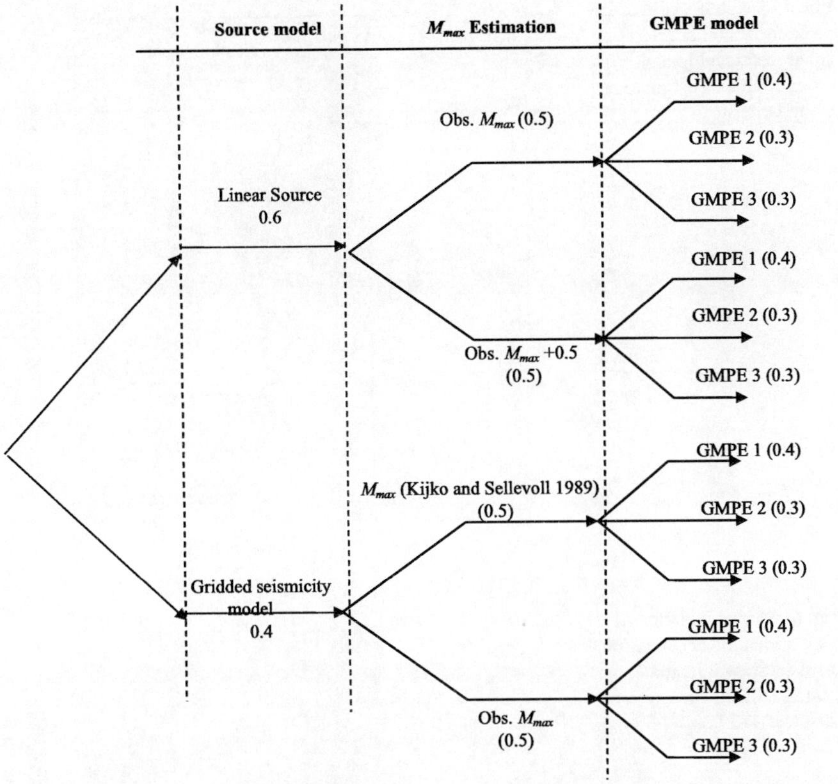

Fig. 4 Parameters and weighting factors adopted in the logic tree for PSHA

input into seismic hazard estimation of a region. The seismic hazard evaluation of the Jaitapur based on a state-of-the-art DSHA and PSHA studies has been performed with different source models and attenuation relations. The most recent knowledge of seismic activity in the region has been used to evaluate the hazard incorporating uncertainty associated with different modeling parameters as well as spatial and temporal uncertainties. The hazard analysis has been performed with currently available data and their best possible scientific interpretation using an appropriate instrument such as the logic tree to explicitly account for epistemic uncertainty by considering alternative models (source models, the maximum magnitude in hazard computations, and ground-motion attenuation relationships).

The hazard maps have been produced for horizontal ground motion at bedrock level. Designs based on ground motion estimated from DSHA will be on the conservative side and hence suitable for critical structures like dams, nuclear power plants, etc. Updated information on the active faults and geodetic information will help in redefining seismogenic sources in the region. Strong ground-motion records of earthquakes in the region would be essential to develop region-specific attenuation

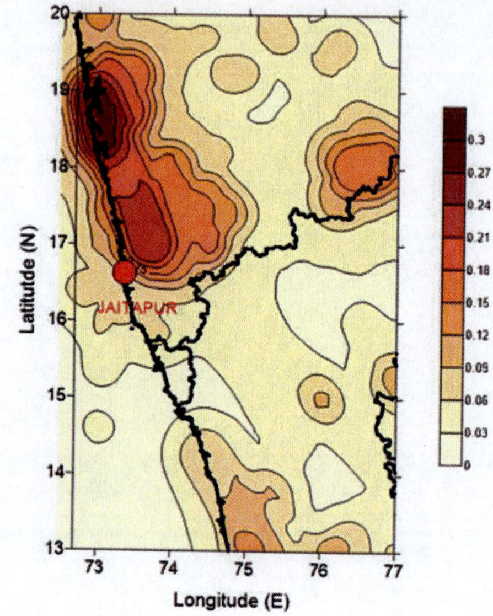

Fig. 5 Spectral variation of peak horizontal acceleration (*g*) at rock level in and around Jaitapur estimated using DSHA

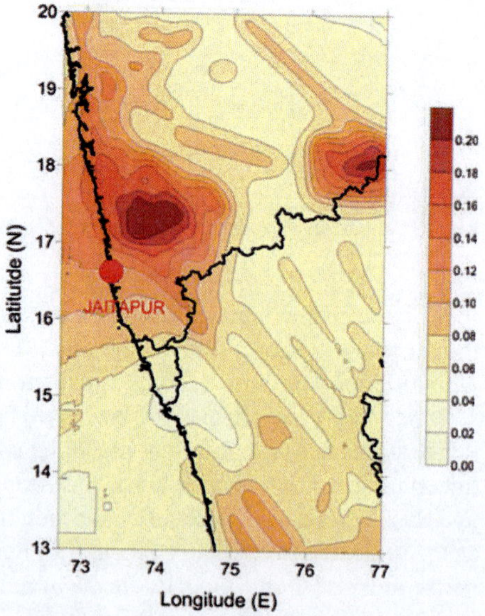

Fig. 6 Spatial variation of peak horizontal acceleration (*g*) at rock level in and around Jaitapur for a return period of 475 years

equations for various tectonic provinces in the region. The updated and reliable information on these factors would go a long way in redefining the seismic hazard of the region in a better way [29].

References

1. Dasgupta S, Pande P, Ganguly D, Iqbal Z, Sanyal K, Venkatraman NV et al (2000) Seismotectonic atlas of India and its environs. Geological Survey of India, Calcutta
2. Frankel A (1995) Mapping seismic hazard in the Central Eastern United States. Seismol Res Lett 66(4):8–21
3. Martin C, Secanell R, Combes P, Lignon G (2002) Preliminary probabilistic seismic hazard assessment of France. In: Proceedings of the 12th ECEE, paper reference 870, Septembre, London, England
4. Woo G (1996) Kernel estimation methods for seismic hazard area source model. Bull Seismol Soc Am 86(2):253–362
5. Wahlstrom E, Grunthal R (2000) Probabilistic seismic hazard assessment (horizontal PGA) for Sweden, Finland and Denmark using different logic tree approaches. Soil Dyn Earthq Eng 20:45–58
6. Lapajne J, Motnikar BS, Zupancic P (2003) Probabilistic seismic hazard assessment methodology for distributed seismicity. Bull Seismol Soc Am 93(6):2502–2515
7. Vilanova SP, Fonseca JFBD (2007) Probabilistic seismic-hazard assessment for Portugal. Bull Seismol Soc Am 93(6):2502–2515
8. Jaiswal K, Sinha R (2007) Probabilistic seismic-hazard estimation for peninsular India. Bull Seismol Soc Am 97(1B):318–330
9. Kalkan E, Gülkan P, Yilmaz N, Çelebi M (2009) Reassessment of probabilistic seismic hazard in the Marmara Region. Bull Seismol Soc Am 99(4):2127–2146
10. Menon A, Ornthammarath T, Corigliano M, Lai CG (2010) Probabilistic seismic hazard macrozonation of Tamil Nadu in Southern India. Bull Seismol Soc Am 100:1320–1341
11. Gutenberg B, Richter CF (1944) Frequency of earthquakes in California. Bull Seismol Soc Am 34:185–188
12. Utsu T (1965) A method for determining the value of b in the formula log N = a - bM showing the magnitude—frequency relation for the earthquakes. Geophys Bull Hokkaido Univ 13:99–103
13. Aki K (1965) Maximum likelihood estimate of b in the formula log N = a-bm and its confidence limits. Bull Earthq Res Instit Univ Tokyo 43:237–239
14. Wiemer S, Wyss M (2000) Minimum magnitude of complete reporting in earthquake catalogs: examples from Alaska, the Western United States, and Japan. Bull Seismol Soc Am 90:859–869
15. Rydelek PA, Sacks IS (1989) Testing the completeness of earthquake catalogs and the hypothesis of self-similarity. Nature 337:251–253
16. Wiemer S (2001) A software package to analyze seismicity: Zmap. Seismol Res Lett 72(2):374–383
17. Kijko A (2004) Estimation of the maximum earthquake magnitude, mmax. Pure Appl Geophys 161:1655–1681
18. Gupta ID (2002) The state of the art in seismic hazard analysis. ISET J Earthq Technol 39(4):311–346
19. Iyengar RN, Ghosh S (2004) Microzonation of earthquake hazard in greater Delhi area. Curr Sci 87:1193–1202
20. Raghu Kanth STG, Iyengar RN (2006) Seismic hazard estimation for Mumbai city. Curr Sci 91(11):1486–1494
21. Bozzoni F, Corigliano M, Lai CG, Salazar W, Scandella L, Zuccolo E, Latchman J, Lynch L, Robertson R (2011) Probabilistic seismic hazard assessment at the Eastern Caribbean Islands. Bull Seismol Soc Am 101(5):2499–2521
22. Raghu Kanth STG, Iyengar RN (2007) Estimation of seismic spectral acceleration in Peninsular India. J Earth Syst Sci 116(3):199–214
23. Atkinson GM, Boore DM (2006) Earthquake ground-motion prediction equations for Eastern North America. Bull Seismol Soc Am 96(6):2181–2205
24. Graizer V (2016) Ground-motion prediction equations for central and eastern North America. Bull Seismol Soc Am 106(4):1600–1612. https://doi.org/10.1785/0120150374

25. Budnitz RJ, Apostolakis G, Boore DM (1997) Recommendations for probabilistic seismic hazard analysis: guidance on uncertainty and use of experts (No. NUREG/CR--6372-Vol. 1; UCRL-ID--122160). Nuclear Regulatory Commission, Washington, DC (United States). Div. of Engineering Technology; Lawrence Livermore National Lab., CA (United States); Electric Power Research Inst., Palo Alto, CA (United States); USDOE, Washington, DC (United States)
26. Bommer JJ, Scherbaum F, Bungum H, Cotton F, Sabetta F, Abrahamson NA (2005) On the use of logic trees for ground-motion prediction equations in seismic hazard analysis. Bull Seismol Soc Am 95:377–389
27. Stepp JC, Wong I, Whitney J, Quittemeyer R, Abrahamson N, Toro G, Youngs R, Coppersmith K, Savy J, Sullivan T (2001) Yucca mountain PSHA project members, probabilistic seismic hazard analyses for ground motions and fault displacements at Yucca Mountain, Nevada. Earthq Spectra 17:113–151
28. BIS-1893 (2002) Indian Standard Criteria for earthquake resistant design of structures, part 1: general provisions and buildings. Bureau of Indian Standards, New Delhi
29. Kijko A, Sellevoll MA (1989) Estimation of earthquake hazard parameters from incomplete data files, part I: utilization of extreme and complete catalogs with different threshold magnitudes. Bull Seismol Soc Am 79:645–654

Liquefaction Potential of Ash Pond Using SPT

Jithin P. Zachariah and Ravi S. Jakka

Abstract In India, thermal power plants produce a considerable amount of ash annually, and out of which, a small proportion is used in industries such as cement, concrete, etc. A significant portion of the produced ash is deposited into onsite storage ponds forming ash ponds, which ultimately leads to the wastage of vast areas. This ash pond occupied land can be wisely utilized for constructing lightweight structures, parking lots, etc., by proper reclamation methods. In seismically active regions, a detailed analysis of such ash ponds is required to ensure the safety of any proposed construction. This paper deals with the assessment of liquefaction potential of a hydraulically deposited ash pond in Delhi region, namely Badarpur ash pond which is fed by deposited ash from Badarpur Thermal Power Station. Standard penetration tests have been conducted in the ash pond and observed low SPT N values indicating the high susceptibility of ash pond to liquefaction. The seismic demand on the ash pond has been determined under an earthquake magnitude of 7.0 with a peak ground acceleration 0.21 g. The result clearly points toward the low liquefaction resistance of ash pond.

Keywords Liquefaction potential · Pond ash · SPT N · CSR · CRR

1 Introduction

About 80% of thermal energy consumed by India is generated by burning coal from different places and ash, being a by-product of these thermal power plants, is produced in a substantial quantity annually. The waste ash thus produced is deposited into onsite storage ponds in the form of ash slurry and leads to the formation of ash ponds. This cause wastage of large area of useful land and trigger critical environmental issues. However, the survival of humans and industries without the aid of this thermal power is unimaginable. This steers to the idea of reusing ash, and ash deposits effectively rather than to stop the production of energy using coal. The

J. P. Zachariah (✉) · R. S. Jakka
Department of Earthquake Engineering, Indian Institute of Technology Roorkee, Roorkee, India
e-mail: rsjakka@gmail.com; jpzachariah@eq.iitr.ac.in

T. G. Sitharam et al. (eds.), *Seismic Hazards and Risk*, Lecture Notes in Civil Engineering 116, https://doi.org/10.1007/978-981-15-9976-7_3

reuse of deposited ash can be executed in large geotechnical earthwork like highway construction, embankment construction, etc. Besides, the area filled with disposed ash can be reclaimed by constructing low load structures, parking slots, parks, etc. Proper studies on engineering aspects of these deposits have to be carried out before attempting any construction activities in order to ensure the safety of the proposed structure.

The characterization of pond ash has been conducted by many researchers in the recent past and produced remarkable conclusions. In ash ponds, the ash slurry contains particles coarser than fly ash and shows a gradual reduction in particle size from inflow to the outflow points, i.e., the deposits at the inflow points are in the range of fine sand, whereas the deposits at outflow points are in the range of silt size [1–4]. Moreover, the pond ash behaves like natural cohesion less soil during compaction which indicates the poor performance of pond ash and ash pond on any type of loading [5]. In earthquake prone areas, the behavior of pond ash under any dynamic loading and susceptibility of pond ash to liquefaction is be taken into account. Badarpur pond ash possess low static and cyclic strength which directly points to its high liquefaction susceptibility in its loose as well as dense condition [6–8]. All these studies show the poor performance of pond ash on loading and recommend detailed liquefaction assessment of pond ash before proceeding to any reuse or reclamation activity.

Still, it is to be noted that these studies are performed in laboratory conditions, not in the site, i.e., the ash pond. In particular, the liquefaction study of Badarpur ash pond using any field test is not reported so far. Hence, an evaluation of liquefaction potential of Badarpur ash pond using measured Standard Penetration Resistance (SPT N) value has been carried out in this study.

2 Site Selection

Electricity demands of the capital city of India is mostly solved by thermal power plants in the region, and the ash produced by burning pulverized coal in these power plants is deposited into nearby ash ponds. The ash pond formed near Badarpur thermal power plant (closed permanently in 2018) is selected for this particular study which adopts the method of wet disposal system to dispose ash slurry into the pond. Location of Badarpur ash pond is shown in Fig. 1. With five operational units, Badarpur ash pond is a huge ash pond of area approximately equals to 3.5 km^2 with an ash deposit of 8–12 m depth. Risk is involved in conducting any type of construction activities in this region as it is situated in the region belonging to earthquake Zone 4 according to Indian standards [9].

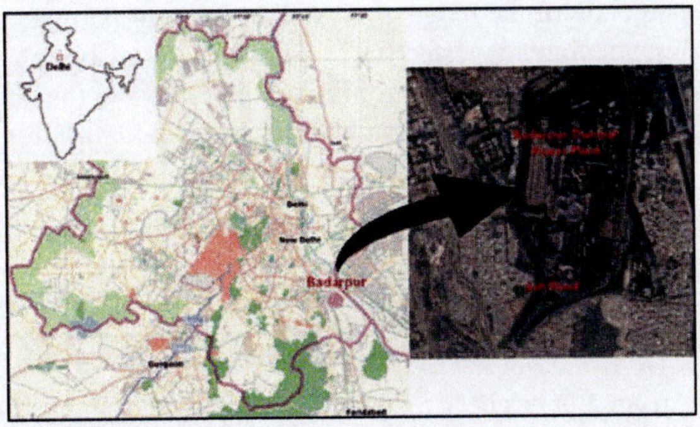

Fig. 1 Location of Badarpur ash pond

3 Geotechnical Characterization of Pond Ash

Index properties of pond ash, such as grain size distribution and Atterberg limits were determined following standard laboratory tests [10, 11]. The characterization has been done for ash samples collected from the ash pond by which the representative samples are extracted from two particular locations of the ash pond. The sample collected from inflow and outflow point of Badarpur ash pond is designated as "*BP-I*" and "*BP-O*," respectively.

3.1 Grain Size Distribution

Grain size distribution curves of samples collected at different locations of the ash pond have been plotted following Indian Standards IS: 2720 (Part 4) -1985 (2006) which is shown in Fig. 2. The gradation curves imply that the samples are well graded. The pond ash samples at outflow points belong to a fine sand range, whereas the inflow

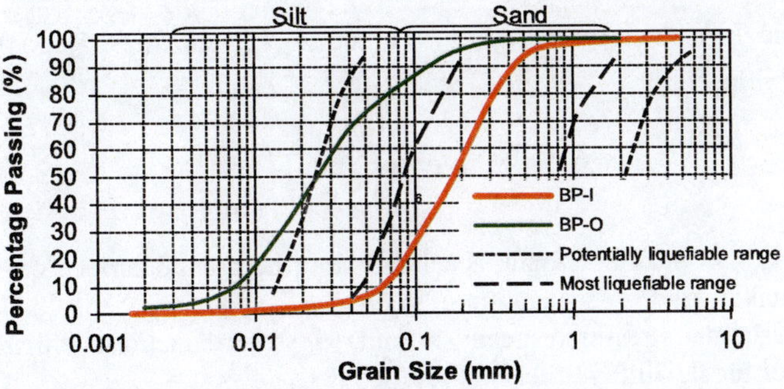

Fig. 2 Grain size distribution curves for the representative samples superimposed on potentially liquefiable ranges [12]

points lie in the range of silts. The range of particle size distribution susceptible to liquefaction suggested by Tsuchida [12] is also plotted. It can be observed that the grain size distribution of inflow ash belongs to the zone susceptible to liquefaction, and the outflow pond ash belongs to the potential liquefaction zone. This indicates the importance of detailed study on liquefaction potential of these materials.

3.2 *Atterberg Limits*

An attempt has been made following Indian Standards IS: 2720 (Part 5) -1985 (2006) using Casagrande's method to determine the liquid limit of the pond ash [11]. But, a consistent number of blows could not be obtained. Also, it was not possible to make threads out of the samples indicating that they are non-plastic.

4 Liquefaction Potential of Non-plastic Fines

Evaluation of liquefaction potential of fine-grained particles is a serious concern from the past. The use of Chinese criteria was widely adopted in those days. Further, studies on fines were reported later, and the behavioral difference of *sand-like* ($PI < 5$) and *clay-like* ($PI > 7$) fines was also considered in the study. The evaluation of liquefaction potential involves the estimation of two parameters: Cyclic Stress Ratio (CSR) and Cyclic Resistance Ratio (CRR). NCEER workshop in 1996 suggested various methods for evaluation of liquefaction potential of soil using different field tests like SPT, CPT, using shear wave velocity, etc., [13].

4.1 *Cyclic Stress Ratio (CSR)*

The seismic demand on soil is expressed in terms of Cyclic Stress Ratio (CSR) and is given in Eq. (1).

$$\mathrm{CSR} = \frac{\tau_{av}}{\sigma_{v0}} = 0.65\left(\frac{a_{\max}}{g}\right)\left(\frac{\sigma_{v0}}{\sigma'_{v0}}\right)r_d \qquad (1)$$

where $a_{\max}$ = peak horizontal acceleration at the ground surface generated by the earthquake; g is the acceleration of gravity; σ_{vo} and σ'_{vo} are total and effective vertical overburden stresses, respectively; and r_d is stress reduction coefficient, which accounts for the flexibility of the soil profile [13].

4.2 Cyclic Resistance Ratio (CRR) Using Measured SPT N Values

Equation (2) gives the expression for evaluating the ability of soil to resist liquefaction, expressed by Cyclic Resistance Ratio (CRR).

$$\mathrm{CRR}_{7.5} = \frac{1}{(34 - (N_1)_{60})} + \frac{(N_1)_{60}}{135} + \frac{50}{(10.(N_1)_{60} + 45)^2} - \frac{1}{200} \tag{2}$$

This equation is valid for $(N_1)_{60} < 30$. For $(N_1)_{60} > 30$, clean granular soils are too dense to liquefy and are classed as non-liquefiable. The presence of fine content in the soil increases the liquefaction potential of the sand. The influence of fine content in the pond ash is incorporated in the study using two parameters α and β. Equivalent clean sand value $(N_1)_{60cs}$ is given by:

$$(N_1)_{60cs} = \alpha + \beta(N_1)_{60} \tag{3}$$

Also, the other corrections in the evaluation of $(N_1)_{60}$ suggested by NCEER workshop are taken into account [13].

4.3 Magnitude Scaling Factors (MSF)

The assessment of CRR value using SPT N values is applicable only to earthquakes with a magnitude of 7.5. For influence scale up or down the curves, which corresponds to other magnitude earthquakes, Magnitude Scaling Factors (MSF) were introduced [13]. Factor of Safety (FOS) shows the influence of MSF on the calculated hazard by:

$$\mathrm{FOS} = (\mathrm{CRR}_{7.5}/\mathrm{CSR})\mathrm{MSF} \tag{4}$$

$\mathrm{CRR}_{7.5}$ is the cyclic resistance ratio for 7.5 magnitude earthquakes; CSR is the cyclic stress ratio. MSF is defined using the relation,

$$\mathrm{MSF} = \frac{10^{2.24}}{M_w^{2.56}} \tag{5}$$

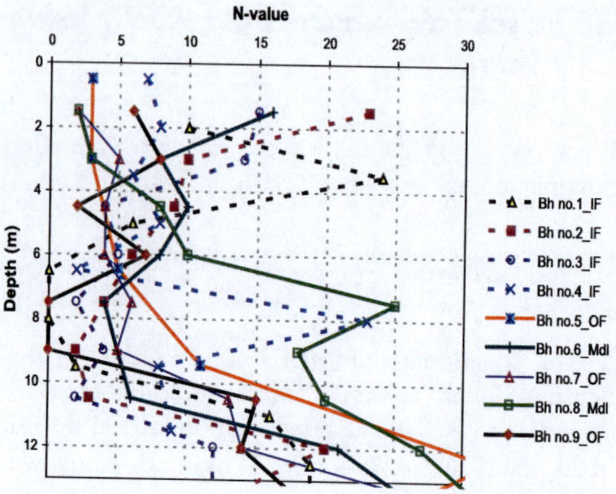

Fig. 3 Measured *SPT-N* values with depth from boreholes drilled at Badarpur ash pond

5 In Situ Testing

Standard penetration tests at different locations of the ash pond were conducted following IS 2131:1981 [14]. Nine boreholes were drilled near the inlet and outlet points of the ash pond and tests conducted at every 1.5 m depth in each of the boreholes to obtain the corresponding "*N*" values. The detailed SPT test results have been plotted in Fig. 3. It can be observed that the measured *N* values of Badarpur ash pond vary in higher rates at a depth of 1.5–12 m. This shows a high variability in the properties of ash along the depth within the ash pond, which indicates the high susceptibility of the ash pond to liquefaction.

6 Liquefaction Potential of the Pond Ash

Liquefaction potential of the particular ash pond has been determined using SPT N values. The value of CSR and CRR at different depth of the deposit is calculated following the above-mentioned procedure. The variation of liquefaction resistance along the depth computed from various boreholes at Badarpur ash pond is shown in Fig. 4 for a postulated earthquake of magnitude, M_w 7.0 and a_{max} of 0.21 g. A decrease in liquefaction resistance after a certain depth is observed due to the fall of *N* values after encountering water table. It is depicted that the ash pond site is very susceptible to liquefaction.

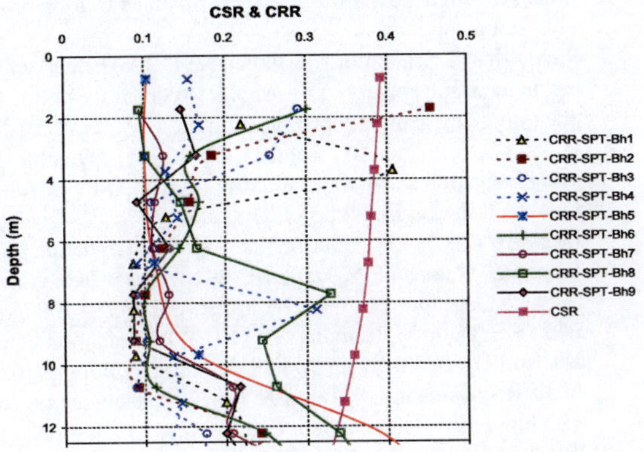

Fig. 4 Variation of CSR and CRR along the depth for different boreholes

7 Conclusions

Evaluation of liquefaction potential of Badarpur ash pond in Delhi region is discussed in this study. Standard Penetration Test (SPT) was conducted using different boreholes drilled near to inlet and outlet points of the ash pond, and corresponding *N* values were determined to calculate the liquefaction potential.

A high variation of SPT N values along the depth of the ash pond is observed indicating change in properties of the deposit with depth. At depths encountering groundwater table, lower *N* values were observed indicating a low resistance of ash pond toward liquefaction. On estimating the liquefaction potential of ash ponds based on the reported values, the ash pond exhibited low liquefaction resistance and is found to be susceptible/vulnerable to liquefaction under the considered earthquake scenario ($M_w = 7.0$ and PGA = 0.21 g).

The present study clearly indicates the need of application of ground improvement techniques to mitigate the liquefaction susceptibility of Badarpur ash pond which is very important before proceeding to any construction activity to ensure the safety of structure and life. This study can be extended using other techniques such as shear wave velocity and cone penetration including laboratory experiments to understand more about the liquefaction behavior of pond ash.

Acknowledgements Work presented here is an extension of the work carried out by the second author during his doctoral research. Authors are thankful to Prof. GV Ramana and Prof. Manoj Datta for their encouragement in carrying out the studies.

References

1. Dayal U, Jain SK, Srivastava AK (1988) Geotechnical investigations of ash properties for dyke construction at korba super thermal power project. In: Report prepared for National Thermal Power Corporation Ltd., New Delhi

2. Datta M, Singh A, Kaniraj R (1996) Spatial variation of ash characteristics in an ash pond, pp 111–119
3. Rajasekhar C, Sridharan A, Pandian NS (1996) Effect of time and concentration of cation on the cation retention cpacity of fly ash. In: Proceedings of 3rd international conference environment planning management, Nagpur, India
4. Jakka RS, Ramaiah BJ, Ramana GV (2011) Dynamic characterization of settled pond ash using measured shear wave velocity (Vs) and SPT-N values: correlation between Vs & N. Int J Geotech Earthq Eng 2:83–97. https://doi.org/10.4018/jgee.2011010106
5. Jakka RS (2007) Liquefaction resistance and dynamic properties of coal ash
6. Jakka RS, Ramana GV, Datta M (2010) Shear behaviour of loose and compacted pond ash. Geotech Geol Eng 28:763–778. https://doi.org/10.1007/s10706-010-9337-1
7. Jakka RS, Datta M, Ramana GV (2010) Liquefaction behaviour of loose and compacted pond ash. Soil Dyn Earthq Eng 30:580–590. https://doi.org/10.1016/j.soildyn.2010.01.015
8. Jakka RS, Ramana GV, Datta M (2011) Seismic slope stability of embankments constructed with pond ash. Geotech Geol Eng 29:821–835. https://doi.org/10.1007/s10706-011-9419-8
9. IS 1893 (Part 1) :2002 (2002) 7.11 Deformations. Earthq Resist Des Struct 1893:27
10. IS 2720 (Part 4)- 1985 (2006) Methods of test for soils -grain size analysis. Bur Indian Stand New Delhi
11. IS 2720 (Part 5) -1985 (2006) Method of test for soils -determination of liquid and plastic limit. Bur Indian Stand New Delhi
12. Tsuchida H (1970) Prediction and countermeasure against the liquefaction in sand deposits. Abstact Sem Port Harb Res Inst 3:1–33
13. Youd TL, Idriss IM, Andrus RD et al (2001) Liquefaction resistance of soils: summary report from the 1996 nceer and 1998 NCEER/NSF workshops on evaluation of liquefaction resistance of soils. J Geotech Geoenviron Eng 127:817–834. https://doi.org/10.1061/(ASCE)1090-0241(2003)129:3(284)
14. IS:2131–1981 Indian Standard method for standard penetration test for soils. Bur Indian Stand New Delhi

Seismic Hazard and Risk Assessment in Maharashtra: A Critical Review

Abdullah Ansari, K. S. Rao, and A. K. Jain

Abstract In peninsula India, the state of Maharashtra has suffered from frequent intra-plate earthquakes. In 1967, the Koyna earthquake caused few hundred deaths and structural damage, and then, in 1993, the Killari earthquake in Latur district caused thousands of deaths and enormous structural damage. There are two basic methods for assessing the seismic ground motion hazard in a particular region or at a specific site, namely deterministic methods and probabilistic methods. In deterministic seismic hazard analysis, a particular earthquake scenario is presumed. On another hand, uncertainties in earthquake location, size, and time are considered in case of probabilistic seismic hazard analysis. Geotechnical and geophysical data-based site characterization helped to divide the area into smaller zones giving a new direction toward microzonation. This paper deals with the compilation of various studies carried out at different region of the Maharashtra based on several approaches. A critical review is presented on each of these methods, highlighting their limitations and suitability of application.

Keywords Seismic hazard · Seismic response · MASW · Liquefaction · Shear wave velocity

1 Introduction

Millions of people are affected by natural hazards every year, and their impact can be calamitous. From the destructions of building to the spread of disease, natural hazards can devastate entire countries overnight. Tsunamis, earthquakes, and typhoons do not just wreak havoc on the land; they also disrupt people's lives in both densely populated cities as well as remote villages and rural areas. Increasing effects of natural hazards viz. landslides, tsunami, floods around the world have a direct social, economic, and environmental impact. In the domain of the Indian sub-continent, the ongoing collision between the Indian plate and the Eurasian plate has resulted in

A. Ansari (✉) · K. S. Rao · A. K. Jain
Department of Civil Engineering, Indian Institute of Technology, Delhi, New Delhi 110016, India
e-mail: cez188391@iitd.ac.in

T. G. Sitharam et al. (eds.), *Seismic Hazards and Risk*, Lecture Notes in Civil Engineering 116, https://doi.org/10.1007/978-981-15-9976-7_4

the formation of the Himalayas in the north and a subduction zone in the southwest along the Sumatra-Andaman trench. Indian sub-continent experienced several devastating catastrophic events, viz. the 1993 Latur earthquake (Mw 6.2) and 2001 Bhuj earthquake (Mw 7.6) in Stable Continental Region (SCR) of Western India; 2005 Muzaffarabad earthquake (Mw 7.6) in Kashmir; 2015 Gorkha earthquake (Mw 7.8) along the active plate boundary in Himalaya; and 2004 Sumatra-Andaman earthquake (Mw 9.3) associated with a transoceanic tsunami in the Indian Ocean [1].

In peninsula India, the state of Maharashtra, along with Gujarat and Madhya Pradesh, has suffered from frequent deadly and damaging earthquakes. The western coastal areas of Maharashtra along with the state capital Mumbai and Pune city have both been strongly shaken in the past in 1618 and 1764, respectively, [2]. Seismologists identified several faults in this region out of which many show evidence of movement during the Holocene epoch. The major Chiplin fault lies along the Sahyadri range running from the mouth of Bombay Harbor to the Sangammeshwar area in Ratnagiri district. While the west coast fault runs along the eastern shore of the Thane Creek in the Mumbai area, along the flanks of the Parsik Hills. The Vidharbha region is intersected by Kadam fault which runs from Bhusawal city into northern Andhra Pradesh. The 1967 Koyna earthquake (Mw 6.6) is the largest known case of Reservoir Induced Seismicity (RIS) in the world. The devastating Latur earthquake (Mw 6.2) of 1993 in southern part of the state has claimed an estimated 11,000 human lives [3].

To study these seismic events, various methods and techniques are used. Deterministic- and probabilistic-based seismic hazard analysis of the earthquake prone source-site area carried out to know about the peak ground acceleration and return interval of the earthquakes in a particular area or site. Geotechnical and geophysical data-based site characterization and ground response analysis help to divide the area into smaller zones giving a new direction toward microzonation. Liquefaction occurs generally due to rapid loading during seismic events, where there is not sufficient time for dissipation of excess pore-water pressures through natural drainage. Liquefaction Potential Index (LPI) is evaluated at each borehole location from the obtained Factors of Safety (FS) to predict the potential of liquefaction to cause damage at the surface level at the site of interest. This paper deals with the compilation of various studies carried out at different region of the Maharashtra based on several approaches. A critical review is presented on each of these methods, highlighting their limitations and suitability of application.

2 Earthquakes in Maharashtra

After 2001 Bhuj earthquake in Gujarat, seismic hazard map of India was updated by the Bureau of Indian Standards (BIS). The latest seismic zonation map of India includes the Beed, Osmanabad, and Latur, along with eastern sections of Ahmednagar, Pune, Satara, and Sangli districts in Zone III. Districts viz. Raigad, Ratnagiri, and Satara lie in Zone IV, where maximum expected intensity of seismic motion is

Table 1 Major earthquakes in Maharashtra after 1950

Date	Region	Location	Earthquake data	
			Magnitude (Mw)	Depth (km)
April 08 1951	Konkan area	18.500 N, 70.800 E	6.0	13
December 13 1957	Koyna area	17.300 N, 73.700 E	6.1	15
December 11 1967	Koyna area	17.450 N, 73.850 E	6.6	27
September 26 1970	Wai area	18.000 N, 74.000 E	5.5	19
September 14 1983	Bhatsa area	19.640 N, 73.540 E	5.2	33
September 30 1993	Latur area	18.090 N, 76.470 E	6.2	10
March 12 2000	Koyna area	17.244 N, 73.707 E	5.3	9
July 27 2003	Sindvani area	21.878 N, 74.341 E	5.1	20
March 14 2005	Koyna area	17.139 N, 73.687 E	5.1	14
September 16 2008	Koyna area	17.289 N, 73.815 E	5.5	11
December 17 2009	Satara area	17.421 N, 73.810 E	5.1	10
July 12 2015	Satara area	17.212 N, 73.75 E	5.3	6
August 19 2017	Karad area	17.194 N, 73.825 E	5.2	8

VIII. The remaining western and northwestern districts of the state fall in Zone III [4, 5]. India Meteorological Department (IMD) under Government of India responsible for monitoring seismic activity in and around the country helping in collection the real-time earthquake data. Following Table 1 gives the detailed description about the major earthquakes event occurred in the Maharashtra region after 1950.

3 Seismic Hazard Assessment

The seismic hazard could be defined as the likelihood of experiencing a specified intensity of any damaging phenomenon at a site or all over a region during a considered period of time. The development of the methodology for analyzing the probability of seismic hazards has originated from the engineering needs for better designs [6]. Seismic hazard assessment is analogous to long-term earthquake prediction [7, 8]. Two basic methodologies used for Seismic Hazard Assessment (SHA) are the Deterministic Seismic Hazard Assessment (DSHA) and the Probabilistic Seismic Hazard Assessment (PSHA) approaches. In the Deterministic Seismic Hazard Assessment (DSHA), the strong ground motion parameters are estimated for the maximum credible earthquake, assumed to occur at the closest possible distance from the site of interest, without considering the likelihood of its occurrence during a specified exposure period. On the other hand, in PSHA, rather than searching for elusive worst-case ground motion, all possible earthquake events and resulting ground motions along with their associated probabilities of occurrence are considered to estimate the level of ground motion intensity exceeded with some tolerably low rate.

3.1 Deterministic Seismic Hazard Analysis (DSHA)

This method of seismic analysis is based on Maximum Credible Earthquake (MCE) and estimates the maximum size of an earthquake located at defined distance from the site. The basic steps involved in the process of deterministic seismic hazard assessment are as follows, as shown in Fig. 1.

a. **Identification of earthquake sources**: These sources could be either clearly understood and defined faults or less well understood and less well defined geologic structures or seismotectonic provinces of many thousands of square kilometers. Earthquake sources could be either clearly understood and defined faults or less well understood and less well defined any geologic structures or seismotectonic region of many thousands of square kilometers. The individual sources could be configured as points, lines, areas, or volumes [6]. The geometry and potential of each earthquake source need to be defined well before jumping to next step.
b. **Evaluation of source-to-site distance**: For each source zone, the shortest distance between the source zone and site of interest is selected.

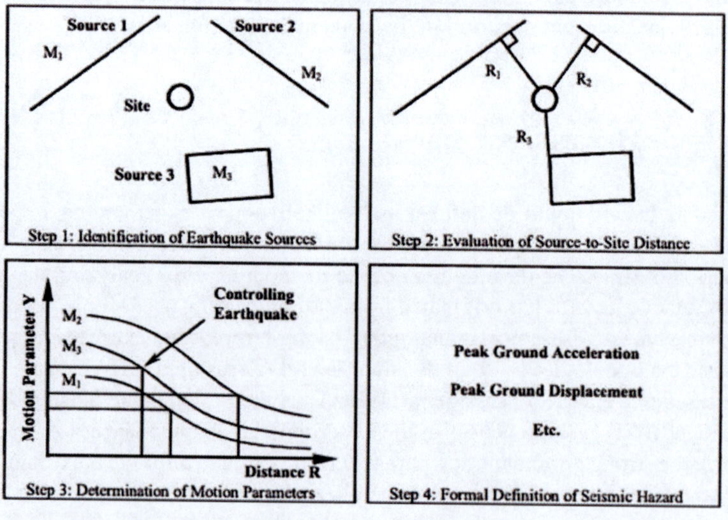

Fig. 1 Steps of deterministic seismic hazard analysis (DSHA) [9]

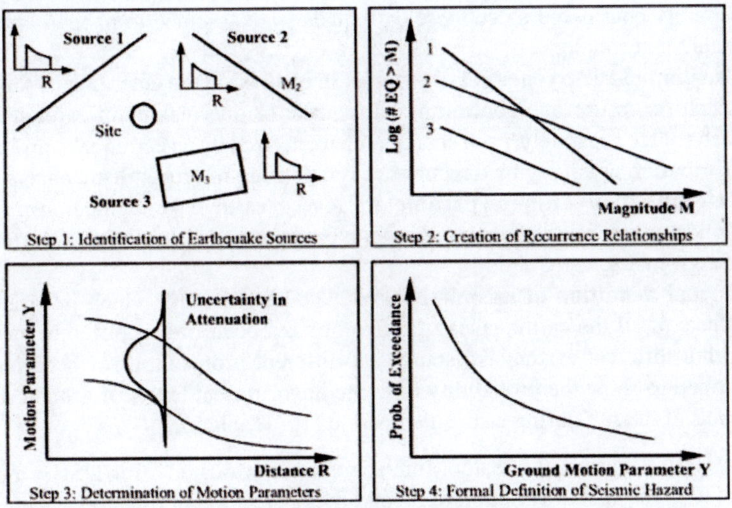

Fig. 2 Steps of probabilistic seismic hazard analysis (PSHA) [11]

c. **Determination of motion parameters**: The earthquake potential of each source specified in the first step should be defined in terms of a maximum earthquake known as either "maximum credible earthquake" which is the maximum earthquake capable of occurring in an area or on a given fault during the current tectonic regime. They are also defined as "maximum expectable earthquake," or "design earthquake." For regional seismicity, two levels of maximum earthquakes could be considered as in the case of the U.S. Nuclear Regulatory Commission: "Safe shutdown earthquake" and "operating basis earthquake" selected for hazard assessment.
d. **Formal definition of seismic hazard**: This can be defined as the hazard at the site of interest represented by specific peak ground motion such as response spectrum ordinate values, describing the earthquake effects [10] (Fig. 2).

3.2 *Probabilistic Seismic Hazard Analysis (PSHA)*

Probabilistic seismic hazard analysis is a rational solution to the different types of dilemmas posed by uncertainties. The first methodology applied to most of the probabilistic seismic hazard analyses was defined by [6] in 1968. The four steps involved in this method are explained below:

a. **Identification of earthquake sources**: The types of the sources may range from small planar faults to large seismotectonic provinces. The source-site distances for each source are calculated considering the spatial uncertainty. The distances from all possible locations to the site within the earthquake source should be certainly considered since these earthquakes are assumed to happen anywhere within that source.

b. **Creation of recurrence relationships**: In contrast to the deterministic analyses which try to use one controlling earthquake or one maximum earthquake for each source separately, each source is characterized by an earthquake probability distribution or simply by a recurrence relationship in probabilistic analyses.
c. **Determination of motion parameters**: A set of earthquake attenuation or ground motion curves relating a ground motion parameter to distance for an earthquake of a given size is required.
d. **Formal definition of seismic hazard**: The hazard curve which integrates the effects of all the earthquakes of different sizes occurring at different locations within different earthquake sources at different probabilities of occurrence is formed to show the probability of exceeding different levels of ground motion levels at the site during a specified period [10] (Table 2).

The seismic hazard of Greater Mumbai region is evaluated using DSHA methodology [12]. The spatial variation of 50 and 84th percentile of Spectral Acceleration (S_a) values shown in Fig. 3 for 0.2 and 1 s period. Seismic hazard for Peak Horizontal Acceleration (PHA) increases steadily from relatively low to high seismicity from the center to southern and northern parts of the Greater Mumbai region. The same hazard pattern is observed for S_a values at short period (0.2 s); however, for long period (1 s), the hazard is observed to be concentrated on southern part of the region

Table 2 Advantages and disadvantages of PSHA and DSHA [9]

Advantages	Limitations
Deterministic Seismic Hazard Analysis (DSHA)	
Based on design maximum earthquake	Frequency or probability of earthquakes is not considered
Gives conservative results and high level of safety	Uncertainties involved in the selection of inputs are not considered
	If active faults are not well defined, then the source-to-site distance is difficult to determine
Probabilistic Seismic Hazard Analysis (PSHA)	
The spatial uncertainty, i.e., randomness in earthquake location is explicitly considered	A range of possible ground motion levels, instead of a unique value, is the result
The temporal uncertainty, i.e., probability of occurrence of earthquakes is included	Deaggregation of the results becomes essential due to the multiple contributions of sources, magnitudes, and distances
Uncertainties involved in the process are incorporated through a logic tree	The use of logic tree in PSHA in itself involves use of the subjective judgment
Risk and cost of a project can be easily optimized by selecting an appropriate return period of earthquake	Inclusion of uncertainties introduces more complexities in the analysis

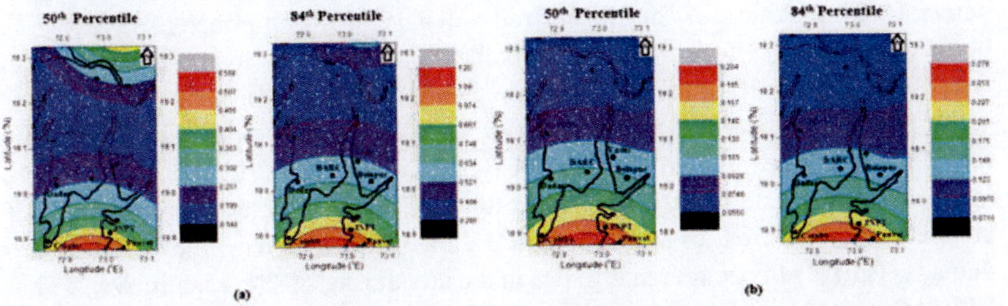

Fig. 3 Spatial variation of 50th percentile and 84th percentile values for **a** 0.2 s, **b** 1 s at bedrock [12]

with a steady decrease toward north side. The PHA values for Greater Mumbai region ranges between 0.076 and 0.328 g.

4 Liquefaction Hazard Assessment

Liquefaction Potential Index (LPI) is a parameter used to represent regional liquefaction potential [13, 14]. LPI at a specific site is computed by integrating the Factors of Safety (FS) along the soil column. A weighting function is added to give more weight to the layers closer to the ground surface. The Liquefaction Potential Index (LPI), proposed by [15, 16] is expressed as follows:

$$\mathrm{LPI} = \int_0^{20} F.w(z)\mathrm{d}z \tag{1}$$

where z is depth of the midpoint of the soil layer), and dz is differential increment of depth. The weighting factor, $w(z)$, and the severity factor, $F(z)$, are calculated as per the following expressions:

$$\begin{aligned} F(z) &= 1 - FS \quad \text{for} \quad FS < 1.0 \\ F(z) &= 0 \quad \text{for} \quad FS \geq 1.0 \\ w(z) &= 10 - 0.5z \quad \text{for} \quad z < 20\ \text{m} \\ w(z) &= 0 \quad \text{for} \quad z > 20\ \text{m} \end{aligned}$$

For the soil profiles with the depth less than 20 m, LPI is calculated using the following expression [4],

$$\mathrm{LPI} = \sum_{i=1}^{n} w_i F_i H_i \tag{2}$$

where H_i is thickness of the discretized soil layers; n is number of layers; F_i is liquefaction severity for ith layer; FS_i is the factor of safety for ith layer; w_i is the weighting factor (=10–0.5z_i); and z_i is the depth of ith layer (m).

To predict the potential of liquefaction causing damage at the surface level of the site of interest evaluated the Liquefaction Potential Index (LPI) at each borehole location from the obtained Factors of Safety (FS) for Greater Mumbai [17]. LPI contour maps prepared for Mumbai region showed that a high degree of liquefaction damages is likely to occur at many sites in the city during severe seismic event. These LPI contour maps help to check the vulnerability of the area against liquefaction. These contour maps can also be used effectively for seismic safety plans and in the seismic hazard mitigation programs followed by seismic microzonation.

5 Seismic Response and Geophysical Investigation

Seismic hazard analysis gives the ground shaking at the bedrock. However, the ground shaking and the associated damage to engineered structures can be strongly influenced by the geology and topography in their vicinities. The influence of local site conditions on the nature of the ground motions and the damage they may cause have been observed and studied [13]. The early work of seismologists studied site amplification shows that they used to assume linear soil behavior, and they rarely considered the soil non-linearity in their assessments of site conditions [7, 18]. On the other hand, the non-linear approach is complicated and cannot model the actual hysteretic stress–strain behavior of cyclically loaded soil. Therefore, the equivalent linear method that modifies the linear approach may be an acceptable estimation of ground response for practical problems of interest [16]. The equivalent linear model provides few parameters that represent typical soil behavior when subjected to cyclic loading. In the equivalent linear method, an iteration procedure is used to make sure the parameters used in the analysis which are compatible with the computed strain level in all the layers (Fig. 4).

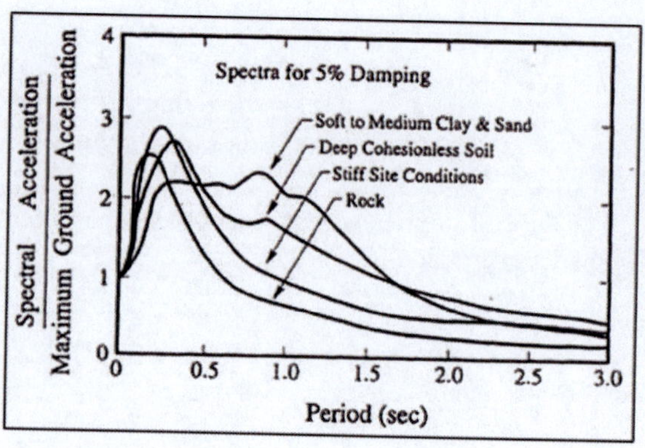

Fig. 4 Spectral shapes for different site conditions [7]

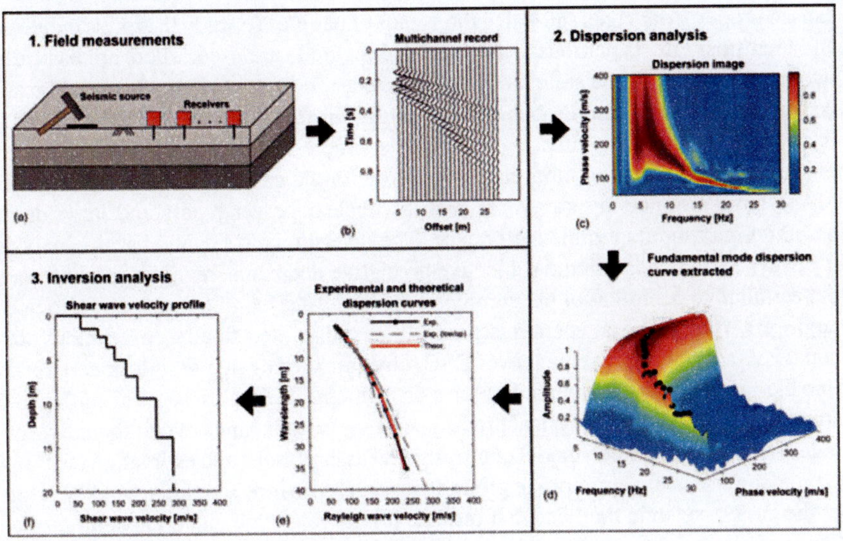

Fig. 5 Procedure followed during MASW survey [22]

Shear wave velocity (V_S) is an essential parameter for evaluating the dynamic properties of soil in the shallow subsurface [8, 19]. Several geophysical methods have been proposed for near-surface characterization and measurement of shear wave velocity by using a great variety of testing configurations, processing techniques, and inversion algorithms. The most widely used technique is Multichannel Analysis of Surface Waves (MASW). The MASW method was first introduced in geophysics by [20, 21]. Following Fig. 5 represents the steps considered while doing geophysical survey using MASW.

Characteristics of strong ground motion and ground response analysis for different sites in Nanded city of the Maharashtra state studied using MASW [20, 23]. Shear wave velocity averaged over top 30 m of the soil is obtained from MASW survey at 60 sites at depths of 20–50 m below the ground surface. The high V_{S30} values (900–1000 m/s) are found toward the northeast and southern part of the city. Around the Chikhli Khurd and Shrinagar areas observed shear wave velocity is 500 m/s. The eastern phase areas are characterized by relatively low V_{S30} values (<200 m/s). They are associated with the Godavari River carrying alluvial deposits composed of brown clay with intercalated bands of sand and gravel.

6 Summary and Conclusion

Maharashtra state suffered many earthquakes in the past few decades including major events like 1967 Koyna earthquakes and 1993 Latur earthquakes. These earthquakes caused property damages and killed thousands of people. To study theses earthquakes and their post effects, various techniques and methods are used. There are basically two methods used for seismic hazard assessment viz. Deterministic Seismic Hazard Assessment (DSHA) and Probabilistic Seismic Hazard Assessment (PSHA). DSHA is an easy and quick method and provides a simple framework for the evaluation of the worst-case ground motions associated with the catastrophic failure damages of the large engineering structures such as nuclear power plants and large dams based on maximum credible earthquake. The deterministic seismic hazard analysis is always associated with the subjective predictive decisions, regarding for instance the earthquake potential, of the combined opinions and judgments of seismologists, engineers, risk analysts, economists, social scientists, and finally government officials. PSHA is an improvement over DSHA by considering all possible uncertainties into the hazard estimation procedure, arising in magnitude scales; types of earthquake sources and its nature of physical processes; using various kinds of earthquake recurrence models and its inter-model or intra-model assumptions and estimation of strong ground motion using appropriate ground motion prediction equations. MASW is one of the easiest seismic methods that provide highly favorable and competent results. Shear wave velocity obtained from MASW survey gives a new direction toward liquefaction hazard assessment and seismic microzonation.

References

1. Ameen AAMM (2018) Modeling the 2004 Andaman-Sumatra tsunami and historical tsunamis from Andaman and Nicobar Island: towards estimation of tsunami hazard along the adjoining areas of Indian Ocean. M.Tech., thesis, Indian Institute of Technology Kanpur, India
2. Dasgupta S, Pande P, Ganguly D, Iqbal Z, Sanyal K, Venkatraman NV, Dasgupta S, Sural B, Harendranath L, Mazumdar K, Sanyal S, Roy K, Das LK, Misra PS, Gupta H (2000) Seismotectonic Atlas of India and its environs. Geol Surv India
3. Ansari A, Satake K, Malik JN (2017) Modelling the 2004 Indian Ocean Tsunami to estimate tsunami heights and its amplitude and to study its effects on coastal areas. ERI Earthquake Conference, University of Tokyo, Japan
4. IS 1893: Part 1.: Indian standard criteria for earthquake resistant design of structures, part—1, general provisions and buildings. Bureau of Indian Standards, New Delhi, India (2002)
5. Raghukanth STG (2011) Seismicity parameters for important urban agglomerations in India. Bull Earthquake Eng 9(5):1361–1386
6. Cornell CA (1968) Engineering seismic risk analysis. Bull Seismol Soc America 58:1583–1606
7. Finn WD, Iai S, Matsunaga Y (1995) The effects of site conditions on ground motions. In: Proc. of 10th ECEE, pp 2607–2612
8. Huang HC, Tseng YS (2002) Characteristics of soil liquefaction using H/V of microtremors in Yuan-Lin area, Taiwan. Terr Atmos Oceanic Sci 13:325–338
9. Kramer SL (1996) Geotechnical earthquake engineering. Prentice Hall Publishers, Englewood Cliffs, New Jersey
10. Reiter L (1990) Earthquake hazard analysis: issues and insights. Columbia University Press, New York, USA
11. Hashash YM, Hook JJ, Schmidt B, John I, Yao C (2001) Seismic design and analysis of underground structures. Tunnelling and underground space technology 16(4):247–293. https://www.sciencedirect.com/science/article/pii/S0886779801000517

12. Desai S, Choudhury D (2014) Deterministic seismic hazard analysis for greater Mumbai, India. ASCE GSP 234:389–398
13. Seed HB, Idriss IM (1971) Simplified procedure for evaluating soil liquefaction potential. J Soil Mech Found 97:1249–1273
14. Seed HB, Idriss IM, Arango I (1983) Evaluation of liquefaction potential using field performance data. J Geotech Eng ASCE 109(3):458–482
15. Iwasaki T, Tokida K, Tatsuoka F, Watanabe S, Yasuda S, Sato H (1982) Microzonation for soil liquefaction potential using simplified methods. In: Proceedings, 3rd international conference on microzonation, Washington D.C., National Science Foundation, pp 1319–1330
16. Luna R, Frost JD (1998) Spatial liquefaction analysis system. J Comput Civil Eng 12:48–56
17. Dixit J, Dewaikar DM, Jangid RS (2012) Assessment of liquefaction potential index for Mumbai city. Nat Hazards Earth Syst Sci 12:2759–2768
18. Faccioli E (1976) A stochastic approach to soil amplification. Bull Seismol Soc Am 66(4):1277–1291
19. Gouveia F, Lopes I, Gomes RC (2016) Deeper V_S profile from joint analysis of Rayleigh wave data. Eng Geol 202:85–98
20. Park CB, Miller RD, Xia J (1999) Multi-channel analysis of surface waves. Geophys J 64(3):800–808
21. Park CB, Miller RD, Rydén N, Xia J, Ivanov J (2005) Combined use of active and passive surface waves. J Environ Eng Geophys 10(3):323–334
22. Olafsdottir EA, Erlingsson S, Bessason B (2018) Tool for analysis of multichannel analysis of surface waves (MASW) field data and evaluation of shear wave velocity profiles of soils. J. Canadian Geotech Eng 55:217–233
23. Subhadra N, Padhy S, Prasad P, Seshunarayana T (2015) Site-specific ground motion simulation and seismic response analysis for microzonation of Nanded City, India. Nat Hazards 78:915–938

Risk Assessment of the Structures Built on Reclaimed Lands in Jabalpur M.P.

Sagar Shrivastava, Sanjay Kumar Verma, and S. D. Pimprikar

Abstract The city Jabalpur falls in zone III of the seismic hazard zone map of India (IS: 1893–2002) (IS: 1893 (2002), Indian Standard Criteria for Earthquake Resistant Design of Structures, Bureau of Indian Standard, 5th Revision.). Earthquakes of intensity up to VI or higher and magnitude of 6.0 or more have occurred in past at Jabalpur. In the present context, an earthquake of M 6.0 (IMD, USGS) occurred at 04 h, 21 Min 30.8 s (IST) on May 22, 1997 (Kanwar in A Geoscientific Study, Geological Survey of India, Special Publication No. 51:i–iv, 2000). This paper describes the risk assessment of the civil engineering structures built on reclaimed lands in Jabalpur. The problems that are likely to arise wherever major civil engineering structures are being planned on lands that have been reclaimed from perennial waterlogged areas or from the areas reclaimed by backfilling the old surface water bodies and old garbage dumps. This practice is being followed in Jabalpur city which once supported 52 major and 84 small water bodies out of which nearly 100 have lost their existence and residential colonies now occupy lands reclaimed from them. As in Jabalpur so also in other towns, this activity is going on without considering the negative effects of using reclaimed land for major constructions. In the present paper, a study of risk assessment on the structure built on reclaimed land has done during the 1997 earthquake at Jabalpur.

Keywords Municipal solid waste (MSW) · SONATA · IMD · NSF · Microzonation

S. Shrivastava (✉)
Gyan Ganga Institute of Technology and Science, Jabalpur, India
e-mail: sagar.shrivastava@gmail.com

S. K. Verma
Indian Geotechnical Society Jabalpur Chapter, Jabalpur, India
e-mail: verma.sanjay090@gmail.com

S. D. Pimprikar
Geological Survey of India, Jabalpur, India
e-mail: pimprikarsad@yahoo.com

T. G. Sitharam et al. (eds.), *Seismic Hazards and Risk*, Lecture Notes in Civil Engineering 116, https://doi.org/10.1007/978-981-15-9976-7_5

1 Introduction

Land reclamation is usually known as the landfill process of creating new land from the ocean, riverbeds, or lake beds. The land reclaimed is known as reclamation ground or landfills. Industrial and commercial developments of the urban areas require a large usable land which could be possible by reclamation. So many advanced countries, e.g., Japan, Singapore, USA, Hong Kong, Netherlands, Dubai (UAE), etc., are obtaining lands by reclaiming seashore. Palm Island in Dubai, The entire East Coast Park at Singapore, Peninsula long beach USA, Hong Kong International airport are the best examples of advanced constructions built on reclaimed lands. These countries are so advanced in technology and well equipped with the modern techniques for land reclamation so that these advanced countries apply the process of reclamation and use the reclaimed lands for modern constructions but in small towns like Jabalpur where advanced techniques of land reclamation are not of such standards so we cannot ignore the disadvantages of wrong landfills. Constructions on improper landfill always invite disaster. This paper presents a field study of losses of structures in the 1997 earthquake, which were built on lands obtained after reclamation or building structures built on edges of the lakes of the town Jabalpur encroached by the residents.

2 Lakes in Jabalpur

Jabalpur is located at 23° 10′ N 79° 57′ E/23. 17° N 79° 95′ E. The Gond kings invested in their vision to make and preserve the lakes as an important source of water. Some of the lakes, viz., Sangram Sagar, Adhartal, Ranital, Maharaj Sagar, Suraj Tal, Imarti Tal, Ganga Sagar, Devtal, Phool Sagar, Soopatal, etc., were built in this period. It is said that at that time more than 130 ponds or lakes were in the town. Now only nearly 36 ponds are remaining [10]. Most of the lakes, e.g., Cherital, Hathital, Bhanwartal, Madhatal, Shreenath ki talaiya, etc., were reclaimed by the garbage waste, municipal solid waste (MSW), and reclaimed areas are encroached by the peoples or developed by the builders and colonizers [8]. A major part of the population settled on these areas by making non-engineered and poor quality houses to live and these houses become sensitive about earthquakes and settlements of ground.

3 Seismicity of Area

The moderate earthquake struck central India on the morning hours of May 22, 1997, in Jabalpur [6]. In addition to these, several earthquakes in the magnitude range 5.0–6.0 occurred in Peninsular India during the last 70 years. The main shock of the Jabalpur earthquake of May 22, 1997 and its aftershocks are also interpreted to have

Fig. 1 Map showing the location of fault and epicenter points in Jabalpur M.P.

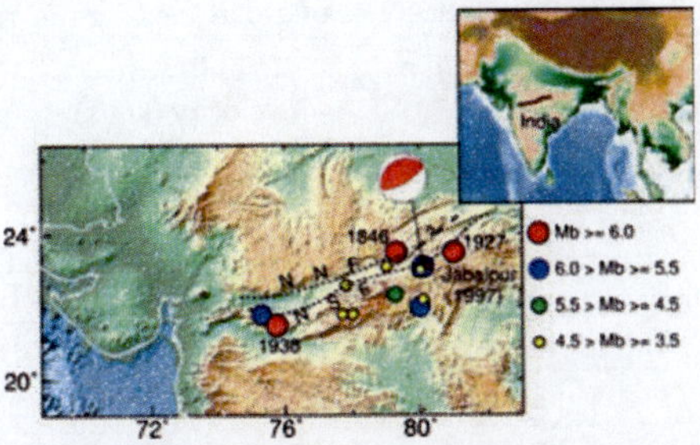

generated as a result of reactivation of the Narmada South Fault (NSF) at the crust-mantle. The epicenter of the 1997 earthquake was at a distance approx. 30 km from the Jabalpur city toward the south-east at Kosamghat [2]. Locations of faults and epicenter in Jabalpur are shown in Fig. 1

4 Intensity Accentuations and Seismic Rigidity

In the present study, the damages accentuations are discussed primarily to be a function of seismic rigidity as originally defined by Reid and subsequently established by Medvedev. The seismic rigidity is the product of the rate of the propagation velocity of longitudinal seismic waves (Vc) and density (σ) [5]. Based on these parameters seismic rigidity value for different grounds in the earthquake-affected area of Jabalpur has been analyzed. On analyzing intensity accentuation for different locations, the physical characteristics of the ground in terms of seismic rigidity have been considered and are shown in Table 1.

5 Assessments of Losses of Structures Built on Reclaimed Lands in Jabalpur

About 100 lakes of the town Jabalpur are reclaimed in the last 50–60 years. The maximum distress due to the 1997 earthquake was experienced in an area of about 400 km^2 which formed the eye of disaster covering the historical town of Jabalpur and surrounding villages. Maximum damages have been observed in the area where the houses were built on reclaimed lands either low lying area reclaimed or the areas which were developed after reclaiming the water bodies [9]. A study was done by authors on such areas affected in the 1997 earthquake and some other structures built on reclaimed land, described below.

Table1 Basic categories of ground and seismic rigidity [5]

S. No	Formations	(Vc) P-wave velocity (km/s)	Density (σ) (gm/cc)	Seismic rigidity (Vc × σ)	Intensity increment factor (after Medvedev, 1965)
1	Reclaimed Fill	0.2–0.6	1.3–1.5	0.26–0.90	2.3–2.9
2	Alluvium	1.8	1.9–2.8	3.42–5.04	1.7–2.8
3	Daccon Basalt	4.5–4.8	2.9–3.0	13.05–14.40	0
4	Lameta Gondwana Sediments	3.8	2.4–2.55	9.12–9.69	0.24–1.8
5	Vindhyan	4.0–4.8	2.5–2.7	10.00–12.96	0.1–0.5
6	Mahakaushal Granite	5.6	2.9	16.24	0
7	Mahakaushal Metasediments	5.9	2.8–2.9	16.52–17.11	0
8	Mahakaushal Volcanics	6.7	3.1	20.77	0

5.1 *Gokalpur Lake*

This is the area developed mainly on the periphery of one of the biggest lakes of the Jabalpur namely Gokalpur Lake. The area having about 2000 houses suffered damages. The earthquake was associated with a rumbling sound with vibrations in the North–South direction.

5.2 *Ranital Area*

This area is located beside Ranital Lake in the central part of the city. In Ranital area lot of structures suffered damages. Shear and tension crack developed in the walls. Himgiri apartment, this four-storied residential building in this area of Jabalpur, tilted on one side and rested against the adjacent building.

5.3 *Gulaua Lake*

The area is developed on the reclaimed landfill and there are more than 1000 houses. In a three-story building, the ground floor damages in the form of open shear-conjugate cracks. Most of the houses were 50–100 years old and made from brick and mud with roofs of wooden frames covered by earthen tiles. Such houses were more vulnerable to earthquake shocks, and as a result, most of such houses have suffered a partial or total collapse [1].

5.4 Overhead Water Tank Beside Gulaua Lake

An overhead water storage tank just beside the Gulaua Lake is a prominent structure. It consists of a 20 m high, 6 m dia. reinforced shaft, with a 6 m high, 8 m dia. water containing superstructure. The tank is founded on fill material. At the time of the earthquake, the tanks were at full capacity and due to dynamic loading following signs of distress were developed.

(i) Three sets of horizontal cracks, 0.30–0.60 m apart, along with a construction joint 0.75 m above the ground, were observed on the western perimeter of the shaft.
(ii) 2 m long hairline cracks with an inclination of 45° emanated from the crack described above [7].

5.5 Gagna Sagar Lake

This area is developed all around the Lake Ganga Sagar. Maximum houses in this area suffered damages severely. A higher grade of damage was noticed along the edges of Ganga Sagar Lake, maximum over thin alluvial cover. The houses over reclaimed lands damage badly.

5.6 Madfaiya Lake and Baksera Talab

There are about 220 houses settled just on the bank of Madfaiya Lake. Maximum houses damage with the cracks of width up to 2 cm. In this locality, the houses situated near ponds suffered more damages. The pronounced damages were observed near Baksera Talab.

5.7 Imarti Talab, Phool Sagar Lake

This area developed beside the lake on land reclamation. All the houses were affected by the earthquake in this area. In the Phool Sagar Lake area, maximum houses were suffered by large damages. So many damages showed in walls, staircase, parapet, etc. The locality lies in an old reclaimed pond area and hence damages were pronounced.

5.8 *Bagha Talab and Supa Tal Lake Area*

These areas are developed beside the Bagha Talab and Supa Tal. The maximum area of this locality settled on the reclamation of the pond area. Maximum houses were affected by major damages. In the Supa Tal area also houses damaged badly.

5.9 *Sanjeevani Nagar Area*

This locality is situated around a big lake known as Shahi Talab. The Geological Survey of India office building was subjected to a varying degree of damages. Though it is a well-constructed structure with columns and beams, filler walls developed cracks at the wall corners. Similarly, the Geological Survey of India colony and Income Tax colony quarters were damaged with various grades. The foundations of the structures have been raised on reclaimed fill [1].

5.10 *Seven-Story Building Collapse Near Cherital Lake in Jabalpur*

A seven-story building recently collapses down on August 17, 2016 in Jabalpur. On preliminary observation, it was found that the building was constructed near Cherital Lake and the site location was reclaimed earlier by garbage and waste material filling.

6 Discussion

On the study of the risk assessment of structures built on reclaimed land in Jabalpur M.P. in earthquake May 22, 1997, and afterward, it has been observed that maximum damages took place in such places where the houses were built on reclaimed lands. Cheri Tal, Madhotal, Bhanwartal, Marhatal, etc., are so many lands reclaimed places in Jabalpur, where damages were at a higher level. Non-engineered land reclamation is always risky because such lands if not properly prepared for construction, possibilities of tilting, and unequal settlements of structures will always be there. The level of disaster may be higher particularly in the earthquake-prone area like Jabalpur.

7 Summary

Since Jabalpur is a city of lakes, so many non-engineered structures were built on these lakes after reclamation. Maximum damages and losses were observed after the May 22, 1997 earthquake in such areas where major non-engineered structures were built on reclaimed lakes areas. The scale and pattern of damages of various parameters of which seismic rigidity of the ground was found to be the primary factor in the present study. Local damage accentuation was due to the low seismic rigidity of the filled ground. Jabalpur is a seismically prone area that falls under zone III and hence all measures should be implemented for safe and stable constructions. An urgent need for a review of seismic zoning, comprehensive seismotectonic evaluation, microzonation of selected sectors, and earthquake awareness programs.

Acknowledgements The authors Sagar Shrivastava, Sanjay Kumar Verma, and S.D.Pimprikar thank the officers and staff of the Geological Survey of India, Jabalpur for providing the necessary literature and data for this study. The authors would like to thanks the professor of Government Model Science College and Jabalpur Engineering College for providing their support in the study of the Geological and Geotechnical aspects of Jabalpur. The authors also thank the India Meteorology Department for providing the required data for this study.

References

1. Chaturvedi RK, Venkatraman NV, Pimprikar SD, Keshava AV, Shrivastava RK et al (2000) Macroseismic survey: the Jabalpur earthquake of May 22, 1997, A Geoscientific Study, Geological Survey of India, Special Publication no. 51:23–161
2. Gupta HK, Chadha RK, Rao MN, Narayana BL, Mandal P, Kumar MR, Kumar N (2000) The Jabalpur earthquake of May 22, 1997. Geol Soc India 50:85–91
3. IS: 1893 (2002) Indian Standard Criteria for Earthquake Resistant Design of Structures, Bureau of Indian Standard, 5th Revision
4. Kanwar SS (2000) Introduction: the Jabalpur earthquake of May 22, 1997, A geoscientific study, Geological Survey of India, Special Publication no. 51:i–iv
5. Mishra PS, Venkatraman NV, Abhinaba R (2000) Geotechnical assessment: the Jabalpur earthquake of May 22, 1997, A geoscientific study, Geological Survey of India, Special Publication no. 51:181–195
6. Rai DC, Narayan JP, Pankaj KA (1997) Jabalpur Earthquake of, May 22, 1997. University of Roorkee, Department of Earthquake Engineering, pp 1–100
7. Soni MK, Roy A, Mishra PS, Tiwari M et al (2000) Macroseismic survey: the Jabalpur earthquake of May 22, 1997. A geoscientific study, Geological Survey of India, Special Publication no. 51:48–49
8. Kumar VS, Saleem A (2016) Effect of earthquake on structures built on reclaimed lands in Jabalpur of India. Int J Civil Eng Technol 7(2):376–386
9. Kumar VS, Saleem A (2016) Need of geotechnical assessments for earthquake resistant structures: a study at Jabalpur. Int Adv Res J Sci Eng Technol Smart Innov Technol Eng Sci SITES, 3(4):31–36. ISSN (Online) 2393–8021 ISSN (Print) 2394–1588
10. Sanjay V (2011) Jabalpur Ke Sarovar Anmol Dharohar, Jal Stroat Sanrakshan Prakoshta, pp 1–46

A Comparative Study on the Mitigation of Seismic Response of Two Adjacent Buildings by Using Viscous Dampers Study

C. L. Mahesh Kumar, K. G. Shwetha, and Shadab Khan

Abstract For a long time, the dampers have been used as a method to mitigate earthquake force in structures. These dampers are traditionally provided inside the buildings. Thereby, in this paper, a different approach is being adapted by carrying out a comparative study between two adjacent RCC buildings of fifteen and ten story buildings in which: (i) No dampers are Provided in Buildings (ii) The dampers are being provided between the two buildings, (iii) dampers are being provided inside the buildings. Further, parameters such as variation in base shear and displacement of the buildings are taken under consideration to draw a proper comparison between the above three cases. The results were obtained for the optimized placement of the damper case after carrying out an extensive trial and error process. Results have shown that connecting the adjacent buildings of different fundamental frequencies by these dampers can effectively reduce the earthquake-induced responses of either building. Providing dampers inside the building gave better results than providing the dampers between the buildings.

Keywords Adjacent connected buildings · Viscous dampers · Seismic response · Displacements · Base shear · Mitigation · Lateral forces

1 Introduction

In this paper, two adjacent RCC buildings of fifteen and ten story buildings are considered and dampers are provided in-between them in case 2 and inside the buildings in case 3 as shown in Figs. 2 and 3, in order to reduce the response of

C. L. Mahesh Kumar (✉) · K. G. Shwetha · S. Khan
Department of Civil Engineering, Nitte Meenakshi Institute of Technology, Yelahanka, Bangalore 560064, India
e-mail: Maheshkumar.cl@nmit.ac.in

K. G. Shwetha
e-mail: Shwetha.kg@nmit.ac.in

S. Khan
e-mail: shadkhan047@gmail.com

T. G. Sitharam et al. (eds.), *Seismic Hazards and Risk*, Lecture Notes in Civil Engineering 116, https://doi.org/10.1007/978-981-15-9976-7_6

this structure by dissipating the seismic energy. From the various studies carried out, it was found that providing dampers can effectively reduce the earthquake-induced responses of building [1–4]. And also reduce both drift-sensitive and acceleration-sensitive damage [5]. The study also showed that linking two buildings allows us to reduce the in-between gap size substantially while structural pounding can be still prevented [6]. Further, it was noted that it is not necessary to provide dampers at all floors but lesser dampers at appropriate locations can also significantly reduce the earthquake response.

The dampers are the devices that divert a portion of forces acting on the structures onto themselves and mitigate those forces by vibration, friction, or movement, etc. In this paper, viscous dampers are used, these devices consist of a viscous fluid which is displaced by being forced through a small opening called as orifice, and hence the work is done by using a part of seismic forces which were acting on the structure. The present paper aims at analyzing seismic response of adjacent buildings with and without being connected by viscous dampers by using response spectrum analysis and also the results obtained between the dampers being provided between the buildings and dampers being provided inside the buildings.

2 Methodology

2.1 Details of Buildings

In this paper, E-Tabs 2016 tool is used for modeling the buildings and E-TABS is an analysis and designing software used in the industry. It stands for "extended three-dimensional analysis of building system." E-TABS is a powerful program that can greatly enhance an engineer's analysis and design capabilities for structures. The details of buildings taken for study are shown in Table 1.

As the pounding effect parameter is not considered in this paper, a safe distance of one meter is considered between the adjacent buildings to satisfy the code provision of IS 1893:2016, clause 7.11.1. T1 is the fifteen-story building and T2 is ten-story building and the plan view of considered buildings is as shown in Fig. 1.

2.2 Damper Data

Stiffness coefficient of the damper is calculated by taking lateral load by displacement for each floor, and then finding an average for all those stiffness's. Damping coefficient is found by considering the buildings to be vibrating freely after the initial excitation, thereby using the formula.

$$C = 2 * m * w * \xi \tag{1}$$

where ξ = (zeta) damping ratio = 1 m = mass of each story.

w = natural frequency of the mode shapes.

Fig. 1 Plan view of fifteen and twenty story buildings

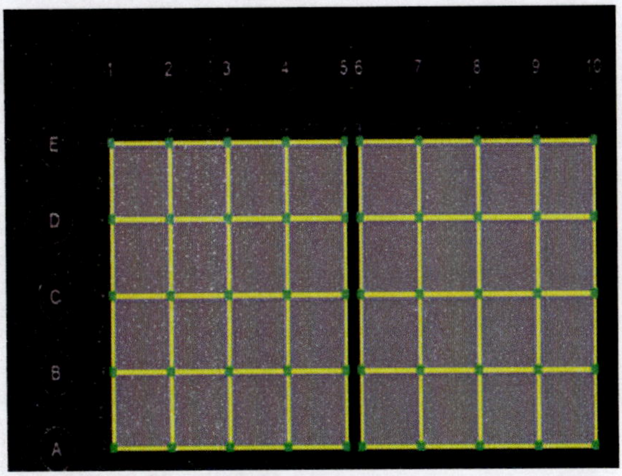

Parameter	Value
Average stiffness coefficient	35,932.3 KN/m
Damping coefficient	19,165.91 KNs/m

2.3 *Placement of Dampers*

In the case two of providing dampers in-between the buildings, dampers were provided in an X bracing and iterations such as dampers between all the floors, alternate floors, etc., were carried out and the optimized placement of dampers was determined. It was found that 24 number of dampers were needed to make the building stable based on code criteria. Four dampers were provided each in 10th, 5th, and 3rd storied level. Also six dampers were provided each in 1st and 2nd storied level.

Therefore, 24 dampers were used in case three as well to establish a proper comparison. The dampers were distributed in the two buildings in 3:2 ratio (that is 14 dampers in T1 and 10 dampers in T2 building), based on the adjacent building height ratio. In case 3 as well, variations such as the dampers being provided in the base floor, two dampers being provided in all floors, etc., were carried out and then providing dampers in inverted v shape in alternate floors was selected as it gave the best result (Figs. 2 and 3).

3 Results and Discussion

The following graphs represent the change in the values of base shear and displacement in the following cases.

No damper case.
Dampers provided in-between the adjacent building (external).
Dampers provided inside the buildings (internal).

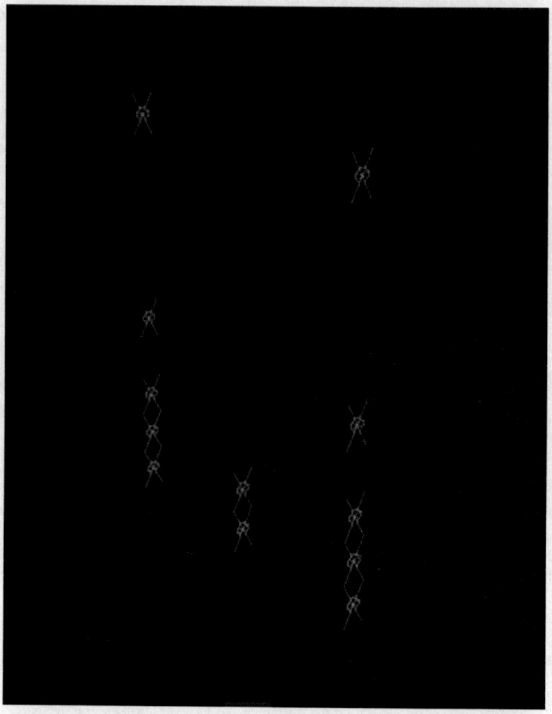

Fig. 2 Dampers provided in-between the buildings (external)

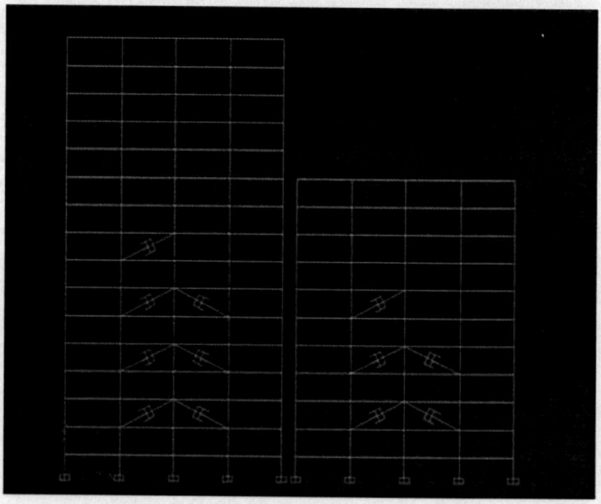

Fig. 3 Dampers provided inside the building (internal)

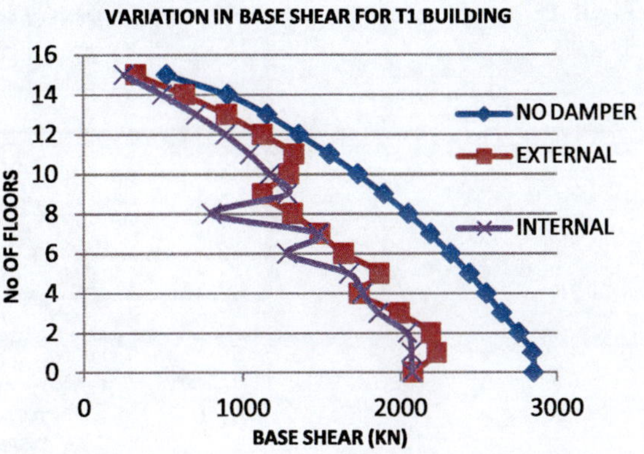

Fig. 4 Base shear for T1 building

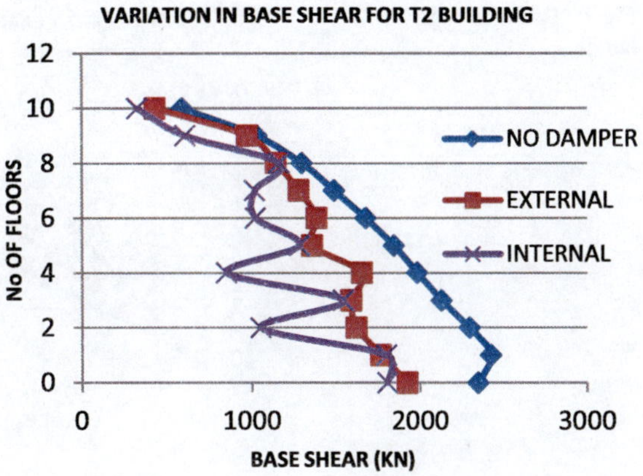

Fig. 5 Base shear for T2 building

3.1 Base Shear

For the T1 building, when compared with no damper case, there is a 26.98% decrease in base shear for case 2, and 27.19% decrease in case 3. For T2 building, when compared with no damper case, there is an 18.17% decrease in base shear for case 2 and 23.19% decrease in case 3. The variations in base shear for T1 and T2 buildings are shown in Figs. 4 and 5.

3.2 Displacements

The displacement value obtained for T1 was 179.27 mm for case 2 and 140.69 mm for case 3. T2 building's displacement value for case 2 was 120.332 mm and for case 3 it was 80.55 mm. For this iteration, the T1 building satisfies the code limitations

Fig. 6 Displacements for T1 building

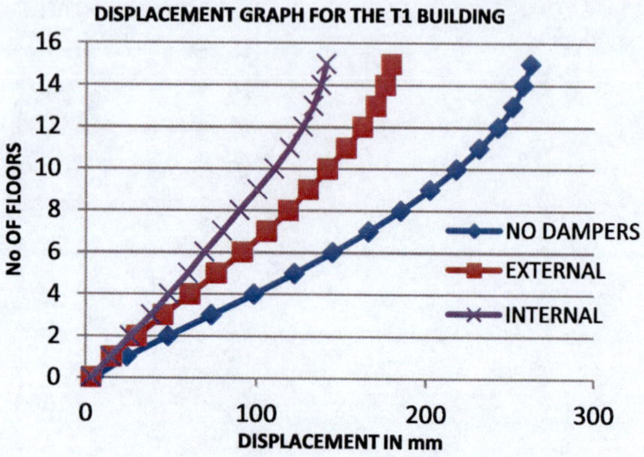

Fig. 7 Displacements for T2 building

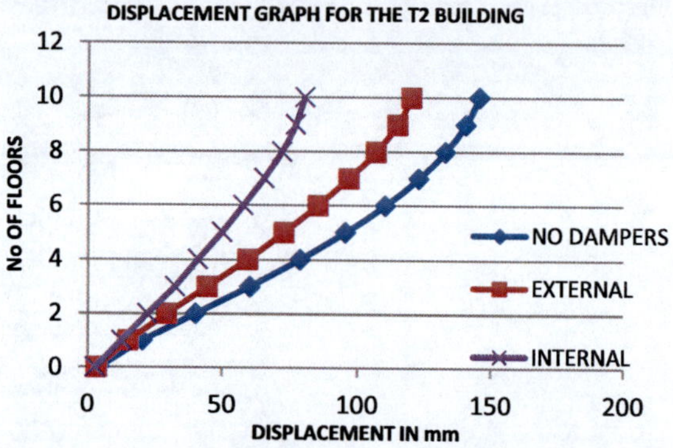

of IS code, whereas the T2 building fails in case 2 by a near 0.3 mm value. The variations in displacments for T1 and T2 buildings is shown in Figs. 6 and 7.

3.3 *Story Velocity*

For the T1 building, when compared with no damper case, there is a 21.52% decrease in story velocity for case 2, and 39.19% decrease in case 3. For T2 building, when compared with no damper case, there is a 35.19% decrease in story velocity for case 2 and 38.12% decrease in case 3. The variations of story velocity for T1 and T2 Buildings are shown in Figs. 8 and 9.

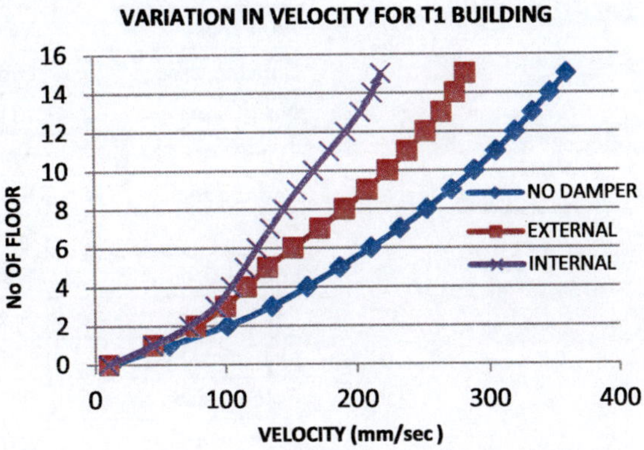

Fig. 8 Story velocity for T1 building

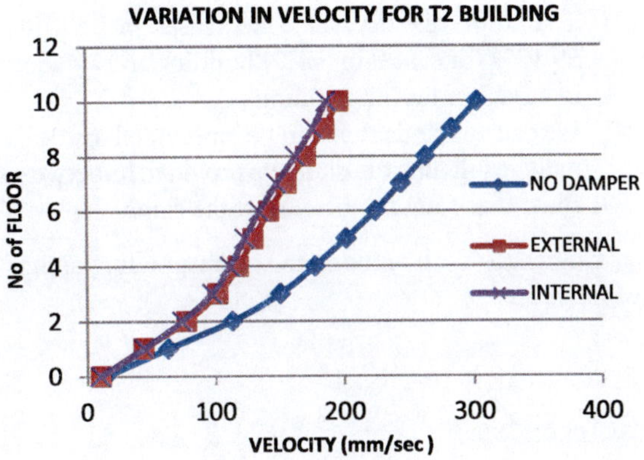

Fig. 9 Story velocity for T2 building

4 Conclusions

From the above graphs, it is clear that dampers provided inside the building give better results than dampers provided in-between the buildings. The results are in agreement with the study carried out that the effects of the viscous damper is lesser on taller building when compared to shorter building [6]. Following inferences can be made from the results above.

(1) The base shear values decrease in both the cases for T1 building and not much variation is noted in the overall reduction which was found to be 27% approximately. For the T2 building, the case 3 gave a 5% more reduction in base shear compared to case 2.

(2) The displacement values decrease in both the iterations, by about 82.93 mm for case 2 and 121.51 mm for case 3 in T1 building and by about 25.38 mm for case 2 and 65.15 mm for case 3 in T2 building.

Table 1 Buildings data

Properties	Values
Structure type	Ordinary moment resisting frame
No. of story	15 story (T1), 10 story (T2)
Typical story height	3 m
Seismic zone	V
Soil	Soft soil
Grade of concrete	M30
Grade of steel (fy)	Fe 500
Slab thickness	175 mm
Column size	500 × 500 mm
Beam size	250 × 500 mm

(3) The story velocity value decreases in both iteration 21.52% for case 2 and 39.19% for case 3 in the T1 building and by about 35.19% in case 2 and 38.12% in case 3 in the T2 building.
(4) We can conclude that dampers provided inside the building give better displacement result than the dampers provided in-between the buildings. But overall base shear was found to be almost the same.

On behalf of all authors, the corresponding author states that there is no conflict of interest.

References

1. Mevada NM, Mevada SV, Patel SB (2018) Seismic response of adjacent buildings connected with non-linear viscous and viscoelastic dampers seismic response of adjacent buildings connected with non-linear viscous and viscoelastic dampers, Nov 2018
2. Uz ME, Hadi MNS (2009) Dynamic analyses of adjacent buildings connected by fluid viscous dampers. WIT Trans Built Environ 104:139–150
3. Yang Z, Xu YL, Lu XL (2003) Experimental seismic study of adjacent buildings with fluid dampers. J Struct Eng 129(2):197–205
4. Qi H, Youlin X (1998) Dynamic analysis of damper-connected adjacent buildings under earthquake. Trans Tianjin Univ 4(2):128–133
5. Del Gobbo GM, Blakeborough A, Williams MS (2018) Improving total-building seismic performance using linear fluid viscous dampers. Bull Earthq Eng 16(9):4249–4272
6. Jankowski R, Mahmoud S (2016) Linking of adjacent three-storey buildings for mitigation of structural pounding during earthquakes. Bull Earthq Eng 14(11):3075–3097
7. Zhao J, Shield C, French C, Posbergh T (2004) 13th world conference on earthquake engineering effect of servovalve/actuator dynamics on displacement controlled testing. Electr Eng 267
8. Bhaskararao AV, Jangid RS (2004) 13th World conference on earthquake engineering seismic response of adjacent buildings connected with dampers. 3143
9. Bureau of Indian Standard (BIS), "Is 1893," Bur Indian Stand, vol 1893, Dec 2016
10. Patil J, Talikoti PRS Effect of gap between building on seismic pounding force, pp 101–108
11. Yaghobzadeh S (2016) Improving the dynamic behavior of adjacent buildings with fluid viscous dampers. Mod Appl Sci 10(12):245

A Parametric Analysis of Adjacent Elevated Service Reservoir with Structural Coupling at Various Locations

C. B. Nayak, S. U. Sayyad, R. R. Khartode, U. T. Jagadale, and S. G. Morkhade

Abstract The idea of linking two adjacent buildings has been effectively applied to diminish the structural response due to strong ground motions. However, some of the linking methods for the elevated service reservoir (ESR) are unsuccessful because of the structural behavior of these structures. During an earthquake, closely spaced high rise ESRs have more chances to collide with each other and hence lead to damage and collapse. In this research work, two types of coupling methods are used to investigate the structural responses considering different staging height of ESR. A nonlinear static analysis (pushover analysis) and nonlinear dynamic (time history analysis) are carried out to check the performance of structure using software SAP 2000 at various locations of ESR. The effectiveness of both coupling methods is examined in terms of base shear, story drift, time period, displacement, velocity, and acceleration. Based on the results, it is concluded that the coupling of ESRs using both the strategy is effective for controlling the seismic response of the structures.

Keywords Coupling · Bracing · Dampers · Analysis · Response · Reduction · SAP2000

1 Introduction

One of the major issues that may lead to failure of ESR is earthquakes. Therefore, the analysis of ESR must be carefully performed so that safety can be assured. During powerful earthquakes, performance of ESR is a significant concern. The ESRs considered lifeline structures as they are used for a basic need like domestic and fire fighting [1–5]. The irregular shape of an ESR, for which most of the mass of liquid, confluent in the upper part of the reservoir makes it more sensitive for the staging of a reservoir to fail especially due to an earthquake [6–10]. In India, structures are constructed very close to each other due to limited land available and unavoidable situations. Hence, to reduce the seismic response of the adjacent high

C. B. Nayak (✉) · S. U. Sayyad · R. R. Khartode · U. T. Jagadale · S. G. Morkhade
Department of Civil Engineering, VPKBIET, Baramati 413133, India
e-mail: cbnnayak@gmail.com

T. G. Sitharam et al. (eds.), *Seismic Hazards and Risk*, Lecture Notes in Civil Engineering 116, https://doi.org/10.1007/978-981-15-9976-7_7

rise and closed structures, different type coupling configurations with different end conditions are used. The concept of coupling in the adjacent building is to allow two dynamically dissimilar structures to exert control forces acts upon one another to reduce the overall response of the system [11–15]. The coupling provided in the adjacent structure may connect only one floor or it may connect to the adjacent floor. A link could be provided at any height and location as per the well-situated structural configurations. In order to inspect the effectiveness of link location on the structural modal properties, seismic analysis of linked ESR is evaluate using bracing and viscous damper using SAP2000. In viscous damper, high flow resistance fluid is used. The velocity of the distortion of a viscous damper is corresponding to force developed. Viscous dampers are alike to shock absorbers in a vehicle that shows the example of the viscous damper, thereby fluid viscous damping is mostly preferred among passive control devices.

From the last several years, many researchers were involved in the analysis and design of the ESR using various methodologies. Some of the literature is reviewed here to know the recent work related to this research work. Kobayashi et al. studied a large-scale seismic test program analyzed using seismic response applied by nonlinear seismic response analysis [16]. Spencer et al. reviewed the rapid recent development of controlled civil structures, incorporation of more modern algorithms and technologies [17]. Chiara et al. studied the damper which is used for building coupling [18]. Patil et al. studied the elevated tank which is used for storage and pressure for a water system. Due to post-seismic functional needs, the safety of the tank is important [19]. Luis et al. studied various coupling for minimizing the frequency of vibrations of the structure [20]. An earthquake and wind analysis of tall structures investigated by Nayak et al. with different shapes [21]. Jagadale et al. studied the seismic response frames with footing taking into account soil–structure interactions [22]. Nayak and Thakare conducted an extensive investigation that consists of an analytical and experimental study on seismic response of the water tank analysis of elevated water tank in Pune district considering Uttarkashi Earthquake and reviewed retrofitting and rehabilitation methods of various types and shapes of ESRs [23–25]. Saha et al. studied the response of long-span truss considering vertical ground motions [26]. Nerkar and Nayak studied the seismic response of ESR by the finite element method [27]. Wadgave and Nayak investigated the performance of tuned mass damper equipped with chimney [28]. Many researchers investigate the different problem of coupled buildings with different approaches. There is no research work was done toward reducing the seismic response of coupled ESRs. Hence, the present research works have been carried out to fulfill this gap.

2 Methodology

The methodology presented in this work is to evaluate the behavior of the structures with the help of coupling provided between two structures. The nonlinear static analysis (pushover analysis) and nonlinear dynamic (time history analysis) are carried

Table 1 Selected ground motions

Sr. no.	Name of ground motion	PGA	Richter magnitude	Place and year	Duration (s)	Scaling factor
1	Bhuj	0.30 g	7.7	2001	75	–

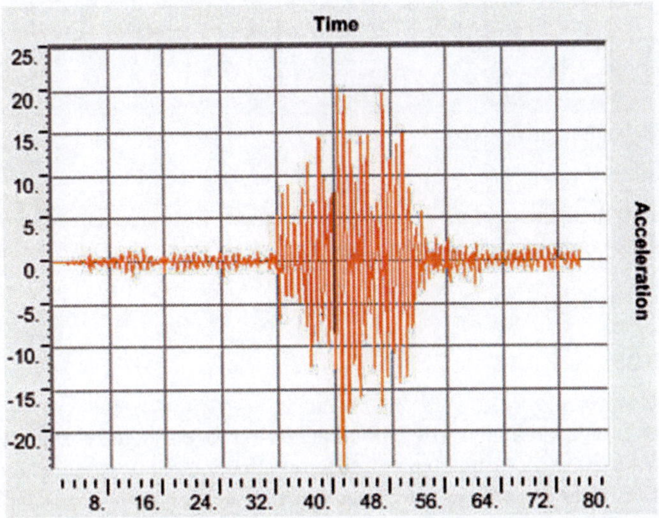

Fig. 1 Acceleration of Bhuj earthquake

out to check the performance of structure using software SAP 2000 [29–33]. The recorded time history is selected as base excitation considering the Bhuj earthquake. The epicenter about 9 km southwest of the village of Chobar in Bhachau taluka of Kutch district of Gujarat, India. Table 1 summarizes the earthquake data, and Fig. 1 shows the acceleration time histories of selected ground motions.

3 Case Study

An RC circular ESR of the capacity of 150 m^3 having a diameter of 8 m and a height of 3.6 m is considered in the present study. It consists of 6 columns of 250 mm diameter supported on RC staging and horizontal bracings of 200 × 350 mm at two levels. The lowest supply level is 10.5 m above ground level with the full water level in the tank. The tank is located in seismic zone III having medium soil conditions. Grade of concrete and steel is M20 and Fe 415, respectively. The density of concrete is 25 kN/m^3. Beam couple link size is 250 × 300 in mm. Damper couple link properties are stiffness 2,000,000 N/m, damping coefficient is 200, and damping exponent is 1.

Fig. 2 Model of ESR in SAP2000

The different mathematical models of closely spaced ESR are analyzed for different coupling link configuration shown in Fig. 2.

4 Results

In this work, nonlinear analysis of ESR is carried out for earthquake zone III using the software SAP 2000. A comparison of different characteristics is made to check the performance of the elevated water tank using different configurations and different conditions as below.

Condition I: Same Capacity and Same Staging Height of ESR

A. **Nonlinear Static Analysis**

Base Shear

In this case, the variation in base shear of structure with a different configuration like without coupling, coupled with beam, and coupled with damper values are presented graphically in Fig. 3.

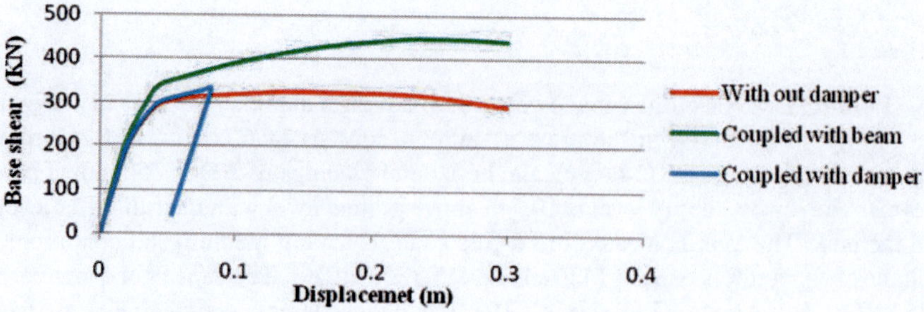

Fig. 3 Pushover capacity curve case

The percentage increase in base shear is 50.15% for coupled with beam and the percentage reduction in base shear is 90.39% for coupled with damper, with respect to without coupling.

Story Drift

The variations in story drift of structure with different configuration without coupling, coupled with beam, and coupled with damper values are presented.

Figure 4 shows the percentage increase in story drift is about 15.23% for coupled with beam and the percentage reduction in story drift is 39.10% for coupled with damper, concerning the base model.

Time Period

The variation in the natural time period of structure with different configurations without coupling, coupled with beam, and coupled with damper values is presented in Fig. 5.

Figure 5 shows the percentage increase in the natural time period is 3.01% for coupled with damper and the percentage reduction in the natural time period is 2.62% for coupled with beam, concerning the base model.

B. Nonlinear Dynamic Analysis

Roof Displacement

The variations in roof displacement of structure with different configuration are presented.

The percentage decreases in roof displacement as shown in Fig. 6 is 21.42 and 32.88% for coupled with beam and coupled with damper, respectively, with respect to the base model.

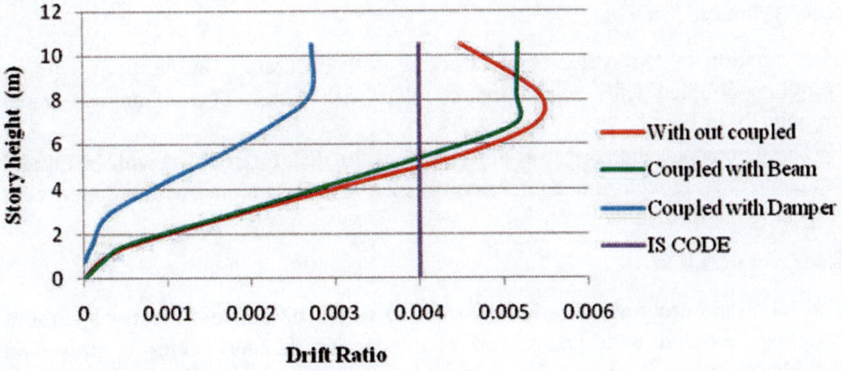

Fig. 4 Story drift curve

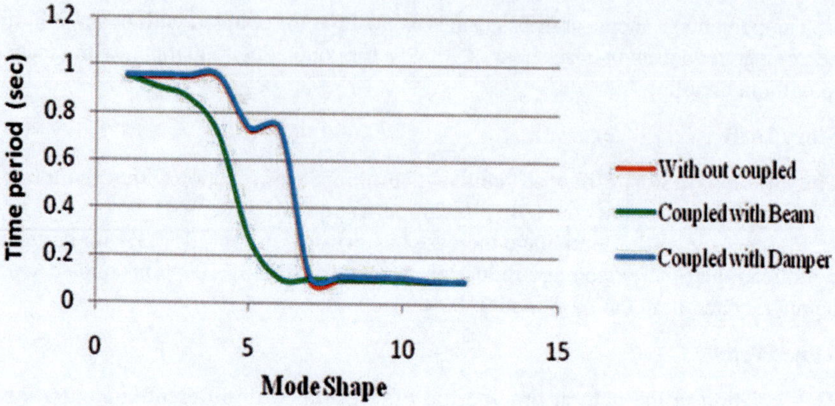

Fig. 5 Time period case

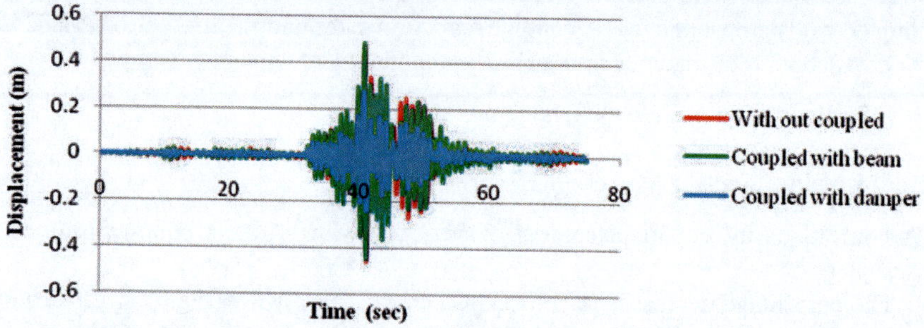

Fig. 6 Roof displacement case

Roof Velocity

The variation in roof velocity of structure with different configurations without coupling, coupled with beam, and coupled with damper values are presented graphically in Fig. 7.

The percentage increase in roof velocity is 30.78% for coupled with beam and the percentage reduction in roof velocity is 21.98% for coupled with damper, with respect to the base model.

Roof Acceleration

The variation in roof acceleration of structure with different configurations without coupling, coupled with beam, and coupled with damper values is presented graphically in Fig. 8.

The percentage reduction in roof acceleration is 1.83 and 31.20% for coupled with beam and coupled with damper, with respect to the base model.

Condition II: Same capacity and different staging height of ESR

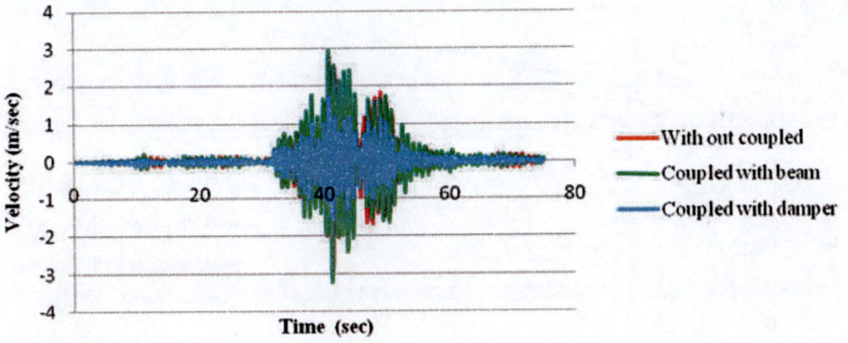

Fig. 7 Roof velocity case

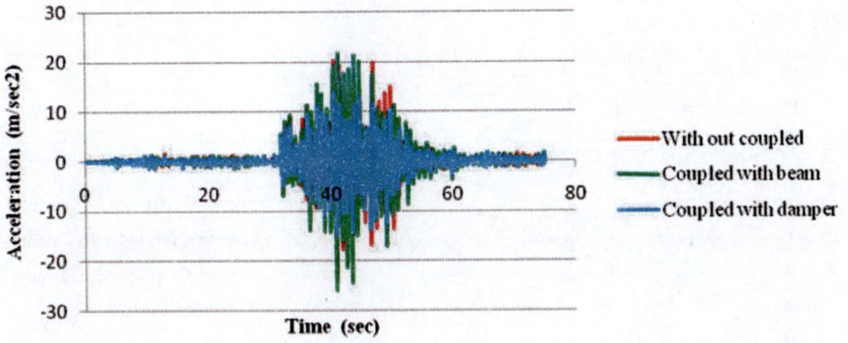

Fig. 8 Roof acceleration case

A. Nonlinear Static Analysis

Base Shear

The variation in base shear of structure with different configuration values is presented in Fig. 9.

The percentage increase in base shear is 49.68% for coupled with beam and the similar base shear for coupled with damper as compared with the base model.

Story Drift

The variation in story drift of structure with different configurations without coupling, coupled with beam, and coupled with damper values is presented graphically in Fig. 10.

The percentage reduction in story drift is 9 and 49.77% for coupled with beam and coupled with damper, with respect to the base model.

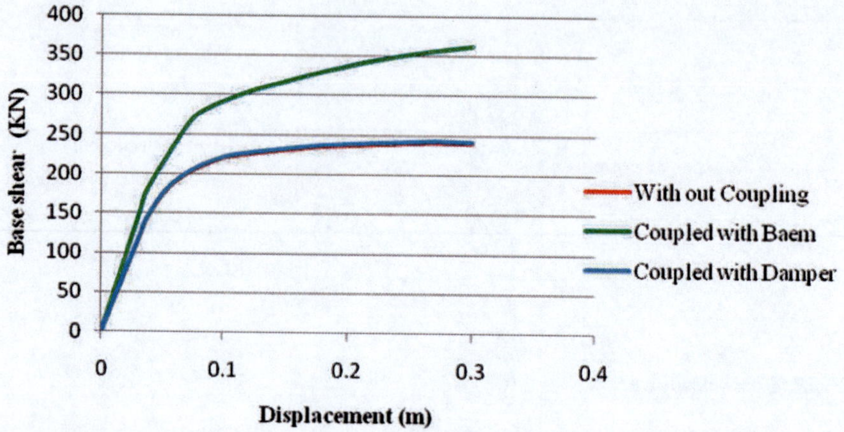

Fig. 9 Pushover capacity

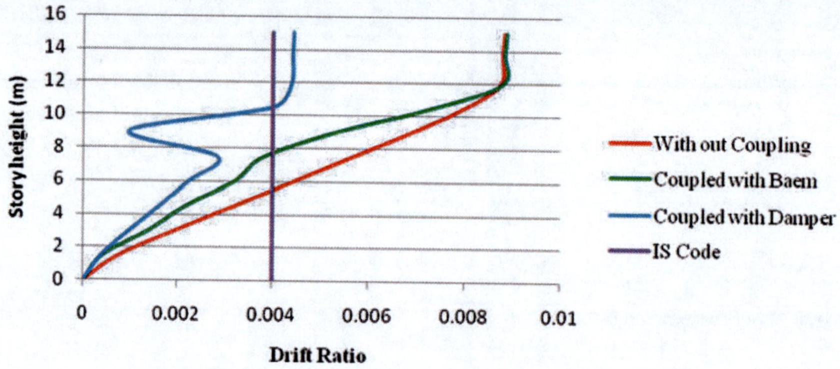

Fig. 10 Story drift

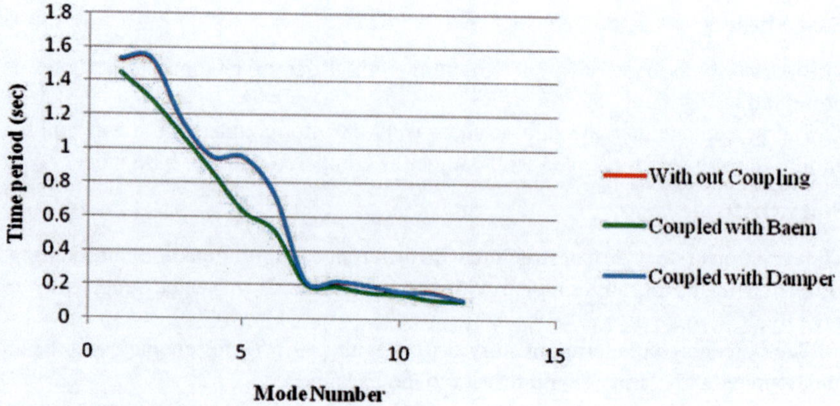

Fig. 11 Time period

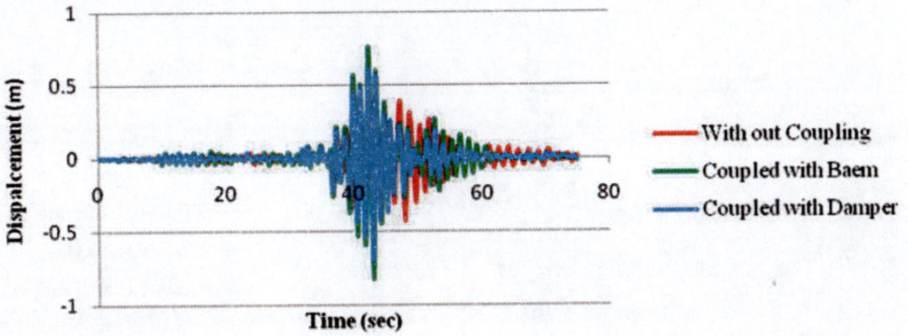

Fig. 12 Roof displacement

Time period

The variations in the natural time period of structure with the different arrangements are presented.

The percentage increase in the natural time period is 4.93% coupled with damper and the percentage reduction in the natural time period is 8.18% for coupled with beam, concerning the base model.

B. **Nonlinear Dynamic Analysis**

Roof Displacement

The variation in roof displacement of structure with different configurations without coupling, coupled with beam, and coupled with damper values is presented graphically in Fig. 12.

The percentage increase in roof displacement is 135.64 and 111.35% for coupled with beam and coupled with damper, with respect to the base model.

Roof Velocity

Figure 13 shows the variation in roof velocity of structure with different configurations.

The percentage increase in roof velocity is 173 and 139% for coupled with beam and coupled with damper with respect to the base model.

Roof Acceleration

The variations in roof acceleration of shown in Fig. 14 with different configurations.

The percentage increase in roof acceleration is 144.97 and 134.02% for coupled with beam and coupled with damper with respect to the base model.

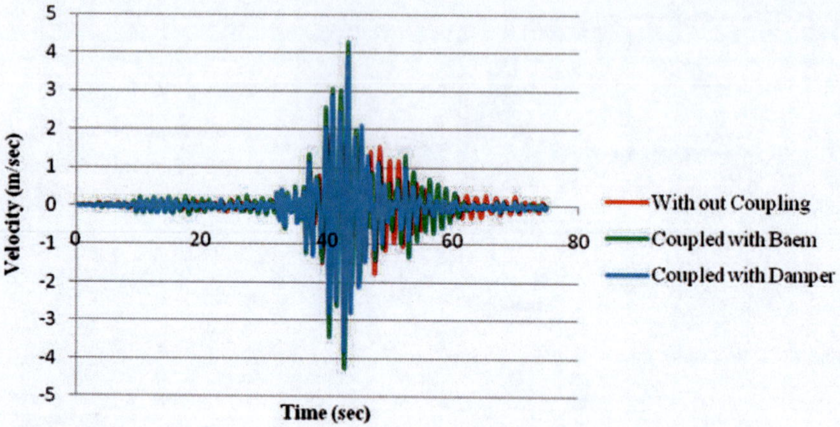

Fig. 13 Roof velocity

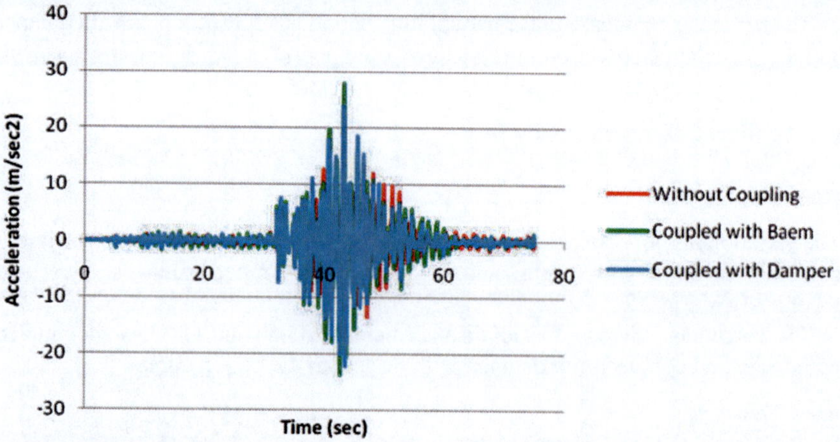

Fig. 14 Roof acceleration

5 Conclusion

In the present work, static and dynamic analysis of ESRs is studied conceding different staging height and two types of coupling systems. The seismic behavior of these ESRs with two case studies of the coupled structure is analyzed using SAP2000. From the above study, the following conclusions are drawn.

Form Condition: I

- From the study conducted in the paper, it is concluded that same height ESR coupled with beam and damper, base shear and story drift is minimum for damper as compared with coupled with beam and base model for static analysis.
- Also, in dynamic analysis, roof displacement, roof velocity, and roof acceleration are minimum for damper coupled structure as compared with coupled beam and base model for dynamic analysis.

Form Condition II:

- The second case study shows a similar base shear capacity of structure for coupled with damper and base model.
- In condition: II dynamic analysis for structure shows a huge difference in roof displacement, roof velocity, and roof acceleration as compared with the base model.
- Form conditions I and II that the coupling method is suitable for the similar height of the structure.
- The energy dissipation of structure is compared with improving performance and comparison data of the structure which has been proved.

References

1. Nayak CB, Thakare SB (2016) Retrofitting and rehabilitation of elevated water tank after investigation of corrosion status by using nondestructive techniques. In: International Congress on Computational Mechanics and Simulation Organized by IIT, Bombay, pp 1244–1247 (2016)
2. Nayak CB, Thakare SB (2016) Investigation of corrosion status in elevated water tank by using nondestructive techniques in Baramati region. In: International conference on construction real estate, infrastructure and project management NICMAR, Pune, pp 1–9
3. Jagadale UT, Nayak CB, Narute G (2018) Structural health monitoring using piezo-ceramics smart material. In: 9th International conference o structural engineering and construction management. University of Paradeniya, Sri Lanka, pp 1–6
4. Morkhade SG, Baswaraj SM, Nayak CB (2019) Comparative study of effect of web openings on the strength capacities of steel beam with trapezoidally corrugated web. Asian J Civil Eng 20(6):1089–1099. https://doi.org/10.1007/s42107-019-00166-6
5. Nayak CB, Thakare SB (2017) Seismic behaviour of existing retrofitted elevated water tank after NDT investigations using SAP 2000. In: The 8th International conference on structural engineering and construction management 2017, Earl's Regency Hotel, Kandy, Sri Lanka, pp 1–6
6. Nayak CB, Vedant, CS (2017) Numerical assessment of probability of detecting defects in civil structures. NICMAR J Constr Manag XXXII(I):28–35
7. Morkhade SG, Kumthekar FP, Nayak CB (2020) Analytical study of steel I-Beam with stepped flanges. Int J Optim Civil Eng 10(2):217–229
8. Jagadale UT, Nayak CB, Mankar A, Thakare SB, Deulkar WN (2020) An Experimental-based python programming for structural health monitoring of non-engineered RC frame. Innov Infrastr Solut 5(1):1–10. https://doi.org/10.1007/s41062-020-0260-x

9. Nayak CB (2020) Experimental and numerical investigation on compressive and flexural behavior of structural steel tubular beams strengthened with AFRP composites. J King Saud Univ Eng Sci. ISSN 1018–3639. https://doi.org/10.1016/j.jksues.2020.02.001
10. Nayak CB, Thorat NS, Thakare SB (2018) Corrosion impact analysis on residual life of structure using cathodic technique and algor simulation software. Eng Struct Technol 10(1):18–26. https://doi.org/10.3846/est.2018.1468
11. Nayak CB, Tade MK, Thakare SB (2017) Strengthing of Beams and Columns using GFRP Bars. Conf Sci Mater Sci Eng 225:1–8. https://doi.org/10.1088/1757-899X/225/1IOP/012144
12. Nayak CB, Jagadale UT, Jadhav KM, Morkhade SG, Kate GK, Thakare SB, Wankhade RL (2020) Experimental, analytical and numerical performance of RC beams with V-shaped reinforcement. Innov Infrastr 6(2):1–13. https://doi.org/10.1007/s41062-020-00363-2
13. Jagtap A, Nayak CB (2020) Corrosion monitoring of RCC structure by using corrosion expansion sensor. In: Pawar P, Ronge B, Balasubramaniam R, Vibhute A, Apte S (eds) Techno-Societal 2018 (2020). Springer, Cham, pp 205–216. https://doi.org/10.1007/978-3-030-16848-3_20s
14. Nayak CB, Jain M, Walke S (2020) Parametric study of dome with and without opening. J Inst Eng (India) Ser A. https://doi.org/10.1007/s40030-020-00447-3
15. Jagadale U, Kharade R, Nayak CB, Deulkar WN (2020) Experimental investigation for damage evaluation of bridges using Piezo-transducers. In: Prashant A, Sachan A, Desai C (eds) Advances in computer methods and geomechanics. Lecture notes in civil engineering, vol 56, pp 319–331. Springer, Singapore. https://doi.org/10.1007/978-981-15-0890-5_27
16. Kobayashi T, Yoshikawa K, Takaoka E, Nakazawa M, Shikama Y (2002) Time history nonlinear earthquake response analysis considering materials and geometrical nonlinearity. Nucl Eng Des 212(1):145–154.https://doi.org/10.1016/S0029-5493(01)00470-8
17. Spencer B, Christenson R, Johnson A, Seto K (2006) Coupled building control considering the effects of building/connector configuration. J Struct Eng 132(6). https://doi.org/10.1061/(ASCE)0733-9445(2006)132:6(853)
18. Chiara P, Andrea B, Alessandra M, Paolo R (2014) Existing structures connected with dampers: state of the art and future developments. In: Second European conference on earthquake engineering and seismology, Istanbul, pp 25–29
19. Patil G, Bhalchandra S (2017) Non-linear time history analysis of elevated water tank. Int J Sci Res Dev 5(8):1–5
20. Luis P, Suzana A, Graciela D (2017) Experimental study of the seismic response of coupled buildings models, X-international conference on structural dynamics. Procedia Eng 199:1767–1772. https://doi.org/10.1016/j.proeng.2017.09.445
21. Nayak CB, Walke S, Kokare S (2019) Optimal structural design of diagrid structure for tall structure. In: International conference on system reliability, quality control, safety, maintenance and management, G. H. Raisoni College of Engineering and Management, pp 263–271. Springer Nature, Singapore. https://doi.org/10.1007/978-981-13-8507-0_39
22. Kare V, Nayak CB, Jagadale UT, Deulkar WN (2019) Earthquake response of 3D frames with strap footing considering soil structure interactions. In: 2nd International conference on advanced technologies for societal applications. In book: Techno-Societal 2018, vol 1, pp 895–904. https://doi.org/10.1007/978-3-030-16848-3_81
23. Nayak CB, Thakare SB (2019) Seismic performance of existing water tank after condition ranking using non-destructive testing. Int J Adv Struct Eng 11:395–410. https://doi.org/10.1007/s40091-019-00241-x
24. Nayak CB, Thakare SB (2017) Analysis and retrofitting of elevated water tank in pune district: by Uttarkashi earthquake. J Eng Technol 6:201–211. https://www.joetsite.com
25. Nayak CB, Thakare SB (2015) A review paper on retrofitting and rehabilitation of elevated water tank. In: International conference on recants developments in engineering and technology, Dr Pauls Engineering College, Tamil Nadu, India, pp 1–7
26. Shaha P, Kamatchi P, Nayak CB (2018) Effect of vertical ground motions on the response of long span roof truss. In: The 16th symposium on earthquake engineering, Department of Earthquake Engineering. IIT Roorkee

27. Nerkar S, Nayak C (2016) Seismic behaviour of elevated storage reservoir by finite element method. Int J Adv Technol Eng Sci 4(1):1188–1197
28. Wadgave P, Nayak C (2016) Vibrational analysis of chimney equipped with strakes and tune mass damper. Int J Adv Technol Eng Sci 4(1):321–333
29. Virkhare D, Vairagade L, Nair V (2015) Pushover analysis of water tank staging. Int Res J Eng Technol 2(4):1786–1805
30. Rofooei F, Attari NK, Rasekh A, Shodja A (2006) Comparison of static and dynamic pushover analysis in assessment of the target displacement. Int J Civil Eng 4(3):212–225
31. IS 1893 (2002) Indian Standard Code Criteria for Earthquake Resistant Design of Structures (Part 1). Bureau of Indian Standard. Manak Bhavan, 9 Bahadur Shah Zafar Marg New Delhi, 110002
32. IS 1893 (2016) Indian Standard Code Criteria for Earthquake Resistant Design of Structures (Part 1). Bureau of Indian Standard. Manak Bhavan, 9 Bahadur Shah Zafar Marg New Delhi, 110002
33. Nayak CB, Narule GN, Surwase H (2021) Structural and cracking of RC T-beams strengthened with BFRP sheets by experimental and analytical investigation. J King Saud Univ Eng Sci. ISSN 1018–3639. In Press

Implementation of Framework for Seismic Resilience Assessment of a R.C. Building

S. Prasanth and Goutam Ghosh

Abstract The need to make the resilient communities was necessary for the developing countries like India due to recent natural and manmade hazards like earthquake, flood, etc. The study has been carried out for a R.C. building (G+10), which was located in India. The vulnerability assessment was made in terms of the probability of exceedance of damage states from slight to collapse limit state as per HAZUS methodology. The resilient structure should have high robustness and rapidity which indicates the minimum damage loss with fast recovery rate respectively after the post-disaster event. The performance and capacity of the building were evaluated by the improved version of static nonlinear analysis given by the Federal Emergency Management Agency (FEMA). FEMA-440 equivalent linearization (EL) is used for the analysis of the buildings for the maximum considered earthquake (MCE) as per the Indian Standard (IS: 1893-2016). The main focus of the study was to address the implementation of the framework for seismic resilience evaluation as per Indian seismic conditions like zone factor, soil conditions, importance factor and response reduction factor. Based on the damage state the repair and replacement cost were estimated which directly related to the direct and indirect losses evaluation due to post-event. The drop-in functionality of the building and the loss of resilience (LOR) was compared with two recovery functions namely trigonometric and exponential.

Keywords Resilient · Recovery · Functionality

1 Introduction

Due to increase in natural disaster over the recent period, there is a need of resilient society towards the post-disaster events. In order to make the structure more resilient,

S. Prasanth (✉) · G. Ghosh
Civil Engineering Department, Motilal Nehru National Institute of Technology Allahabad, Prayagraj 211004, India
e-mail: prasanth@mnnit.ac.in

G. Ghosh
e-mail: goutam@mnnit.ac.in

T. G. Sitharam et al. (eds.), *Seismic Hazards and Risk*, Lecture Notes in Civil Engineering 116, https://doi.org/10.1007/978-981-15-9976-7_8

the concept of resilience-based design emerges in recent times. This concept is based on how the society should be well prepared to withstand the post-disaster events in terms of "recovery" with optimum cost. The implementation of resilience concept was in practice in recent days and it requires lots of economic and social data which will have direct impact on the estimation of the resilience. Two major factors like recovery time and loss estimation play a major role in resilience quantification of the structure. The framework to evaluate the resilience was proposed by many researches. The framework to estimate the resilience for the hospital community system based on damage state of the structure was proposed with different rehabilitation strategy [1]. In addition to above for the loss estimation, the theoretical equation was proposed [2]. The assessment of damage state of building in terms of its functionality to evaluate the seismic resilience capacity based on performance-based methodology was proposed [3]. The global damage state was assessed for the multi-span bridge network by based on HAZUS methodology which was followed by the estimation of loss curves with respective restoration time using Monte Carlo simulation [4]. The new classification structure functionality into three functional states such as unsafe to occupy (Nocc), safe to occupy but not functional (OccLoss) and fully functional (OccFull) was proposed which is based on inspection and closure probabilities after the disaster events [5]. Using performance-based design, the damage loss estimation of eight-storey OMRF building and loss of resilience were found based on functionality of the building and the enhancement of resilience capacity using FRP technique was reported [6]. The increase in the seismic performance of RC ordinary moment resisting frames (OMRFs) which was retrofitted by FRPs and steel X-braces was noted due to the ductility effect [7]. A three-storey frame that was strengthened with FRP around its columns was investigated for its resilience capacity and the increase in strength of the columns with marginally increasing their stiffness was noted [8]. The general framework for the optimal resilience and cost-based prioritization of interventions was reported [3]. Based on the service-level different type of recovery functions was developed with the recovery indicator for assessing the rapidity [9]. Based on above review, very few researches were done in terms of infrastructural resilience and only limited work has been identified in India. Since Indian code is silent over the seismic resilience of the structures, this study shows how to implement the framework to evaluate the resilience capacity of the building based on Indian Standard (IS: 1893-2016). The work comprises of vulnerability assessment in terms of fragility curves which will be followed by the estimation of the seismic resilience capacity of the structure using two recovery functions.

2 Concept of Resilience-Based Design

Resilience has many definitions given by many authors. The tendency of the structure to withstand and to recover back with optimum time and cost is termed as resilience. Resilience is based on two main parameter such as rapidity and robustness. Rapidity implies the ability of the structure to recover back in timely manner based on its

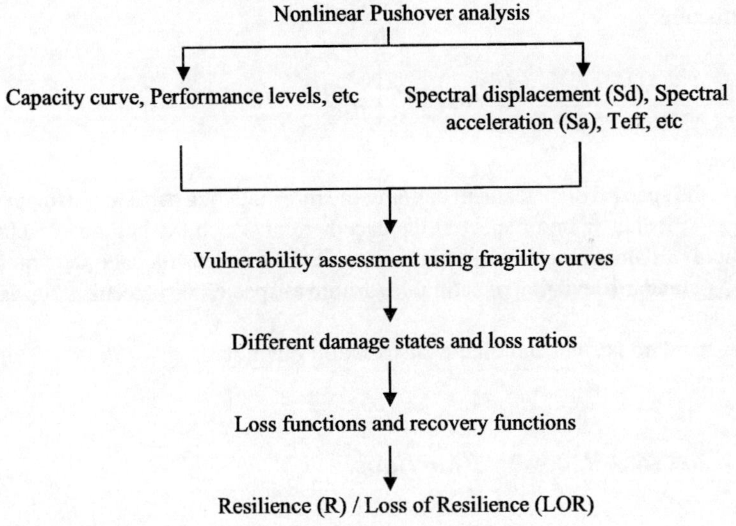

Fig. 1 Flowchart to evaluate the seismic resilience of the structure

functional needs, whereas robustness shows the capacity of the structure to withstand the effect of post-disaster event. The rapidity of the structure was measured in terms of "recovery functions" and the robustness were in terms of "loss function".

$$\text{Resilience}(R) = \int \frac{Q(t)}{T_{LC}} \text{d}t \tag{1}$$

where

$$Q(t) = 1 - \{L(I, T_{\text{RE}}) \times [H(t - t_{\text{OE}}) - H(t - (t_{\text{OE}} + T_{\text{RE}}))] \times f_{\text{rec}}(t, t_{\text{OE}}, T_{\text{RE}})\} \tag{2}$$

$L(I, T_{\text{RE}})$ is the loss function, $f_{\text{rec}}(t, t_{\text{OE}}, T_{\text{RE}})$ is the recovery function, T_{RE} is the recovery time after the occurrence of the event, t_{OE} is the initial time of the event occurred, T_{LC} is the control period for the estimation of resilience (Fig. 1).

2.1 Vulnerability Assessment

The nonlinear pushover analysis was done to find the performance level of the building. The maximum considered earthquake of PGA = 0.36 g based on IS:1893-2016 was considered for the vulnerability assessment. The HAZUS methodology was used to assess the vulnerability of the building by means of developing fragility curves in terms on spectral displacement (S_{d}) as it is resembles the real behaviour of the structure.

$$P\left({}^{\text{ds}}/_{S_\text{d}}\right) = \phi\left[\frac{1}{\beta_{\text{ds}}}\ln\left({}^{S_\text{d}}/_{S_{\text{d,ds}}}\right)\right] \tag{3}$$

where

S_d—the spectral displacement defining the threshold of a particular damage state.

$S_{\text{d,ds}}$—median value of spectral displacement at which the building reaches the threshold of damage states and can be calculated by various damage state models.

β_{ds}—standard deviation of natural logarithm of spectral displacement for damage state (ds).

ϕ—standard normal cumulative distribution function.

2.2 *Loss and Recovery Functions*

All the direct (L_D) and indirect losses (L_{ID}) comes under the loss function. Direct losses includes repair and rehabilitation cost of the damaged structural elements, whereas indirect losses is due to unfunctionality of the structure such as occupancy state. The direct loss was found in terms of dimensionless cost term L_D.

$$L_\text{D} = \sum P_\text{E}(\text{DS} = K) \times r_\text{K} \tag{4}$$

Due to lack of details on the indirect loss like causalities, relocation time, etc., the indirect losses are ignored, and hence, in this study, the attention was given to the direct loss that includes the damage of the structural elements.

The path of post-event restoration of the building was described in terms of recovery function (f_{rec}). In this study, the variation in the resilience was observed using trigonometric and exponential recovery functions [1].

Exponential function:

$$f_{\text{rec}}(t, t_{\text{OE}}, T_{\text{RE}}) = \exp\left[-\frac{(t - t_{\text{OE}})(\ln 200)}{T_{\text{RE}}}\right] \tag{5}$$

Trigonometric function:

$$f_{\text{rec}}(t, t_{\text{OE}}, T_{\text{RE}}) = 0.5\left\{1 + \cos\left[\Pi\frac{(t - t_{\text{OE}})}{T_{\text{RE}}}\right]\right\} \tag{6}$$

3 Reference Case Study: Mid-Rise Building [10]

A mid-rise building of G+10 storey was taken for the study [10]. The building has a plan area of 18 × 18 m as shown in Fig. 2 with total height of 44 m. Based on IS: 1893-2016 [11], the importance factor (*I*) and response reduction factor (*R*) were taken as 1.0 and 5.0, respectively. The medium soil (Type-B as per IS: 1893-2016) conditions and a peak ground acceleration (PGA) of 0.36 g were considered for the study. Since the pushover analysis has not been introduced in the Indian standards, the response spectrum curve was converted into Ca and Cv from ATC-40 [12] which was used in the analysis. The values of Ca and Cv for respective soil type and zone conditions were found and as shown in Table 1. The performance level of the building such as immediate occupancy (IO), life safety (LS) and collapse prevention (CP) under flexure was found by performing nonlinear pushover analysis using SAP2000 V20 based on the hinge formation as per ATC-40 and FEMA-356 [13] . The yield and ultimate displacement were found by bi-linearization of the capacity curve plotted between base shear and the displacement.

The pushover analysis was performed based on FEMA-440 equivalent linearization method [14] and the results were shown in Table 2. To assess the vulnerability of the structure, the fragility curve was develop in terms of spectral displacements of the structure (Fig. 3). The percentage of cumulative probability of exceedance of damage states such from slight to collapse was shown in Table 3.

The percentage of probability of exceedance at each damage state is quite less as structure remains in IO level. This is due to the ductile capacity of the structure. To estimate the resilience capacity of the structure, the rapidity and robustness have to

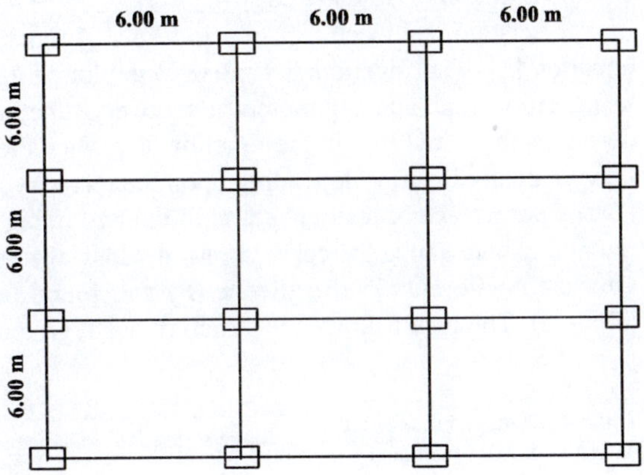

Fig. 2 Typical plan view

Table 1 Ca and Cv for zone V and soil type II

Importance factor (I)	Response reduction factor (R)	Ca	Cv
1.0	5	0.072	0.098

Table 2 Pushover results with importance factor (I) = 1.0

S. no.	Response reduction factor (R)	Performance point (kN)	Displacement (mm)	Spectral displacement (mm)	Ductility demand
1	5	2056.12	54.849	43.392	1.291

Table 3 Percentage of cumulative damage probability of exceedance of damage state

Damage	Cumulative percentage (%)
Slight	9.47
Moderate	3.20
Extreme	0.66
Collapse	0.11

be considered. As mentioned earlier, the robustness was found as loss function which indicates the direct economic loss ratios. In order to estimate the loss function, the direct economic losses were found using equation [7] from discrete damage probabilities. The damage ratio (rK) was taken from HAZUS-MR4 technical manual [15] and direct damage loss ratio was shown in Table 4. The rapidity of the structure in terms of recovery functions used. In this study, two recovery functions such as exponential and trigonometric functions were used and their effect in variation of resilience capacity was noted. In order to understand the implementation of framework to evaluate the seismic resilience, the recovery time was taken as 65 days which is based on the type and functionality need of the structure. The time of occurrence of event was taken as 50th day of the control period which depends on the designer and the availability of resources [6].

The seismic resilience capacity in terms of functionality $Q(t)$ was found using equation [1]. The functionality curve was plotted against the control time period using exponential and trigonometric recovery functions (Figs. 4 and 5). The graph describes the initial loss in the functionality when the event occurs at 50th day and how gradually it varies depending upon the recovery function used.

The percentage of resilience was found by finding the area under the curve. The remaining area above the curve shows the loss of resilience (LOR) of the structure after the post-event. The resilience (R) was found for the two recovery functions (Table 5). The result shows the structure has high resilience capacity which is due

Table 4 Damage ratios based on HAZUS-MR4 technical manual for buildings

Damage state (ds)	Damage ratio (rK)	Direct damage loss ratio (L_D)
Slight	0.10	0.022
Moderate	0.40	
Extreme	0.80	
Collapse	1.00	

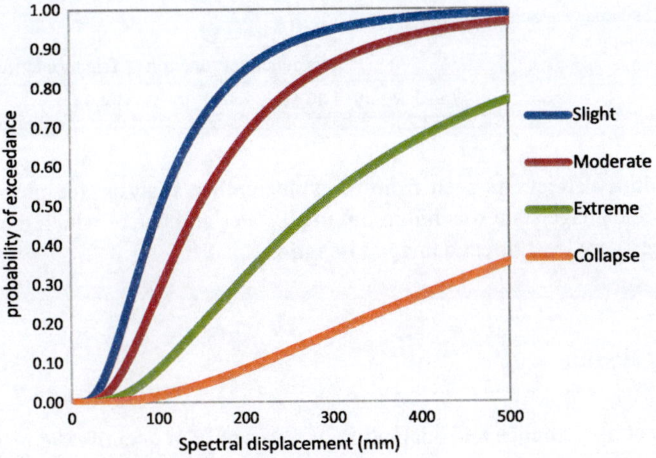

Fig. 3 Fragility curve based on spectral displacement

Fig. 4 Functionality curves using exponential recovery function

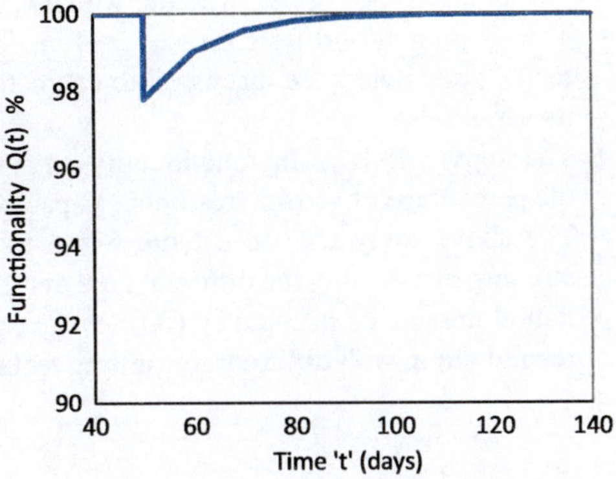

Fig. 5 Functionality using trigonometric recovery function

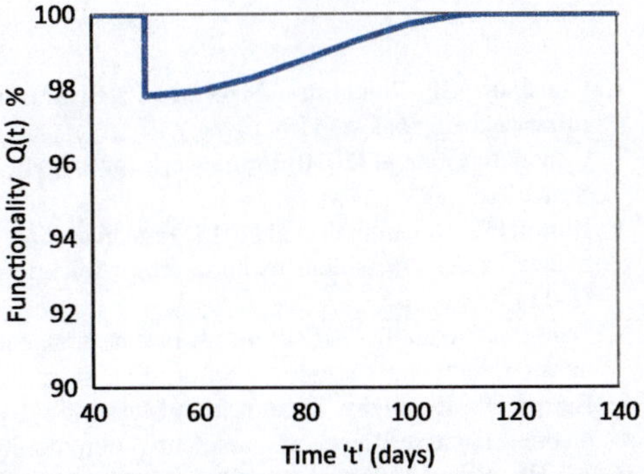

Table 5 Resilience capacity ($T_{RE} = 65$ days)

I/R values	Resilience (%)	
	Exponential function	Trigonometric function
$I = 1\ R = 5$	99.59	98.98

to the ductility effect. As seen from the vulnerability analysis, the probability of exceedance damage state was minimum in all states and due to which the structure experiences very least direct damage loss ratios.

4 Conclusion

The concept of resilience and adaptation of framework to quantify the resilience of the structure based on Indian site conditions was focused in this study.

- The nonlinear pushover analysis based on FEMA-440 was focused to estimate the damage ratios of the structure which is used to find the loss functions.
- The adaptation of different recovery functions such as exponential and trigonometric functions were discussed to estimate the seismic resilience capacity and its variations.
- The steps to develop the functionality curve was explained which was used to find the percentage of seismic resilience capacity of the structure.
- The above study will be extended in a focus to increase the resilience of the building by adopting the different type of rehabilitation strategies such as rehabilitation immediate occupancy (IO) level, rehabilitation life safety (LS) level and rebuild along with different retrofitting techniques.

References

1. Cimellaro GP, Reinhorn AM et al (2010) Seismic resilience of a hospital system. Struct Infrastruct Eng 6(1–2):127–144
2. Cimellaro GP et al (2010) Framework for analytical quantification of disaster resilience. Eng Struct 32:3639–3649
3. Burton HV, Deierlein G et al (2015) Framework for incorporating probabilistic building performance in the assessment of community seismic resilience. J Struct Eng 142(8):C4015007 (1–11)
4. Gehl P, D'Ayala D (2018) System loss assessment of bridge networks accounting for multi-hazard interactions. Struct Infrastruct Eng. https://doi.org/10.1080/15732479.2018.1434671
5. Burton HV, Deierlein G et al (2017) Measuring the impact of enhanced building performance on the seismic resilience of a residential community. Earthquake Spectra 33(4):1347–1367
6. Ali HS, Saeed MS et al (2016) A preliminary case study of resilience and performance of rehabiliated buildings subjected to earthquakes. Earthquakes Struct 11(6):967–982

7. Mahini SS, Hadigheh SA, Setunge S (2015) Seismic resilience of retrofitted reinforced concrete buildings. In: Proceedings of the second international conference on performance-based and life-cycle structural engineering (PLSE 2015). Brisbane, Australia
8. Zou XK, Teng JG, De Lorenzis L, Xia SH (2007) Optimal performance-based design of FRP jackets for seismic retrofit of reinforced concrete frames. Compos B Eng 38(5–6):584–597
9. Shinozuka M, Chang SE et al (2003) Resilience of integrated power and water systems. Multidiscipl Center Earthquake Eng Res 155:22–43
10. Prasanth S, Ghosh G (2019) Seismic vulnerability assessment of a reinforced concrete building located in India. Int J Innov Technol Explor Eng 8(11S):310–314
11. IS:1893 (2016) Criteria for earthquake resistance design of structures, (part-1)
12. ATC-40 (1996) Seismic evaluation and retrofit of reinforced concrete buildings: Applied Technology Council (1996)
13. FEMA-356 (2000) Pre-standard and Commentary for Seismic Rehabilitation of Buildings, Federal Emergency Management Council, Washington DC, USA
14. FEMA-440 (2005) Improvement of nonlinear static seismic analysis procedures. Federal Emergency Management Agency
15. HAZUS, MR4 Technical Manual (2003) Multihazard Loss Estimation Methodology. Department of homeland society, Washington DC, USA

Seismic Source Characterization for Amaravati Capital Region, Andhra Pradesh, India

M. Madhusudhan Reddy, K. Rajasekhara Reddy, and G. Kalyan Kumar

Abstract The majority of Indian cities having populated one million or more fall under seismic zones III, IV and V according to the seismic microzonation manual (2011). However, the expected damages and loss of lives due to earthquakes, quite different within the specified seismic zone, mostly the level of damage during the earthquakes depend on the regional geological, geotechnical, geomorphological characteristics and the size of the earthquake influenced by different seismic sources like faults, lineaments and shear zones. The prime objective of the current study is to develop the seismotectonic map to the proposed capital region of Andhra Pradesh (AP) state, India. The selected area for capital to the state of Andhra Pradesh has situated in zone III as per the seismic code IS 1893:2016. In the present study to understand the seismicity and to identify different seismic sources, the study area extended about 300 km radial distance from Velgapudi (16° 52′ N, 80° 51′ E) village as a centre. A total number of 219 events and 22 seismic sources (17 faults, two lineaments and three shear zones) identified with the study area. The spatial distribution of the seismicity shown in the seismotectonic map with the size of 0.5 Mw interval starting from 3.0 Mw is followed by different seismic sources.

Keywords Amaravati capital region · Andhra Pradesh · Seismicity · Seismic sources

1 Introduction

Among all different types of natural hazards, the hazard due to earthquakes is capable of creating severe damages to the structures and loss of human lives. Most of the casualties take place during the earthquakes due to the rapid urbanization

M. Madhusudhan Reddy (✉) · K. Rajasekhara Reddy
Department of Civil Engineering, Koneru Lakshmaiah Education Foundation, Green Fields, Vaddeswaram, Andhra Pradesh, India
e-mail: Madhu.mrc@gmail.com

G. Kalyan Kumar
Department of Civil Engineering, National Institute of Technology Warangal, Warangal, Telangana, India

T. G. Sitharam et al. (eds.), *Seismic Hazards and Risk*, Lecture Notes in Civil Engineering 116, https://doi.org/10.1007/978-981-15-9976-7_9

and misleading the land-use guidelines while construction. India has witnessed some major earthquakes like Assam (Mw 8.7) in 1897, Kangra-Himachal Pradesh (Mw 8.6) in 1905, Bihar-Nepal (Mw 8.4) in 1934, Assam-Tibet (Mw 8.7) in 1950, Uttarkashi-Uttaranchal (Mw 6.6) in 1991, Latur-Maharashtra (Mw 6.3) in 1993, Jabalpur-Madhya Pradesh (Mw 6.0) in 1997, Chamoli-Uttarakhand (Mw 6.8) in 1999, Bhuj-Gujarat and Muzaffarabad-Kashmir (Mw 7.6) in 2001 and 2005 [1,2]. As per the Bureau of Indian Standard (BIS: 1893–2016), complete Indian lands have been divided into four seismic zones (zone II, III, IV and V) based on the peak ground acceleration (PGA) and comprehensive intensity scale (CIS-64) [1]. Most of the recent earthquakes caused severe structural damage and resulted in the loss of human lives in urban areas of India [3]. Unfortunately, natural hazards are unpredictable and unavoidable in nature. Hence, the microzonation studies will help in reducing property loss during earthquakes. After the Bhuj earthquake (Mw 7.6) in 2001, the Government of India under the ministry of the earth science department has decided to conduct seismic microzonation studies all over India. It has been recognized and mentioned in the seismic code IS 1893:2002 that not even a single part of Indian landmass is safe against seismic hazard, and according to the seismic microzonation manual (2011), it has been recognized that a total of twenty-seven Indian cities fall under the vulnerable zone due to the earthquake. Unfortunately, the level of damage is not possible to predict with available zone PGA values because the damage level completely depends on the geology, and the conflicting seismic response may change location to location and region to region for the same size of the earthquake, even if there are small changes in geology of the site or region [1]. The Chamoli (1999) and Bhuj (2001) earthquakes are the best example in this perspective. The epicentre of Chamoli and Bhuj earthquakes is situated approximately 250 km away from the Delhi and Ahmedabad cities, resulted in moderate to medium failures in the structures of both cities. Hence, the identification of seismic sources and characterization of the past seismicity are the most important steps in the evaluation of microzonation studies.

In the present study, an attempt is made to identify different seismic sources and spatial distribution of seismicity for the Amaravati capital region of Andhra Pradesh state. To characterize the seismicity, the study is extended about 300 km radial distance from the Velgapudi village (situated approximately centre to the proposed capital region) with the latitude of 16° 53′ 09″ N and longitude of 80° 51′ 02″ E shown in Fig. 1. The selected study area has covering a total of five Indian states approximately. According to the seismic hazard map of India, the selected circular region fall in seismic zone III and II in Peninsular India (PI) and classified as low- to medium-risk zone from the earthquakes. However, the PI states also experienced with major earthquakes like Coimbatore (MW 6.3) in 1900, Koyana (MW 6.3) in 1967, Ongole (MW 5.4) in 1967, Bhadrachalam (MW 5.7) in 1969, Killari (MW 6.4) in 1993, Latur (MW 6.1) in 1993, Jabalpur (MW 5.7) in 1997 and Bhuj (MW 7.7) in 2001, whereas in the eastern part of the South India shield and Bay of Bengal region, the seismic activity is being increased due to the subduction of Burma plate towards the Bay of Bengal and resulting in the development of new faults and reactivation of older faults [4, 5]. As per the Indian seismic code (IS 1893:2016), 60% of Indian land

Fig. 1 Location map of the study area from India map

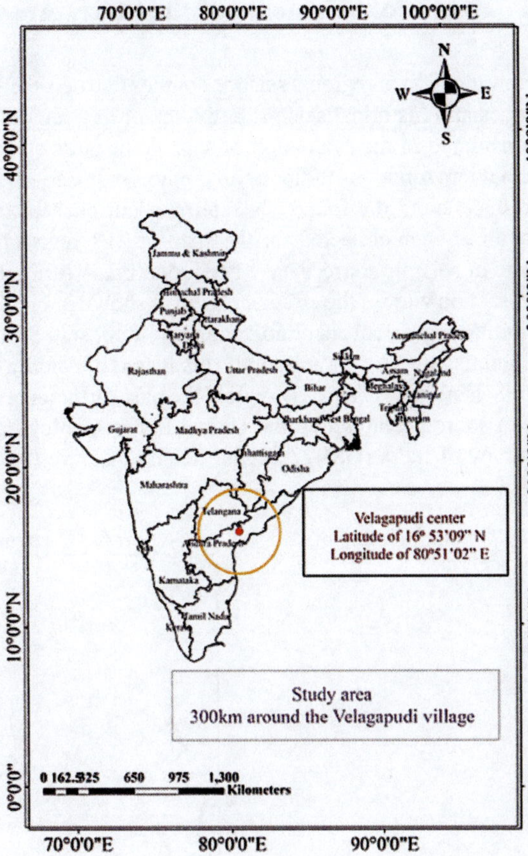

falls under vulnerable to earthquake, in which 26% of landfall in zone III. Amaravati capital region is one of the largest populated urban cities in the world with its suburbs Vijayawada and Guntur being the 3rd and 24th most densely populated cities in the world. Geographically, the capital region of Andhra Pradesh is situated between the Guntur and Krishna districts of Andhra Pradesh. Topographically, Andhra Pradesh's new capital region has very close relations with Vijayawada city (Krishna district). As per the seismic microzonation manual (2011) and as per the BIS (1893:2016), the Vijayawada city and adjoining areas are recognized as low- to medium-risk zone under seismic zone III with PGA value of 0.16 g.

2 Details of Seismicity of the Study Area

Identification of regional seismic sources like neo-tectonics (faults/lineaments), shear zones and historical seismic activities of any region is the most important features considered in the evaluation of seismic hazards analysis, ground response analysis and microzonation studies of any region in future [6]. However, in the present study, to understand the frequency rate of occurrence of the earthquakes, the seismicity data has been collected for the span of 219 years (1800 AD to 2019 AD), and all type of seismic sources have been identified within the selected circular region, the extraction view of the circular region as shown in Fig. 2 with the state boundaries. The circular region covers almost four adjoining states (southwest parts of Maharashtra, Chhattisgarh, Odisha and south Telangana) of Andhra Pradesh. In order to understand the seismicity of the study region, the past earthquake details collected from different data source centres like the United States Geological Survey (USGS), Geological Survey of India (GSI), Amateur Seismic Centre (ASC) and Indian Meteorological

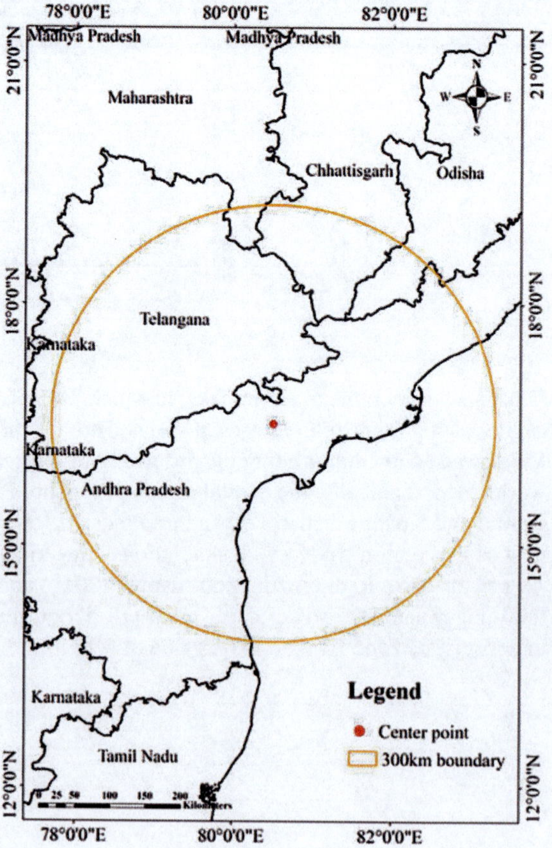

Fig. 2 Extraction view of the study area

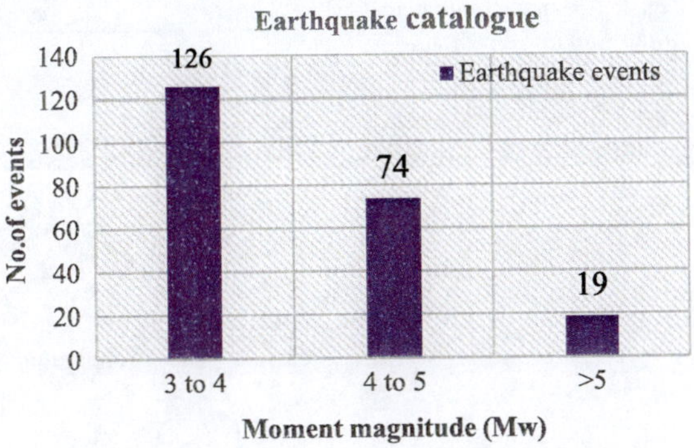

Fig. 3 Statistics of the earthquake catalogue

Department (IMD). The collected earthquake catalogue is a combination of both pre-instrumental and instrumental periods.

Due to the lack of consistency in seismicity measuring devices, the collected earthquake data contains the information about earthquake size in different scales including intensity (*I*), surface-wave magnitude (Ms), body-wave magnitude (Mb), local magnitude (ML) and moment magnitude (Mw). In order to avoid the confusion in earthquake size, all different sizes of earthquakes are converted into moment magnitude (Mw). First, to convert the body-wave and surface-wave magnitudes into moment magnitude, the empirical relationships given by the Scordilis (2006) have been used [7, 8], whereas Scordilis (2006) not suggested any empirical relation to converting the local magnitude (ML) into Mw. However, to convert such magnitudes into Mw, the equation suggested by Heaton et al. (1986) has been used. The earthquake data belongs to the pre-instrumental era (intensity-scale values), converted into moment magnitude using the relationship ($Mw = (2/3)\, I + 1$); here, "*I*" represents intensity scale [9]. After declustering, earthquake catalogue has been prepared with 219 main events starting from 3.0 Mw [10, 11]. Figure 3 shows the statistics between the cumulative earthquakes to the magnitude range. Another most important step in seismic hazard evaluation is identification faults. In the present study, data of different seismic sources has been collected from the Seismotectonic Altals of India (SEISAT 2000). The Geological Society of India has prepared 43 seismotectonic maps on 1:1 million scales and published with the name Seismotectonic Atlas of India and its environs (2000) [12]. To develop the seismotectonic map for the Amaravati capital region, suitable seismotectonic maps are considered (total four sheets), and separate shapefiles for each seismic source were created after georeferencing using Geographic Information System (GIS) software. To prepare the seismotectonic map, all different shapefiles superimposed on a single layer and the seismotectonic map are developed with the 17 faults, two lineaments and three shear zones as shown in Fig. 4 [13].

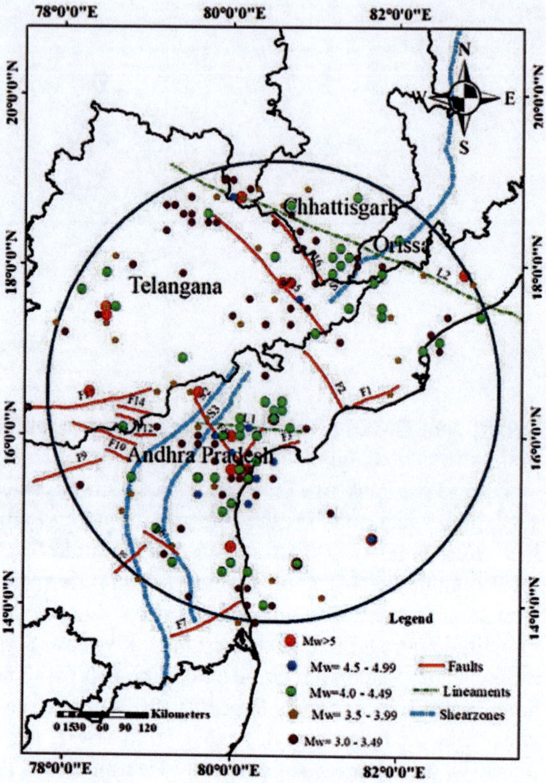

Fig. 4 Seismotectonic map of the Amaravati capital region

3 Results and Discussion

Seismically, Andhra Pradesh state located in peninsular India and has been categorized as low- to moderate-risk zone. However, the state has experienced many noticeable earthquakes in the past. In the current study, the earthquake catalogue has been prepared with 219 main events starting from 3.0 Mw. The strongest earthquake of this region recorded in the Kichanapalle-Gollagudem area with 5.7 Mw in 1969, and this event caused considerable damage in Bhadrachalam and adjoining areas, and it is also referred to as the Bhadrachalam earthquake. The focus of the Bhadrachalam earthquake (5.7 Mw) lies at depth of 25 km and located roughly 197 km from east-southeast of Kakinada and 209 km to the Vishakhapatnam south. The earthquake catalogue contains a total of 82 events between the magnitude range of $3.0 \leq Mw \leq 3.49$, 44 events between $3.5 \leq Mw \leq 3.99$, 62 events between $4.0 \leq Mw \leq 4.49$, 12 events between $4.5 \leq Mw \leq 4.99$ and 19 events $Mw \geq 5.0$. The adjoining seismicity of a particular fault is treated as associated with the same fault. In such a case it has been identified, the Gundlakamma fault prominent active fault in this region because

60 plus events associated with the Gundlakamma fault. To prepare the seismotectonic map of the proposed capital region, the details of fault, lineaments and shear zones are collected from Seismotectonic Atlas of India and its environs (SEISAT 2000). The seismotectonic map is prepared with 219 events and 21 seismic sources. The details of seismic sources like starting, ending latitudes and longitudes with the length of the particular seismic source are summarized in Table 1. Out of all different faults, the Nekkantivagu fault is the shortest fault and Kolleru lake fault is the longest fault in this seismic region. Other major seismic sources like Karempudi-Nakirekallu lineament, Musi lineament, Nallamala, Cuddapah eastern margin and Sileru shear zones identified in this region.

Table 1 Details of the seismic source which present within the 300 km circular region (SEISAT 2000)

Id. no.	Name of the seismic source	Starting		Ending		Length in km
		Latitude	Longitude	Latitude	Longitude	
F1	Vasishta-Godavari fault	18.3	91	18.2	90.6	43.7
F2	Kolleru lake fault	18.1	90.5	20	89.3	246.0
F3	Nizampatnam-Nagayalanka fault	17.7	89.9	17.6	89.6	33.7
F4	Addanki-Nujividu fault	17.5	89.1	17.8	89.4	46.1
F5	Gundlakamma fault	17.3	89.3	18.2	88.5	131.1
F6	Badvel fault	16.2	88.3	16.5	87.8	62.9
F7	Karakambadi-Swarnamukhi fault	15.6	89.1	15.1	88.2	111.4
F8	Papaghani fault	16.3	87.9	16	87.5	54.0
F9	Gani-Kalva fault	17.4	87.3	17.1	86.4	101.2
F10	Atmakur fault	17.4	87.6	17.6	87.3	38.8
F11	Bhavanasi river fault	17.7	87.6	17.9	87.1	57.4
F12	Rudravagu fault	17.8	88	17.8	87.7	31.8
F13	Nekkantivagu fault	18	87.7	18	87.5	21.2
F14	Nallavagu fault	18.8	87.9	18.1	87.5	88.5
F15	Raichur-Nagarkurnool fault	18.4	87.9	18.1	86.1	193.0
F16	Kinnerasani-Godavari fault	19.6	89.8	20.7	88.3	198.7
F17	Godavari valley fault	19.8	90.1	20.8	89.3	139.0
L1	Karempudi-Nakirekallu lineament	16.6	80.8	16.1	79.5	139.0
L2	Musi River lineament	17.2	83.7	19.14	79.2	520.3
S1	Sileru shear	17.5	82.7	20.7	81.16	392.3
S2	Nallamala shear	16.2	80	12.9	79.1	354.8
S3	Cuddapah eastern margin shear	16.7	80.2	13.7	79.5	340.4

It has been concluded that the seismotectonic map of the Amaravati region can be used as basic information in further seismic hazard studies like an evaluation of seismic hazard analysis and microzonation studies. The seismotectonic details can also useful to estimate the Mmax of this region in the feature.

References

1. Mohanty WK, Walling MY (2008) Seismic hazard in megacity Kolkata, India. Nat Hazards 47(1):39–54
2. Rao KS, Satyam DN (2007) Liquefaction studies for seismic microzonation of Delhi region. Curr Sci 92(5):646–654
3. Ganapathy GP, Rajawat AS (2012) Evaluation of liquefaction potential hazard of Chennai city, India: using geological and geomorphological characteristics. Nat Hazards 64(2):1717–1729
4. Reddy PR, Chandrakala K (2004) Seismicity in and around Ongole, Andhra Pradesh-an appraisal. J Indian Geophys Union 8(2):143–146
5. Satyam ND, Towhata I (2016) Site-specific ground response analysis and liquefaction assessment of Vijayawada city (India). Nat Hazards 81(2):705–724
6. Anbazhagan P, Sitharam TG (2008) Seismic microzonation of Bangalore, India. J Earth Syst Sci 117(2):833–852
7. Scordilis EM (2006) Empirical global relations converting M_S and m_b to moment magnitude. J Seismolog 10(2):225–236
8. Vipin KS, Anbazhagan P, Sitharam TG (2009) Estimation of peak ground acceleration and spectral acceleration for South India with local site effects: probabilistic approach. Nat Hazards Earth Syst Sci 9(3):865–878
9. Scordilis EM (2005) Globally valid relations converting Ms, mb and MJMA to Mw. In: Meeting on earthquake monitoring and seismic hazard mitigation in Balkan Countries, NATO ARW, Borovetz, Bulgaria, pp 11–17
10. Gardner JK, Knopoff L (1974) Is the sequence of earthquakes in Southern California, with aftershocks removed, Poissonian? Bull Seismol Soc Am 64(5):1363–1367
11. Uhrhammer RA (1986) Characteristics of northern and central California seismicity. Earthquake Notes 57(1):21
12. Geological Survey of India, Dasgupta S, Narula PL, Acharyya SK, Banerjee J (2000) Seismotectonic atlas of India and its environs. Geol Survey India 34
13. Reddy M, Konda RR, Kumar GK, Asadi SS (2020) Site characterization and evaluation of seismic sources for Amaravati region. Int J Geotech Earthquake Eng 11(1):71–86

Seismic Performance Assessment of Reinforced Concrete Building with Precast Shear Wall

Pijush Shil, Shubham Singhal, Ajay Chourasia, and Ravindranatha

Abstract To avoid exceptional seismic effects on structure during the severe earthquake, it is desirable to increase the lateral load resistant capacity of the structure, which may be done by providing reinforced concrete shear walls in the building. In recent times, precast RC shear walls are more prevalent in practice, where is the understanding of seismic behavior of such walls has not new studied adequately. In the present study, proposed building models are G+5 story reinforced concrete (RC) frame building with and without precast RC shear wall which were modeled in SAP 2000. Non-linear pushover analysis and incremental dynamic analysis are performed to assess and compare the seismic performance of both the systems. Fragility curves are developed for different damage states, i.e., slight, moderate, extensive, and complete damage based on HAZUS methodology. Fragility curves gave the probability of damage at the corresponding damage state, which is useful in seismic assessment of precast RC buildings with precast shear wall.

Keywords Precast shear wall · Pushover analysis · Incremental dynamic analysis · Seismic collapse assessment · Fragility curves · Shear link

1 Introduction

Historical record shown that the effect of earthquake is impendent and devastating. It caused great destruction of life and economy. During an earthquake, structure experiences horizontal distortion. Due to which different kind of structural damage occurs such as cracks, drift, deformation, crushing, and buckling [1, 2]. Day by

P. Shil (✉)
Manipal Institute of Technology, MAHE Manipal, Manipal 576104, India
e-mail: piyushshil2014@gmail.com

S. Singhal · A. Chourasia
CSIR—Central Building Research Institute, Roorkee 247667, India

Academy of Scientific and Innovative Research (AcSIR), Ghaziabad 201002, India

Ravindranatha
Manipal Institute of Technology, MAHE Manipal, Manipal 576104, India

T. G. Sitharam et al. (eds.), *Seismic Hazards and Risk*, Lecture Notes in Civil Engineering 116, https://doi.org/10.1007/978-981-15-9976-7_10

day world population growth is increasing, which creates pressure to construction industry. Therefore, speedy and quality construction technique need to develop. Many researcher presently they study about precast concrete prefabricated structures and introduced this allover world as a substitute of conventional cast-in-situ concrete because of its quality and rate of construction speed [3, 4]. Research-based study those are related with precast concrete, very much focus on structural component during an earthquake and performance of various structural joint. RC shear wall act as a vertical cantilever [5]; it has large lateral stiffness and transfer the lateral load to foundation, due to which RC shear wall widely used as a lateral load resisting component in buildings specially in seismic zone area [6, 7]. By using proper materials and RC shear wall, it will be possible to prevent severe damage caused by earthquake even though structural element such as beam and column are in poor condition [8, 9]. Analytical approach has been used to described response of proposed building structure. According to ATC40, Hazus technical manual, FEMA356 [10, 11] guideline developed fragility curves using pushover analysis as well as incremental dynamic analysis (IDA) results. Although, it requires several steps to perform IDA under suitable ground motion. (1) choosing suitable ground motion, (2) using appropriate algorithms, (3) introducing proper interpolation, (4) summarization, (5) Behavior of structure in limit state, (6) probabilistic seismic hazard assessment (PSHA) [12–14]. To evaluate the performance of structure, at first need to find performance point, where response spectrum curve and capacity spectrum curve are intersecting to each other for selected ground motion. The objectives of this study are as follows

- To embellish IDA and pushover procedure.
- To illustrate fragility curve, arise by IDA and push over analysis.
- To find performance of precast shear wall building, while joints of precast panels are modeled using shear link.

2 Methodology

2.1 Model and Ground Motion Records

Two models are prepared in SAP2000 v20 computer software. Type-1 model is reinforced concrete building without shear wall and Type-2 with precast shear wall. All geometrical and structural details are shown in Table 1. Design of considering structural model done as per guideline suggest by IS 1893-2016. To model type-2 structure in SAP2000, considered shear link as precast joint [8, 9, 15]. The model which is without precast shear wall has fundamental time period of first mode which is $T_1 = 0.7143$, and model with precast shear wall has fundamental time period of first mode is $T_1 = 0.437$. In addition to conduct incremental dynamic analysis (IDA) need to set target spectra and collect suitable ground motion. In this study 6, number of suitable ground motion historical data has been selected from PEER database those have magnetite in between 6.5

Table 1 Architectural and structural details

<table>
<tr><th>Details</th><th colspan="2">Structure without precast shear wall</th><th colspan="2">Structure with precast shear wall</th></tr>
<tr><td>Number of stories</td><td colspan="2">G+5</td><td colspan="2">G+5</td></tr>
<tr><td>Height of building in meter</td><td colspan="2">20</td><td colspan="2">20</td></tr>
<tr><td>Maximum length along transverse direction (X-axis) (m)</td><td colspan="2">25.180</td><td colspan="2">25.180</td></tr>
<tr><td>Maximum length along longitudinal direction (Y-axis) (m)</td><td colspan="2">9.5</td><td colspan="2">9.5</td></tr>
<tr><td rowspan="2">Beams size in mm</td><td>B_1, B_2, B_3, B_4, B_6</td><td>230 × 500</td><td>B_1, B_2, B_3, B_4, B_6</td><td>230 × 500</td></tr>
<tr><td>B_5</td><td>300 × 500</td><td>B_5</td><td>300 × 500</td></tr>
<tr><td rowspan="2">Column size in mm</td><td>C_1, C_2, C_3, C_8</td><td>450 × 450</td><td>C_1, C_2, C_3, C_4</td><td>450 × 450</td></tr>
<tr><td>C_5, C_6, C_7</td><td>450 × 600</td><td>C_5, C_6, C_7</td><td>450 × 600</td></tr>
<tr><td rowspan="2">Loading</td><td>Dead load</td><td>Self-weight
Floor finish</td><td>Dead load</td><td>Self-weight
Floor finish</td></tr>
<tr><td>Live load</td><td>3.5 kn/m^2</td><td>Live load</td><td>3.5 kn/m^2</td></tr>
<tr><td>Software used</td><td colspan="2">SAP2000 v20, Autocad, SeismoSignals, SeismoMatch</td><td colspan="2">SAP2000v20, Autocad, SeimoSignals, SeismoMatch</td></tr>
<tr><td>Ground motion</td><td colspan="2">Selected from PEER</td><td colspan="2">Selected from PEER</td></tr>
</table>

and 7 and moderate distance shown in Table 2. After that scaled them with proper scaling factor and performed number of time-history analysis systematically for each recorded data. To construct IDA curve need to considered intensity measured (IM) and damage measured (DM) [1]. In present study, spectral acceleration at fundamental time period for 5% damping, Sa $(T_1, 5\%)$ considered as an IM and inter-story drift ratio considered as a DM.

Table 2 Selected ground motion

Sl. no.	RSN no.	Event	Station	Magnitude	PGA (g)
1	12	Kern county	LA—Hollywood Stor FF	7.36	0.0422
2	15	Kern county	Taft Lincoln school	7.36	0.1588
3	68	San Fernando	LA—Hollywood Stor FF	6.61	0.2247
4	164	Imperial Valley-06	Cerro Prieto	6.53	0.1682
5	280	Trinidad	Rio Dell Overpass FF	7.2	0.0277
6	286	Irpinia_Italy-01	Bisaccia	6.9	0.09553

2.2 Hazus Method

Hazus manual gives appropriate guidelines to prepared fragility curve as per data has been collected, which is involves by lognormal functions and helps to explain probability of structure reaching or exceeding various structural and non-structural damage states such as slight, moderate, extreme, and collapse state for target spectral displacement. Fragility curve shows uncertainty, variability, and collapse probability of structure integrated with capacity curve and selected ground motion [16]. Results collected from pushover analysis, means peak lateral drift and corresponding lateral force or base shear, both are needed to convert in terms of spectral displacement and spectral acceleration to developed fragility curve corresponding to any spectral displacement. If P represent the probability of damage states for a particular spectral displacement, then

$$P\left[\frac{ds}{S_d}\right] = \emptyset\left[\frac{1}{\beta_{ds}}\ln\left(\frac{S_d}{\overline{S}_{d,ds}}\right)\right] \tag{1}$$

where $\overline{\mathbf{S}}_{\mathbf{d,ds}}$= Threshold spectral displacement for particular damage state, *ds*.

β_{ds} = Standard deviation of the natural logarithm of spectral displacement for particular damage state, *ds*. It represents the combined variability of capacity property, demand, and uncertainty in damage state threshold.

$\emptyset$= the standard normal cumulative distribution function.

S_d = Spectral displacement of the building.

Development of fragility curves from analysis result is based on spectral displacement defined by Barbat et al. [17] (Table 3). Median spectral displacement as per type-1 and type-2 structure analysis results, yield displacement (S_{dy}), and ultimate spectral displacement (S_{du}) values shown in Table 4.

Table 3 Damage state definition as per Barbat et al. (2006)

Damage state	Spectral displacement
Slight damage	$0.7\ S_{dy}$
Moderate damage	S_{dy}
Extreme damage	$S_{dy} + 0.25(S_{du}-S_{dy})$
Complete damage	S_{du}

Table 4 Median damage state threshold for the RC school building from simulation

Description of building	Capacity control points		Damage state threshold values $S_{d,ds}$			
	S_{dy}	S_{du}	Slight	Moderate	Extreme	Collapse
Type-1	27	71	18.9	27	38	71
Type-2	21	53	14.7784	21.112	29.112	53.112

Table 5 Damage state variability beta for fragility β_{ds}

Damage states	Slight	Moderate	Extreme	Collapse
β_{ds} for type-1	0.70	0.74	0.86	0.98
β_{ds} for type-2	0.66	0.73	0.72	0.73

Beta represents values of variability for different damage state with respect to three different parameters such as capacity curve $\beta_{C,}$ demand spectrum variability β_{D}, and threshold of each damage state β_{Tds}. As per HAZUS guideline, beta values can be directly calculated from Tables 5.2, 6.5, 6.6, and 5.7 of HAZUS manual. In present study, proposed structural building comes under medium rise building and high seismic code design considered (Table 5).

After finding all required perimeter substitute in above Eq. (1) to get the probability for different damage state at any specific spectral displacement which is called fragility curve.

3 Non-linear Static and Non-linear Dynamic Analysis

To accomplish the present study, IDA and pushover analysis are carried out in one G+5 building structure which is design for seismic Zone V, importance factor 1 and medium type soil located in Nepal guideline followed by IS1893-2016.

3.1 Non-linear Static Analysis

Pushover analysis has been accomplished referred by FEMA 356, ATC 40 using SAP2000 v20. Geometrical details of both the structure are shown in Fig. 1.

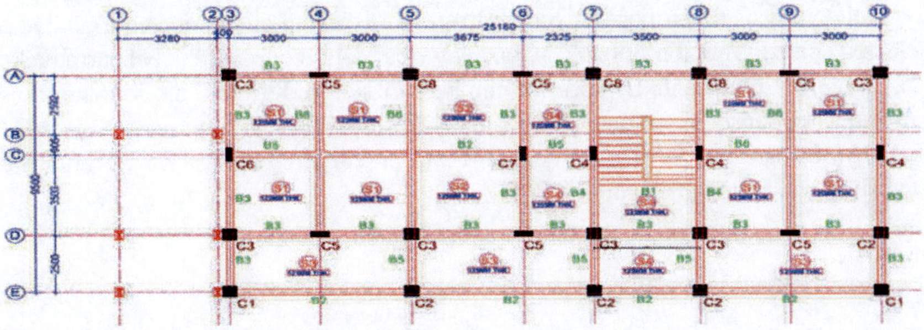

Fig. 1 Plan of proposed building structure (type-1 and type-2)

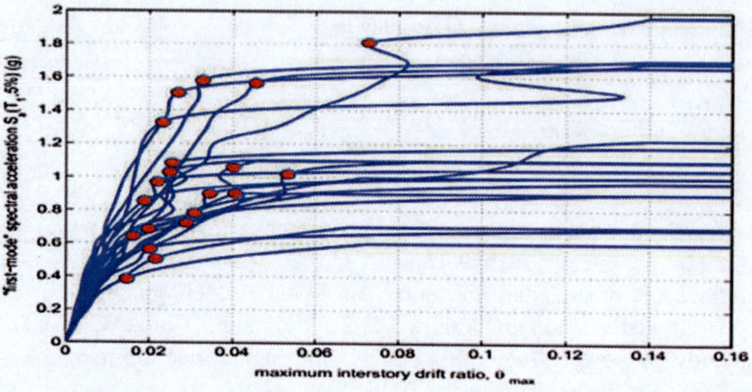

Fig. 2 IDA for different earthquake ground motion [12]

ATC-40 established relationship between base shear and peak displacement is shown in Fig. 2 to explain the performance levels of structure against lateral force. Figure shows point A is origin; B yielding point of structural members; C shows ultimate strength of structural members; line AB indicates elastic behavior structural member because its linearity characteristics; line BC indicates plasticity characteristics, but BC divided by three different levels, such as immediate occupancy (IO), life safety (LS), and collapse prevention (CP). Residual strength of members is shown by point D. CD line indicates maximum ductile capacity of structural members as a result ductile failure of longitudinal bars. Point E represents ultimate failure.

3.2 *Incremental Dynamic Analysis*

Estimation of mean annual frequency (MEF) and global dynamic instability of structural system is challenging task; to meet all these needs, incremental dynamic analysis is recent developed analytical method (referred by FEMA P-695). In which method dynamic analysis of proposed structural model is carried out subjected to suitable selected scaled ground motion (GM) as historical data recorded for real earthquake [12]. Figure 2 represents IDA for 20 numbers of selected ground motion, and dots are shows collapse preventions (CP) limits. To make IDA, it requires to select IM and DM. Considered S_a (T_1, 5%) as a IM and maximum inter-story drift ratio as a DM shown in following figure.

4 Results and Discussion

4.1 Non-linear Static Analysis (Pushover Analysis)

4.1.1 Structural Model Without Shear Wall (Type-1) Analysis Results Are as Following

The pushover curve for proposed structural model RC building without shear wall is shown in Fig. 3. Initially, up to a certain stage curve is linear which indicates elastic behavior of structural member such as beam and column. With increasing of pushing load structural member comes under inelastic range. As a result, curve changed its slope and continuously reducing. Figure 4 shows target displacement of 127 mm and base shear 4220 kN.

Fig. 3 Pushover curve (Push-X)

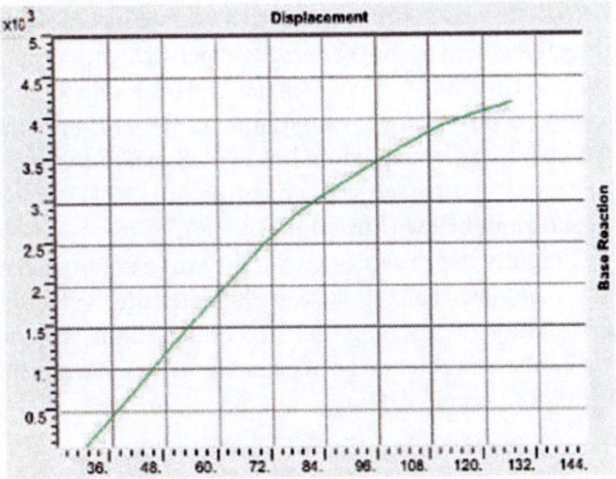

Fig. 4 Performance point

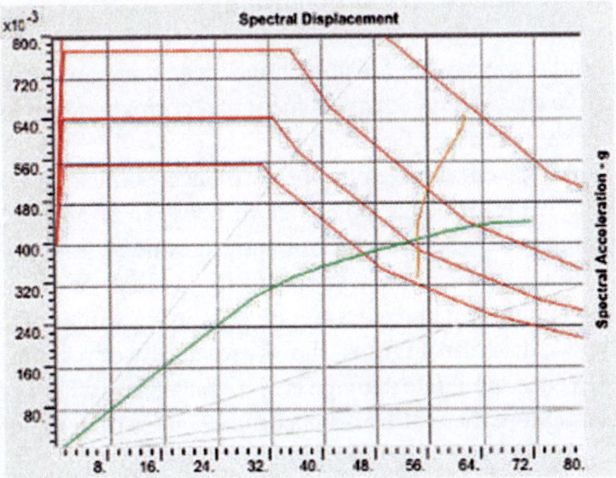

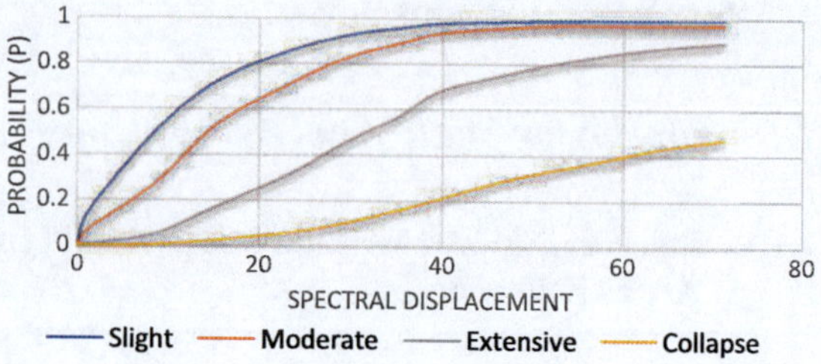

Fig. 5 Fragility curve based on pushover analysis (type-1)

From acceleration displacement response spectrum (ADRS) curve which is formed by spectral displacement (S_d) and spectral acceleration (S_a) gives performance point, where demand curve and capacity curve intersect to each other as shown in Fig. 4. According to the graph, both the curve intersects in between B and C. Therefore, some strength and stiffness are reduced by structural members which are beyond retrofitting required limit. The coordinate of performance point (S_a, S_d) is (0.409 g, 54 mm) as per Fig. 4 shown in below. Similarly, due to push along *Y* direction, fined pushover curve and performance point. But to simplifying the present study push along critical axis that means along *X* direction only considered.

Based on pushover curve and ADRS curve, collected data and solved Eq. (1) which has explained before in HAZUS methodology. To find probability of reaching or exceeding damage states mention in HAZUS manual constructed fragility curve as shown in below (Fig. 5).

Fragility curve shows provability to reach or exceed slight, moderate, extensive, and complete damage state with increasing values of spectral displacement (S_d). Provability of reaching and exceeding slight and moderate damage is more than 90%, where is percentage of extensive damage which is more than 80% but maximum values of complete damage not more than 50%.

4.1.2 Structural Model with Precast Shear Wall (Type-2) Analysis Results Are as Following

Similar ways, after finishing non-linear static analysis for type-2 structure, pushover curve shows target displacement and corresponding load as 63 mm and 10.58 × 10^3 kN shown in Fig. 6.

ADRS curve gives the performance point (S_a, S_d) as (0.784, 37.386) shown in Fig. 7. Performance point lies in between B and C ,indicates structural condition which is beyond retrofitting required limit.

Fragility curve shows provability to reach or exceed slight, moderate, extensive, and complete damage state with increasing values of spectral displacement (S_d). For type-2 structure, Figure 8 shows provability of reaching and exceeding slight damage 90%, and moderate damage is more than 80%, where percentage of extensive damage is not more than 80% also maximum values of complete damage not more than 50%.

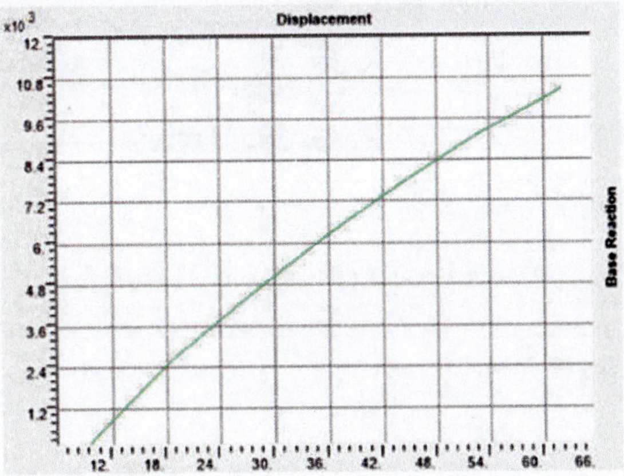

Fig. 6 Pushover curve (Push-X)

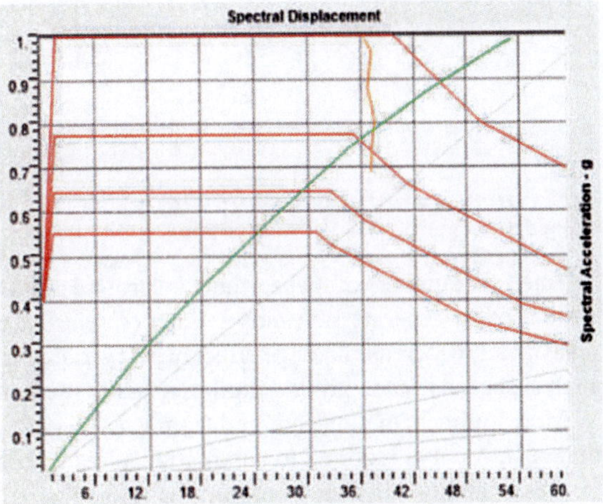

Fig. 7 Performance point

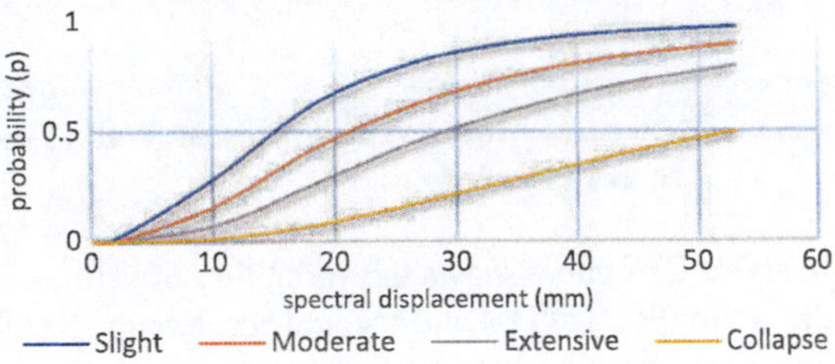

Fig. 8 Fragility curve based on pushover analysis (type-2)

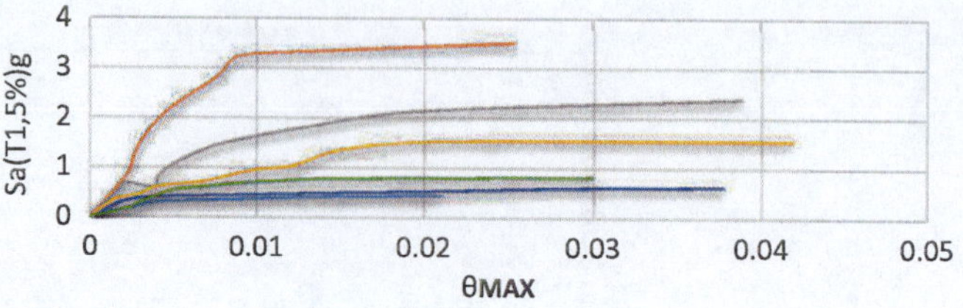

Fig. 9 IDA curve of all six selected ground motion

Fig. 10 Mean IDA curve

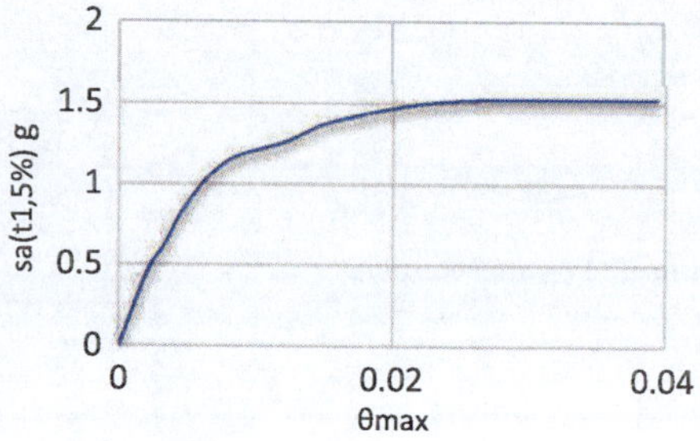

4.2 *Incremental Dynamic Analysis (IDA)*

Selected six numbers of real earthquake ground motion database mentioned in Table 2 are used to generate IDA curve. Each of selected ground motion multiplying by scale factor to increase damage effect in every trial of analysis. IDA is time consuming analysis because each ground motion consists by numbers of trials.

More numbers of data gathered from various ground motion during the construction of IDA curve required to summarize all data according to our study objectives. In present study, all data from various GM records summarized individually and prepared individual IDA curve shown in Fig. 9, then summarized into mean of all data and constructed mean IDA curve shown in Fig. 10.

4.2.1 Structural Model Without Shear Wall (Type-1) Analysis Results Are as Following

In above IDA curve shows the variability of various recorded data in terms of peak drift ratio (θ_{max}) as DM and spectral acceleration Sa (T_1, 5%) as IM. As the intensity increases, structure gradually transforms from linear stage to non-linear stage for each recorded data. Each record represented by response sequence number (RSN) as mention in figure. Near the collapse point with small changes of IM, extensive change of DM has been occurred. As a result, slope of IDA curve is decrease and converted to flatline.

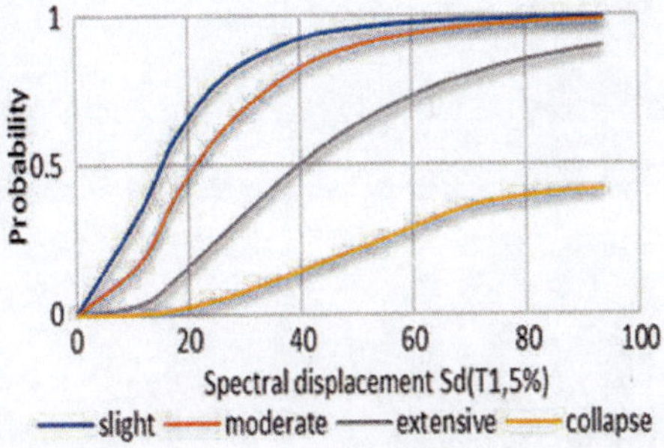

Fig. 11 Fragility curve developed by IDA

Figure 10 represents mean IDA curve to proper understanding of structural behavior under various real earthquake ground motion which is prepared by summarizing large number of analysis result data and properly quantify them. Curve shows non-linearity response of type-1 structure.

As mention in before, fragility curve shows probability to reaching or exceeding damage stages (slight, moderate, extensive, collapse) with increasing of DM (Fig. 11). For type-1 structure, Fig. 11 shows provability of reaching and exceeding slight damage 99%, and moderate damage is more than 90%, where percentage of extensive damage is not more than 90% also probability of complete damage below 45% but more than 40%.

4.2.2 Structural Model with Precast Shear Wall (Type-2) Analysis Results Are as Following

Followed by similar procedure as type-1 structure, constructed IDA curve for type-2 structure shown in Fig. 12 which shows the variability of various recorded data in terms of peak drift ratio (θ_{max}) as DM and spectral acceleration Sa (T_1, 5%) as IM.

Figure 13 represents mean IDA curve to proper understanding of type-2 structural behavior under various real earthquake ground motion. Curve constructed followed by similar procedure as type-1 structure. Mean curve shows non-linear response of type-2.

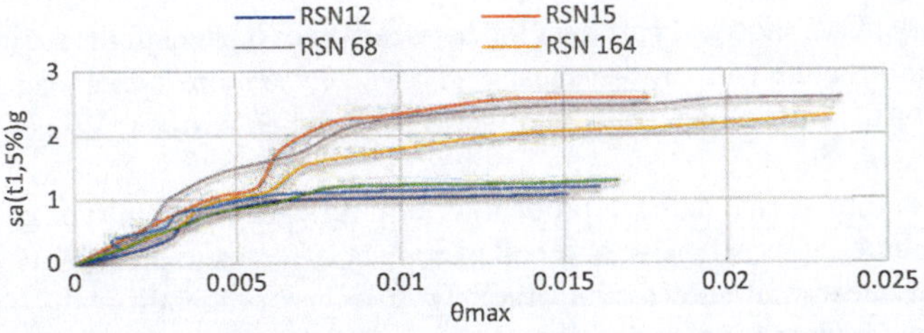

Fig. 12 IDA curve of all six selected ground motion

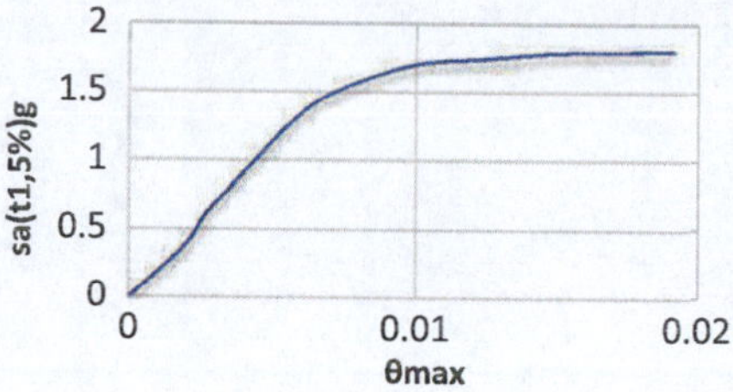

Fig. 13 Mean IDA curve

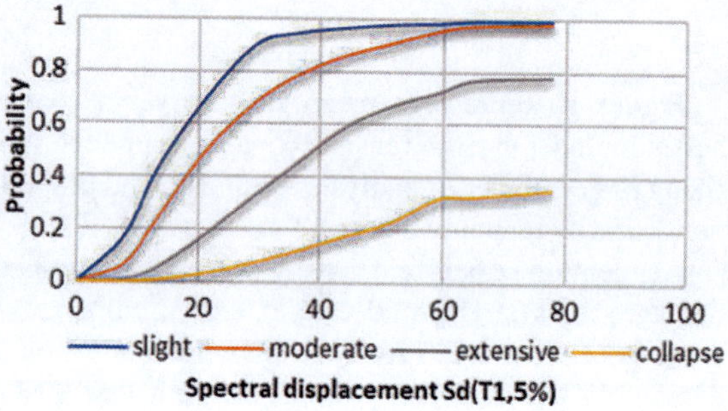

Fig. 14 Fragility curve developed by IDA

For type-2 structure, Fig. 14 shows provability of reaching and exceeding slight damage 99%, and moderate damage is more than 90%, where percentage of extensive damage is not more than 80%; also probability of complete damage is not more than 40%.

5 Conclusion

Two G+5 story reinforced concrete (RC) frame building, one without shear wall (type-1) and another one with precast RC shear wall (type-2), modeled in SAP 2000. To assess seismic collapse performance of proposed structures, performed non-linear static analysis and incremental dynamic analysis and developed fragility curve for both the models of structure. Conclusion of the present study is as follows.

- Fragility curve shows, if participation between precast panel joints are negligible, then the structure with precast shear wall have 20.7% more lateral load resistance capacity than the structure without shear wall, when structure comes under its non-linear stage.
- Slight and moderate damage probability almost same for both type of structure but in the case of extensive and collapse stage of damage probability, structure without shear wall shows comparatively 16% less probability which indicates after the yielding stage energy decapitation capacity of type-2 structure is more.

- For the same type of structure at its fundamental time period (T_1), during incremental dynamic analysis (IDA) response of structure is more; hence, dynamic analysis is more sensitive during the assessment of seismic performance of any structure.
- Therefore, to fulfill the need of durable, safe against earthquake, economical, and speedy construction, this concept is very much useful (specially in earthquake zone area) for construction industry.

References

1. Moridani K, Khodayari R (2013) Seismic performance assessment uses incremental dynamic analysis. J Basic Appl Sci Res 3(8):757–764
2. Chopra AK (2003) Theory and applications to earthquake engineering. Dyn Struct 45:34–54
3. Lu X, Wu H, Zhou Y (2017) Seismic collapse assessment of self-centering hybrid precast walls and conventional reinforced concrete walls. Struct Concr 18(6):938–949
4. Ozkul TA, Kurtbeyoglu A, Borekci M, Zengin B, Kocak A (2019) Effect of shear wall on seismic performance of RC frame buildings. Eng Fail Anal 100:60–75
5. Jamnani HH, Amiri JV, Rajabnejad H (2018) Energy distribution in RC shear wall-frame structures subject to repeated earthquakes. Soil Dyn Earthq Eng 107:116–128
6. Vetr MG, Shirali NM, Ghamari A (2016) Seismic resistance of hybrid shear wall (HSW) systems. J Constr Steel Res 116:247–270
7. Christidis KI, Trezos KG (2017) Experimental investigation of existing non-conforming RC shear walls. Eng Struct 140:26–38
8. Biswal A, Prasad AM, Sengupta AK (2019) Study of shear behavior of grouted vertical joints between precast concrete wall panels under direct shear loading. Struct Concr 20(2):564–582
9. Zhai X, Zhang X, Cao C, Hu W (2019) Study on seismic performance of precast fabricated RC shear wall with opening filling. Constr Build Mater 214:539–556
10. Council BSS (2000) Prestandard and commentary for the seismic rehabilitation of buildings. Report FEMA-356, Washington, DC
11. Federal Emergency Management Agency (2000) Prestandard and commentary for the seismic rehabilitation of buildings. American Society of Civil Engineers (ASCE)
12. Vamvatsikos D, Cornell CA (2002) Incremental dynamic analysis. Earthquake Eng Struct Dynam 31(3):491–514
13. Baker JW (2015) Efficient analytical fragility function fitting using dynamic structural analysis. Earthquake Spectra 31(1):579–599
14. Shinozuka M, Kim HK, Kim SH (2000) Nonlinear static procedure for fragility curve development. J Eng Mech 126(12):1287–1295
15. Ramadan T, Ghobarah A (1995) Analytical model for shear-link behavior. J Struct Eng 121(11):1574–1580
16. Kappos AJ, Panagopoulos G (2010) Fragility curves for reinforced concrete buildings in Greece. Struct Infrastruct Eng 6(1–2):39–53
17. Barbat AH, Lagomarsino S, Pujades LG (2002) Vulnerability assessment of dwelling buildings. projects: REN 2001-2418-C04-01 and REN2002-03365/RIES Universitat Politdecnica de Catalunya, Barcelona, Spain, University of Genoa,Genoa, Italy
18. Honarparast S, Chaallal O (2019) Non-linear time history analysis of reinforced concrete coupled shear walls: Comparison of old design, modern design and retrofitted with externally bonded CFRP composites. Eng Struct 185:353–365
19. Vamvatsikos D, Jalayer F, Cornell CA (2003) Application of incremental dynamic analysis to an RC-structure. In: Proceedings of the FIB symposium on concrete structures in seismic regions, May 2003, pp 75–86

Seismic Fragility of Reinforced Concrete Moment Resisting Frames Designed by Direct Displacement-Based Design and Performance-Based Plastic Design Methods

Sejal Purvang Dalal and Purvang Dalal

Abstract Present study assesses the seismic fragility of a set of 4, 8, 12, and 15 storied reinforced concrete (RC) moment resisting frames which were analysed and designed by direct displacement-based design (DDBD) method and performance-based plastic design (PBPD) method. The seismic loads were inculcated in both the methods as per the Bureau of Indian Standards specifications. The seismic performance evaluation of the frames was done by nonlinear static analysis. Damage state thresholds were defined in terms of yield displacement and ultimate displacement to estimate the seismic fragility. Although the seismic performance of both the methods in terms of strength and ductility was found satisfactory, it was observed that the frames designed by DDBD method frames show a lower seismic fragility over PBPD method.

Keywords Seismic fragility · Direct displacement-based design · Performance-based plastic design · Reinforced concrete moment resisting frames · Seismic performance

1 Introduction

The displacement-based design procedures are known to have better control on structural deformation and damage as they have direct relationship with structural deformations. The direct displacement-based design (DDBD) and performance-based plastic design (PBPD) methods are the two most popular displacement-based seismic design methods. These two methods have been developed by including the inelastic response of the structure which is often reflected during the whole design process especially in the calculation and distribution of the design seismic loads. The required design forces and moments in the beams and columns are calculated for potential

S. P. Dalal (✉)
Sardar Vallabhbhai Patel Institute of Technology, Vasad, Gujarat, India
e-mail: sejalpurvang@gmail.com

P. Dalal
Dharmsihn Desai Institute of Technology, Nadiad, Gujarat, India

T. G. Sitharam et al. (eds.), *Seismic Hazards and Risk*, Lecture Notes in Civil Engineering 116, https://doi.org/10.1007/978-981-15-9976-7_11

hinge locations in case of framed structures. Comparison of both the methods is often based on their seismic performance calculated by probabilistic or deterministic approach (under static or dynamic loading). Seismic fragility is defined as the probability of a response (of a particular structure) exceeding a critical level when subjected to a specified seismic loading. Simply, it is the probability of failure of a structure under seismic loading and it is extremely helpful in estimating the expected damage in structure. In case of low to mid rise buildings, it is a general practice to estimate the fragility based on the evaluation results calculated by simplified seismic evaluation procedures such as capacity spectrum method. In this study, the seismic fragility of a set of 4, 8, 12, and 15 storied reinforced concrete (RC) moment resisting frames designed by both the methods has been compared.

2 The Study Frames

The front views of 4, 8, 12, and 15 storied 2D RC study frames located in seismic zone 5 (PGA of 0.36 g) with an importance factor 1 and resting on a medium soil site are shown in Fig. 1. The seismic mass of the roof floor is 81.58 t (equal to 800 kN seismic weight), first floor is 101.5 t (equal to 996.2 kN seismic weight), and for all other floors is 97.87 t (equal to 959.3 kN seismic weight). The yield stress of steel is taken as 415 N/mm^2 and modulus of elasticity as 200,000 N/mm^2, respectively. The value of yield strain is considered as 0.0023. M25 concrete grade has been used. The

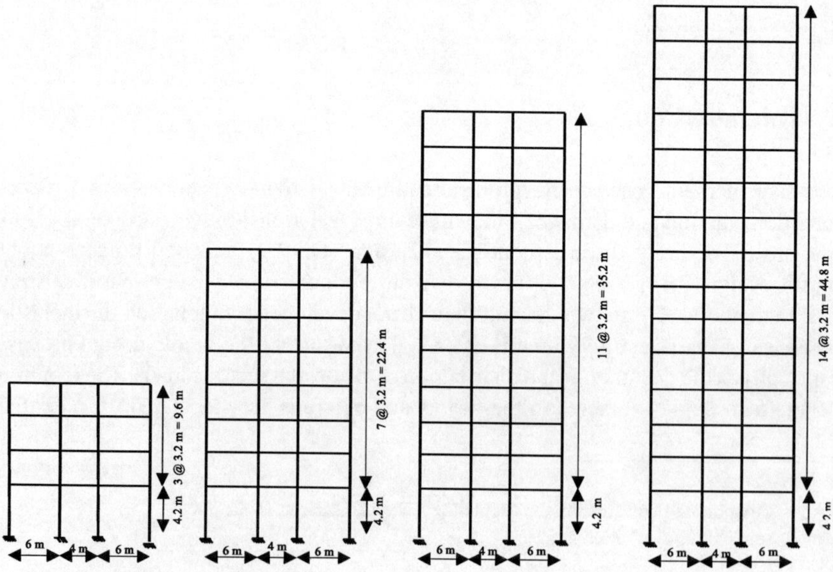

Fig. 1 Front view of the study frame

natural time period of the frames has been calculated as per [7] which works out to be 0.54 s, 0.59 s, 1.17 s, and 1.38 s for 4, 8, 12, and 15 storied frame, respectively.

2.1 *Design Base Shear and Analysis of Frames*

Both the methods start by using a specific acceptable level of displacement or drift to meet performance objective. As defined by ASCE [1], the performance objectives are immediate occupancy (IO), life safety (LS) and collapse prevention (CP) for which the permissible drift is 0.01, 0.02, and 0.04, respectively, for RC frames. As per [7] which follows a force-based design, the permissible yield drift is limited to 0.004 at design load level but it also permits a reduction of seismic loads by applying the response reduction factor (R). These response reduction factor values (Table 9 of [7]) depend on the perceived seismic damage performance characterized by ductile or brittle deformations, and for RC frames, its value is 5. Therefore, by multiplying the permissible drift (0.004) to the R value (5), the design displacement of 0.02 is selected in the study.

In DDBD method, the original structure is substituted with an equivalent SDOF system with a secant (or effective) stiffness at maximum response and the design base shear is calculated as product of this effective stiffness and the design displacement [11]. PBPD method calculates base shear by equating the work needed to push the structure monotonically up to the target drift to the corresponding energy demand of an equivalent SDOF oscillator [9]. The pinched hysteretic behaviour and P-Delta effects were also included in calculation of design base shear in this study. The seismic parameters as per [7] have been inculcated in both methods as suggested and presented in [12]. The design base shear (Table 1) thus calculated is well above the minimum specified requirements of [7].

This base shear is distributed along the height of the frame while considering the higher mode effects in both the methods. This is done by distributing the base shear with respect to the roof level. The DDBD method gives 10% allocation exclusively to the roof level and remaining 90% is distributed in proportion to mass and displacement at discretized mass locations (i.e. at other floor levels) for MDOF systems [11]. The PBPD method uses the distribution scheme as suggested by Chao et al. [3]

Table 1 Design seismic load of study frames

Frame details	DDBD method	PBPD method		
	Design base shear V_B (kN)	V_B/W (%)	Design base Shear V_B (kN)	V_B/W (%)
4 storey	130.36	3.4	150.3	4.05
8 storey	302.41	4.0	229	3.05
12 storey	412.9	3.6	308.9	2.71
15 storey	545.85	3.8	366.6	2.56

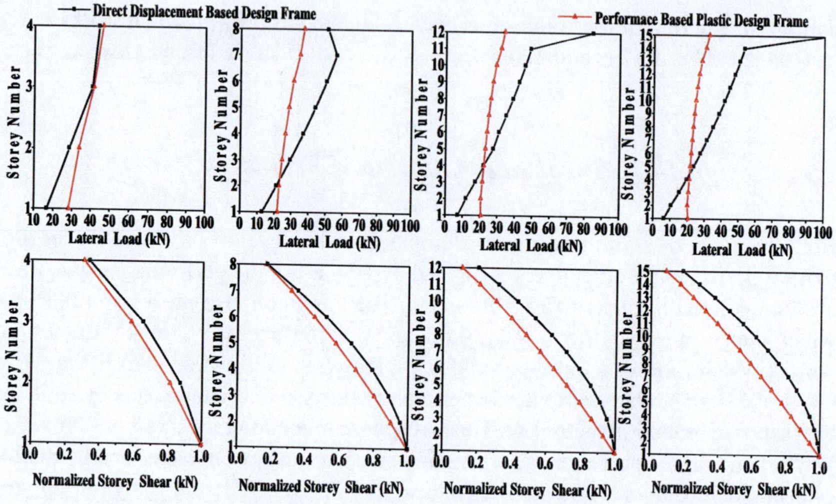

Fig. 2 Distribution of lateral loads and normalized storey shears in the study frames

which gives a for a more uniform distribution of interstorey drift, is more rational and gives a much better prediction of inelastic seismic demands at global as well as at element levels. The lateral load and normalized storey shear for both the methods are presented in Fig. 2.

All the study frames were then analysed for gravity and seismic loads as per [11] for the DDBD method and [10] for the PBPD method. In the DDBD method, the moments at centerlines of columns for each end of the beam are calculated after obtaining storey shear forces in beams by using statically admissible distribution of internal forces. The column moments are calculated based on the assumption that the point of contra flexure will be at a distance of 0.7 times height of the lower storey. In the PBPD method, plastic hinges are assumed to be developed in beams (away from face of column) and at the base of columns of the first storey, and consequently, the beam moments are calculated by plastic analysis method. The columns are analysed with the help of equilibrium equations of free body diagram by applying the expected beam end moments and lateral forces. The calculated beam and column design moments are shown in Fig. 3.

It is observed that the values of base shear and lateral loads calculated by DDBD method are higher compared to PBPD frames except for the four storied (lower height) frame. Notably, the storey shears are more evenly distributed in PBPD method rather than DDBD method. Accordingly, the same pattern is observed in beam moments. The strong column weak beam hypothesis is clearly perceived in both the methods. DDBD frames arrive with larger values of column moments for all the frames.

Due to seismic loading in the lateral direction along with gravity loading, all the beams and columns are designed as beam columns. All the members have then been

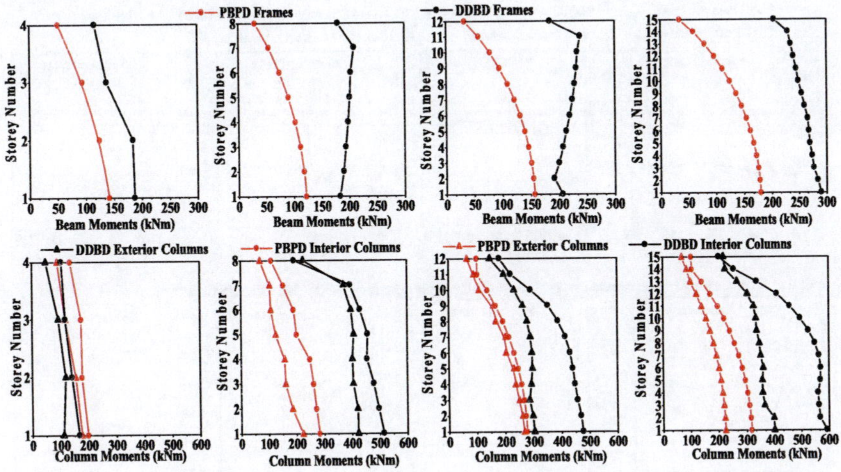

Fig. 3 Beam moments and column moments in the study frames

designed for limit state of collapse and checked for limit states of serviceability as per [6, 8] for both the methods and the detailed design can be retrieved from [12].

3 Nonlinear Static Analysis

Nonlinear static pushover analysis as per [1, 5] using the ETABS software was carried out to obtain the pushover curve and performance points (Fig. 4) of the study frames by applying the seismic loading of Fig. 2. Lumped plasticity models with the hinge properties, as defined in [1], are used by considering the flexural inelastic mechanisms only. Conforming reinforcements have been considered to assign the plastic rotations for the beams and columns. The performance point (intersection of capacity and demand curve) is obtained considering bilinear force–displacement relationship.

The push over curves (base shear versus monitored displacement) and capacity demand curves for the study frames are presented in Fig. 4. For both the methods, the ultimate strength of all the frames remains higher than the one for which they have been designed. As the pushover curve is in the linear zone at the design base shear value for all the frames, it shows that in all the frames, the first hinge is formed only after it exceeds the design base shear value. The immediate occupancy (IO), life safety (LS), and collapse prevention (CP) stage as observed in the software are also shown on the pushover curve. The demand curve as per Indian Code is used to obtain the performance point. The performance point as seen on the capacity demand curve is in terms of spectral displacement and is converted to displacement by standard procedures and placed on the pushover curve. The performance point of

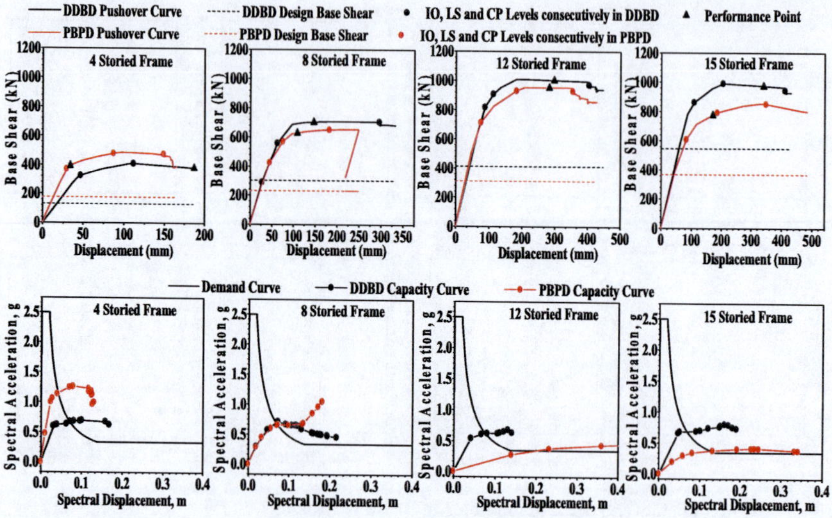

Fig. 4 Pushover curves and capacity demand curves of the study frames

Table 2 Ductility demands of the study frames

Frame details	DDBD method			PBPD method		
	d_u (mm)	d_y (mm)	μ	d_u (mm)	d_y (mm)	μ
4 storey	187.1	52.3	3.57	33.96	15.23	2.23
8 storey	146.4	48.43	3.02	108.5	47.59	2.27
12 storey	297.3	93.84	3.16	282.4	75.8	3.72
15 storey	339.2	98.3	3.45	180.4	70.91	2.54

all the DDBD and PBPD frames is below collapse prevention (CP) performance level as seen from the pushover curve. The yield displacement (d_y), ultimate displacement (d_u), and ductility demand (μ) thus obtained are presented in Table 2. They are used to define the damage states.

4 Seismic Fragility Assessment

The fragility curves in this study give the probability that the expected global damage of the study frames exceeds a given damage state as a function of spectral displacement (Sd). The damage state thresholds used in this study are quantified as per the suggestions of [2] and are presented in Table 3.

For each damage state, the corresponding fragility curve is defined by plotting the probability of exceedance $P[d \geq d_{si}]$ as ordinate and the spectral displacement

Table 3 Damage states thresholds

S. no.	Damage state	Damage state thresholds ($\overline{\mathrm{Sd}}_{\mathrm{si}}$)
1.	Sd_1 (slight)	0.7 d_y
2.	Sd_2 (moderate)	$\mathrm{d_y}$
3.	Sd_3 (severe)	0.75 d_y + 0.25 d_u
4.	Sd_4 (complete)	d_u

(Sd) (of the performance point) as abscissa. The fragility curves follow a lognormal probability distribution and are defined as following probability density function.

$$P\left[\frac{\mathrm{ds}_i}{\mathrm{Sd}}\right] = \phi\left[\frac{1}{\beta_{\mathrm{dsi}}}\ln\left(\frac{\mathrm{Sd}}{\overline{\mathrm{Sd}}_{\mathrm{dsi}}}\right)\right] \quad (1)$$

where

$\mathrm{d_{si}}$ is the given damage state.
$\mathrm{Sd_{si}}$ is the threshold spectral displacement.
β_{dsi} is the standard deviation of the natural logarithm of the displacement threshold and it indicates uncertainties in capacity curves, damage levels, modelling errors, etc. Its value is taken as 0.7 [4].
Φ is the standard normal cumulative distribution function.

The probability of exceedance of the frames to slight, moderate, extreme, and complete damage is presented in Fig. 5. The complete fragility curves are presented in Fig. 6. The four storied DDBD frame has been designed for a lower value of seismic loads (Table 1) and column moments (Fig. 4) and hence is more ductile as compared to the PBPD frame. The d_u and d_y values of DDBD frame are quite higher compared to PBPD frame (Table 2) and they show lesser probability to damage (Fig. 5). The range of damage chances of the PBPD frames is till the spectral displacement of nearly 0.033 m, and at the same time, it has 50% probability of complete damage at this spectral displacement of 0.033 m. Beyond this, the ultimate displacement is reached. Whereas the DDBD frame has probability of complete damage as 11% at 0.033 m. It has nearly 50% probability of complete damage at spectral displacement of 0.18 m. The eight-, twelve-, and fifteen storied DDBD frames were designed for higher seismic base shear and column moments as compared to PBPD frames. But they show lesser probability of damage at different damage levels as shown in Fig. 5. DDBD frames show less vulnerability to damage as shown in Fig. 6.

5 Conclusions

The seismic fragility of a set of 4, 8, 12, and 15 storied reinforced concrete (RC) moment resisting frames which were analysed and designed by DDBD and PBPD method by incorporating the Indian standard code is assessed in this study. As already

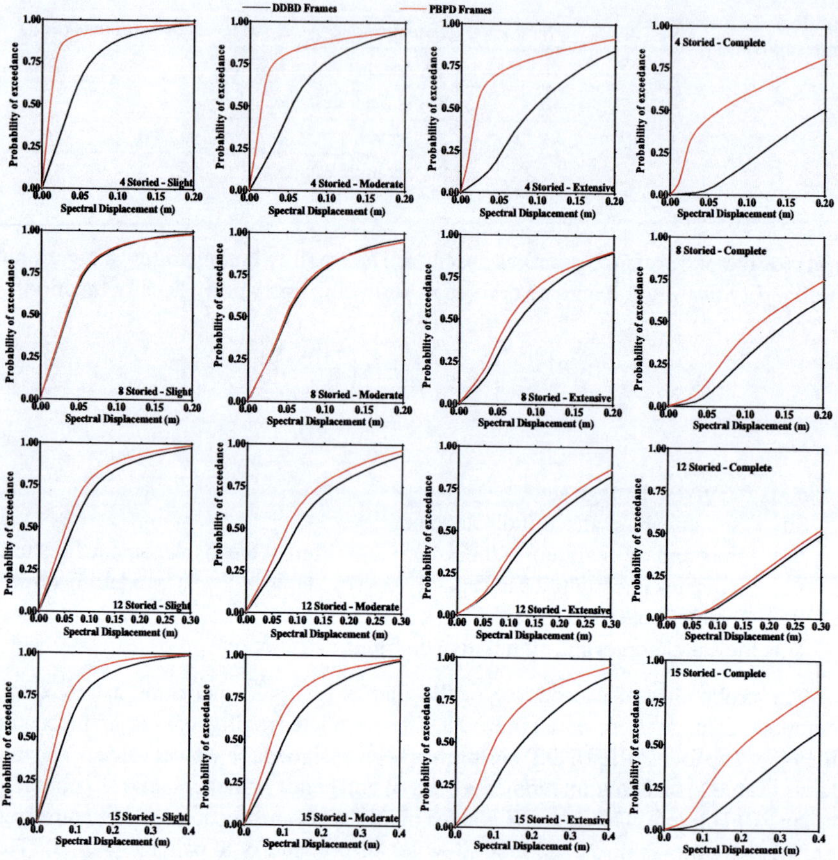

Fig. 5 Comparison of probability of exceedance of slight, moderate, extensive, and complete damage of study frames

concluded by Qammer et al. [12], both the study frames have shown satisfactory performance in terms of drift and capacity. It is further observed in this study that the DDBD frames show a lower seismic fragility over PBPD frames. There is a larger difference in yield and ultimate displacement values in DDBD frames due to which it covers a broader range of spectral displacement. This also indicates higher ductility demand. The DDBD frames have a higher value of ultimate displacement which makes it less vulnerable to damage.

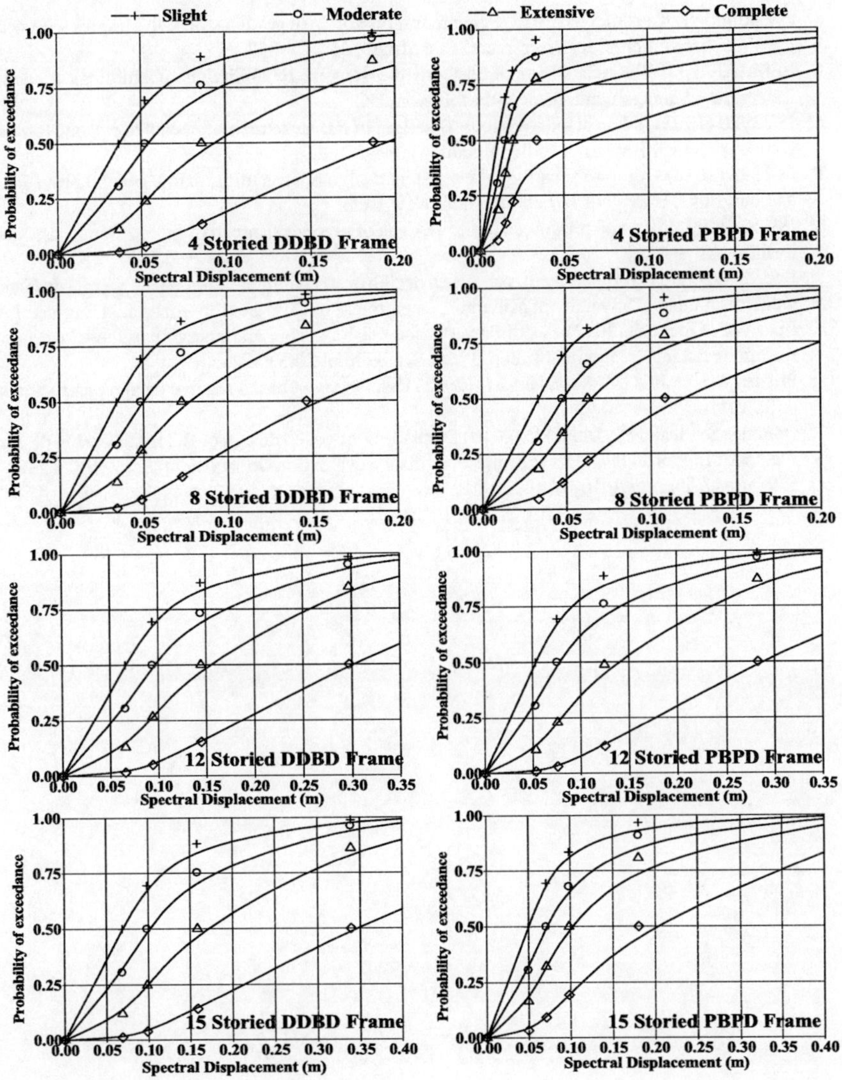

Fig. 6 Fragility curves of the study frames

References

1. ASCE (2006) Seismic rehabilitation of existing buildings. American Society of Civil Engineers, Reston
2. Barbat AH, Pujades LG, Lantada N (2008) Seismic damage evaluation in urban areas using the capacity spectrum method: application to Barcelona. Soil Dyn Earthq Eng 28:851–865
3. Chao SH, Goel SC, Lee SS (2007) A seismic design lateral force distribution based on inelastic state of structures. Earthq Spectra 23(3):547–569

4. Choudhury T, Kaushik HB (2018) Seismic fragility of open ground storey RC frames with wall openings for vulnerability assessment. Eng Struct 155:345–357
5. FEMA 273 (1997) NEHRP guidelines for the seismic rehabilitaion of buildings. Federal Emergency Management Agency, Washington, DC
6. IS 13920 (1993) Indian standard ductile detailing of RC structures subjected to seismic forces-code of practice. Bureau of Indian Standards, New Delhi
7. IS 1893 (2016) Criteria for earthquake resistant design of structures, part-1 general provisions and buildings. Bureau of Indian Standard, New Delhi
8. IS 456 (2000) Plain and reinforced concrete-code of practice. Bureau of Indian Standards, New Delhi
9. Lee SS, Goel SC (2001) Performance-based design of steel moment frames using target drift and yield mechanism. University of Michigan, Department of Civil and Environmental Engineering
10. Liao WC, Goel SC (2012) Performance based plastic design and energy based evaluation of seismic resistant RC moment frame. J Mar Sci Technol 20(3):304–310
11. Pristley MJ, Calvi GM, Kowalsky MJ (2007) Displacement based seismic design of structures. IUSS Press, Pavia
12. Qammer SS, Dalal SP, Dalal PD (2019) Displacement-based design of RC frames using design spectra of Indian code and its seismic performance evaluation. J Inst (India) Ser A 100(3):367–379. https://doi.org/10.1007/s40030-019-00

Seismic Fragility Assessment of Existing 9 Storey Reinforced Concrete (RC) Buildings in Patna, India

Pratyush Kumar and Avik Samanta

Abstract Precise and accurate estimation of building's response is the demand of the time. Since the response of any building is always associated with inherent uncertainty, its consideration is utmost for the preciseness of their vulnerability. The uncertainties can be due to two reasons, (1) randomness in ground motion, also termed as aleatoric uncertainty, and (2) variability in structural modelling (epistemic uncertainty). The present work focuses on obtaining the variability functions and fragility curves by considering those uncertainties for 9 storey RC buildings in Patna, India, which is a region of high seismicity. Variability due to randomness in ground motion was determined by generating several artificial ground motions (comprising both near as well as far-field). Therefore, extensive study of the seismotectonic map of the region was required to determine the nearby faults, past earthquakes and other random parameters contributing variability. Structural variability, on the other hand, is generally attributed to the uncertainty in the estimation of material, geometrical properties, of which the building is made and at the same time due to the uncertainty in design details. Therefore, their consideration is vital in fragility analysis. Material strength, storey height, plan typology and design methodologies were considered as the random parameters in this study. The selection of these parameters was purely based on the extensive survey of buildings in the region. Considering the several combinations of random parameters, the buildings were analysed using non-linear static (NLS) pushover analysis to obtain pushover curves. These pushover curves were converted to capacity curves considering the dynamic characteristic of the first mode of vibration. The obtained capacity curves were then utilised to derive the variability in the capacity spectrum corresponding to particular damage states. Eventually, lognormal variability function and fragility curves for 9 storey representative buildings were obtained considering both the uncertainties. The obtained fragility curves corresponding to particular damage states give a measure for the vulnerability of buildings in that region.

P. Kumar (✉) · A. Samanta
Department of Civil Engineering, Indian Institute of Technology, Patna, India
e-mail: ce18d017@smail.iitm.ac.in

T. G. Sitharam et al. (eds.), *Seismic Hazards and Risk*, Lecture Notes in Civil Engineering 116, https://doi.org/10.1007/978-981-15-9976-7_12

Keywords Aleatory and epistemic uncertainty · Capacity spectrum · Ground motion · Pushover analysis · Variability function · Fragility curve · Vulnerability assessment

1 Introduction

Seismic fragility of structures is the need of the time. The Indian state of Bihar has numerous record of earthquakes tremor. 1934, Bihar earthquake was one of the worst earthquakes in India's history. The earthquake caused widespread damage to life and property in the region of Bihar and Nepal. As per the Bureau of Indian standards [6], the region of Bihar lies in seismic zone IV and V. The capital city Patna, which is the interest of study, however, lies in seismic zone IV. High seismic hazard, together with significant exposure and large population density enhances the need for the vulnerability assessment in case of Bihar. Keeping in mind, the severity of earthquakes in the region, this research was aimed at evaluating the vulnerability of 9 storeyed RC buildings which are prevalently found in the Patna region.

Seismic vulnerability is a measure of the susceptibility of building to damage for a given intensity of ground shaking. It is required to assess the social and economic impact of the earthquakes in any region. Owing to the necessity of vulnerability assessment of the region like Patna, the fragility analysis of 9 storeyed buildings was carried out following an analytical approach, particularly the capacity spectrum method. It is known that the precise assessment of any structural parameters linked with the seismic response is associated with the inherent uncertainty. The uncertainties are attributed to randomness (aleatory) and lack of modelling knowledge (epistemic). Therefore, their consideration is vital for the precise assessment of vulnerability. Several researchers have recognised the importance of considering those uncertainties for accurate estimation of the building's fragility. Kwon and Elnashai [18] recognised the role of material and ground motion uncertainty on the vulnerability curves of RC structures. The effect of material uncertainty viz. concrete compressive strength, steel yield strength, along with the ground motions uncertainty on structural response, were studied. Yu et al. [25] mainly studied the effect of structural uncertainty due to material and geometrical parameters. Their study quantified the lognormal variability in capacity (β_c) corresponding to different damage states.

Subsequently, Ellingwood et al. [11] studied the variability due to uncertainty in structural modelling and ground motions. Uncertainty parameter viz. peak intensity, time varying amplitude, frequency content and strong motion duration of ground motions were included as random variables for computing variability due to ground motion records.

For estimating the variability associated with uncertainties in structural parameters, non-linear static (NLS) pushover analysis was used which is a powerful technique for the seismic performance of any structure; however, it has some pros and cons [17], which is attributed to the assumption under which the method can be applied. The two major drawbacks of NLS include, (1) The unimodal loading profile does not take into account the effect of higher modes that may have significant impact on the structural response in some situation, (2) The time invariant or so-called static load profile neglects the changes in dynamic characteristic of structure which is due to the progressive plasticization of structural element and consequent loss of stiffness.

Several works were done by researchers in the past to estimate the vulnerability of buildings in a particular region. Polese et al. [22] performed the vulnerability assessment of gravity load designed RC buildings of Arenaella district in Naples, Italy. The study comprised an extensive survey of buildings in the region to obtain the statistical parameters of the existing and pre-code designed buildings followed by the simulation, design, and analysis of those buildings based on the capacity spectrum method.

Since the estimation of variability function for a specific region requires details of the buildings of that particular region; therefore, field survey is needed to fetch the details of buildings typology, configuration, construction materials, and design practices which are the key contributors to structural variability. For Patna region, a field survey was done by Sriwastav [24] and Hussain [16]. From the survey, the material and geometrical properties of 9 storey buildings were obtained. Therefore, in the present study, buildings were modelled in such a way through which the maximum representation of buildings of the study area could be made.

Besides, structural parameters extensive study of the seismotectonic map of the region is required for the development of region-specific ground motions. Several studies [1, 21] comprising the details of the past earthquake in the region, their magnitude, epicentral distance, soil condition, and nearby fault, were utilised in the development of ground motion for the Patna region. Eventually, the obtained ground motions were used for the estimation of demand variability.

Thence, the main objective of the present study is to determine the lognormal variability functions considering aleatoric and epistemic uncertainty and utilising those functions for the fragility assessment of nine-storeyed RC buildings for seismically active Patna region.

2 Ground Motion Uncertainty

Uncertainty in estimation of ground motions for any particular region renders variability in demand spectra. Therefore, a thorough study of the seismotectonic map of the region is required to estimate the variability arising due to this uncertainty. In this study, seismic hazard map developed by Anbazhagan et al. [1] and NDMA [21] was extensively utilised to obtain various details of seismic features like faults, shear zone and past earthquakes observed around 500 km radius of the study area. Both near-field and far-field ground motions were generated considering the active faults as well as the past earthquake around 500 km of the region.

For the development of far-field ground motion, a distance greater than 200 km, and moment magnitude of 7.5 or more was considered. Furthermore, for near field ground motion, the epicentral distance between 30 and 45 km and a moment magnitude of less than 6.5 was assumed [1]. Focal depth for moment magnitude between 5 and 6 varies between 5 and 75 km. However, for moment magnitude greater than 6, focal depth varies between 10 and 75 km. For the worst-case scenario, focal depth of 10 km was assumed for an epicentral distance of 200 km or more. Thereby, the focal depth of 5 km was taken for a moment magnitude less than 6 and 10 km for moment magnitude greater than 6 and beyond the epicentral distance of 200 km. For the study area, the shear wave velocity ranges between 73 and 385 m/s [7], thereby soil type was suitably identified based on the shear wave velocity. The criterion for the selection of the above-mentioned parameters was based on the extensive study of the seismotectonic and hazard map of the region. The details of the study region as well as the active faults is illustrated in Fig. 1.

Moment magnitude, epicentral distance, shear wave velocity ($v_{s,30}$) and focal depth of the past earthquakes around the study region were considered for the development of ground motions using strong ground motion simulation programme SGMSv5 [14] to generate synthetic time histories. Consequently, 162 near and far-field ground motions were generated considering those parameters. The ground motions so generated were utilised for the development of response spectra. The variability in the response spectra corresponding to the particular fundamental period of building yields the required demand variability function.

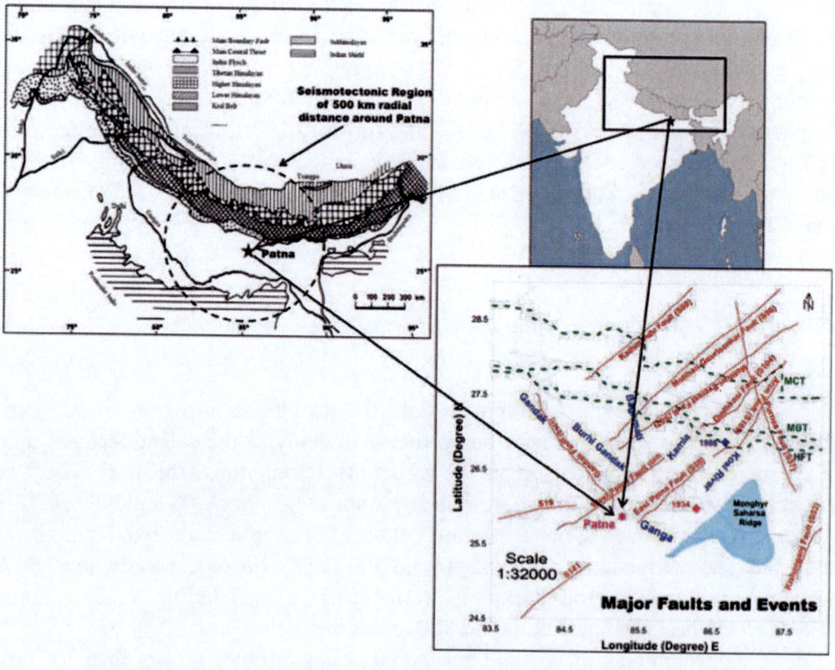

Fig. 1 Study area of Patna along with seismotectonic faults based on Anbazhagan et al. [1]

3 Modelling Uncertainty

Uncertainty in geometrical, design detailing, and mechanical properties of a material are the major source of modelling uncertainty. These parameters were considered as random variables. For the preciseness in structural modelling, the field survey data [16, 24] comprising the details of an extensive sample of different storied buildings was extensively followed. For the estimation of variability function, uncertainty in structural parameters contributing variability needs to be considered. Among the structural parameters, random variables considered include characteristic strength of concrete, yield strength of steel, loading, plan (e.g., *L* shaped, rectangular and square shaped building) and storey height (Table 1).

The number of buildings considered in the analysis depends solely on the data obtained from field survey performed on the 9 storeyed.

Considering the various building parameters, pushover analysis was performed following HAZUS methodology [13, 15] on 51 representative 9 storyed buildings (the details of which have been mentioned in Sect. 4, building properties and configurations) of each configurations following inverted triangular load distribution (Eq. (1)). The selection of the 51 representative building was based on the extensive survey of buildings in the region. This pushover load resembling a triangular pattern was applied at the nodes of a different storey of bare, URM infilled and open ground storey (OGS) buildings in both the directions. Thereby, 153 buildings were considered and analysed following this methodology for the 9 storey and particular design philosophy.

$$F_{xi} = \frac{m_i h_i}{\sum_{i=1}^{n} m_i h_i} \quad (1)$$

where F_{xi} is the lateral load at *i*th floor, m_i denote the mass of *i*th floor,h_i is the height of the *i*th floor from the ground level and *n* corresponds to the number of floors. The mass of the floor was assumed to be lumped at the respective floor. In this study for computationally simpler analysis, floor was considered rigid.

The total lognormal variability in corresponding damage state (β_{ds}) is modelled by the combination of three contributors (capacity variability, demand spectrum variability, and damage state uncertainty) following HAZUS methodology [15],

Table 1 Description of structural and material parameters

Parameters	Material properties considered
Characteristic strength of concrete	M20, M25, and M30
Grade of steel reinforcement	Fe415 and Fe500
Storey height	3 and 3.2 m
Building shape	Square, rectangle, *L* shaped
Building type	Bare frame, infilled frame, open-ground storey

Table 2 Damage states definition

Damage grade	Damage state	Median spectral displacement
D_{s1}	Slight	$S_{d,1} = 0.7d_y$
D_{s2}	Moderate	$S_{d,2} = 1.5d_y$
D_{s3}	Extensive	$S_{d,3} = 0.5(d_y + d_u)$
D_{s4}	Collapse	$S_{d,4} = d_u$

$$\beta_{ds} = \left\{\sqrt{\text{CONV}(\beta_C, \beta_D)^2 + \beta_{T,ds}^2}\right\} \tag{2}$$

where β_C is the lognormal variability in the capacity spectrum, β_D is the lognormal variability in demand spectrum and $\beta_{T,ds}$ is the lognormal variability in the threshold of damage states. However, for the sake of simplicity in the computation of lognormal function, the equation used by [23] was utilised in the present study. The equation indicates the independency or no correlation between capacity and demand variability, thereby resulting in a conservative estimation of desired vulnerability function.

The conditional probability of exceeding or in specified damage state, *ds*, given the spectral displacement, $s_{d,}$ is mentioned in Eq. (3) [15].

$$P(ds/S_d) = \Phi\left[\frac{1}{\beta_{ds}} \ln\left(\frac{S_d}{\overline{S}_{d,ds}}\right)\right] \tag{3}$$

where $\overline{S}_{d,ds}$ is the median value of spectral displacement corresponding to damage states, *ds*, β_{ds} is the lognormal variability in the threshold of damage states, and Φ is standard normal cumulative distribution function. The median spectral displacement at which building reaches the specified damage states ($\overline{s}_{d,ds}$) is required for the fragility assessment. The value recommended by Lagomarsino and Giovinazzi [19] was taken in this study (Table 2). The value $S_{d,1}$ to $S_{d,4}$ corresponds to slight, moderate, extensive and collapse damage states, respectively. The idealised elasto-plastic capacity curve was obtained from the obtained force–displacement curve [19]. The four points of the curve correspond to the four damage states which are either a function of yield displacement (d_y) or ultimate displacement (d_u) or both.

4 Buildings Properties and Configurations in Patna

For the estimation of variability function, which is associated with capacity, a field survey done by Hussain [16] was followed extensively. So, this study includes building configurations, construction material, storey height, plan irregularities as buildings parameters for uncertainty analysis. This study mainly includes square, rectangle and *L* shaped building configurations, which are typically practised in the region (Fig. 2). The behaviour and capacity of the building get changed with

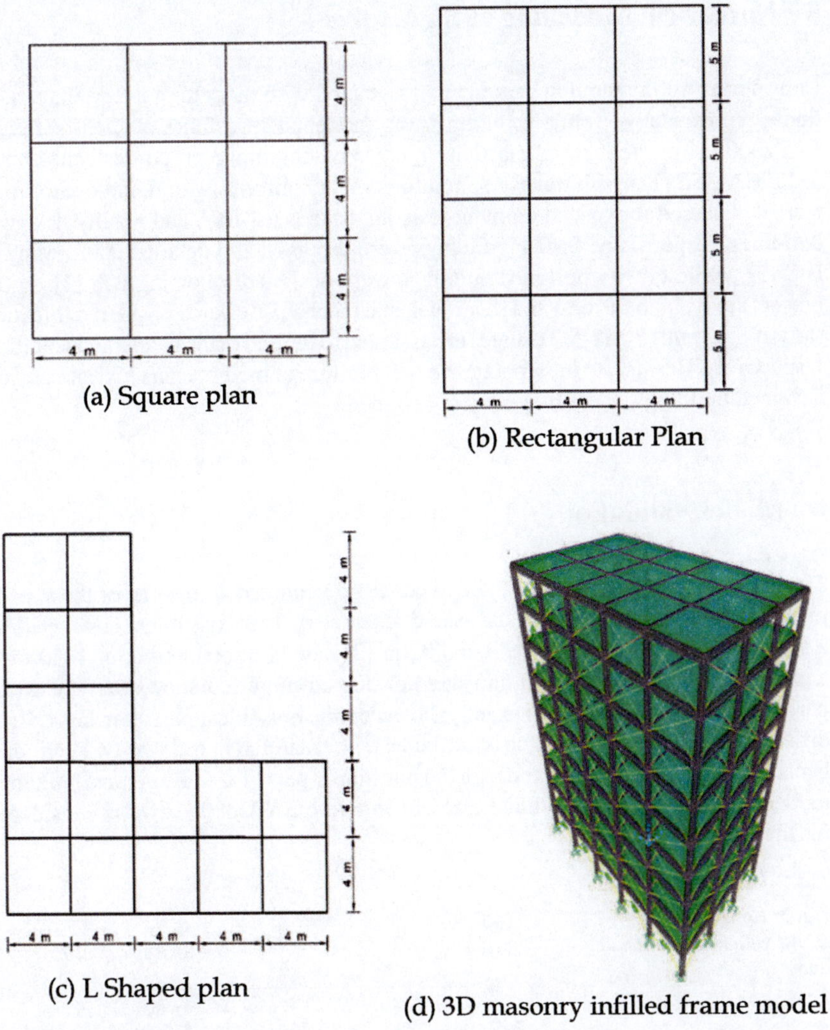

Fig. 2 Various building's configurations for 9 storey building

configuration. The change can be due to additional torsional mechanism as well as due to modification in building stiffness. It was also observed [16] that the study area mostly comprises buildings of storey height between 3 and 3.2 m. Therefore, the storey height of the building was varied to account for the variability in their structural capacity.

5 Numerical Modelling of Structures

Considering all the ramdon parameters associated with modelling uncertainty, non-linear representative frame (building) was modelled in a finite element software SAP2000 [10]. The lumped plasticity models having hinge properties conforming to FEMA-356 [12] with transverse reinforcement conforming and non-conforming were considered for special moment resisting frames (SMRF) and gravity designed buildings, respectively. SMRF buildings were designed and detailed conforming to BIS [3], while gravity designed buildings were designed following BIS [4]. In the present study, both flexural and shear inelastic mechanism were considered. Flexural (M3) hinges and (P-M2-M3) hinges are assigned to the end of the beams and columns, respectively. Besides, to incorporate the shear failure of members in OGS and gravity designed buildings, shear hinges were assigned.

6 Model Validation

The buildings modelled in SAP2000 needs to be validated with some of the experimental or analytical results for the sake of exactness. The experiment conducted on a scaled portal RC frame by Anil and Altin [2] having a total height of 1050 mm, and a total bay width of 1500 mm was considered. Figure 3 shows that the linear part of both the graph matches exactly; however, the non-linear part showing a slight difference (less than 10%) which is attributed to the material non-linearity. Therefore, because of the marginal difference in the non-linear part of the two curves, buildings modelled and analysed in a finite element software SAP2000 [10] was considered for the rest of the study.

Fig. 3 Comparison of results with an experimental study

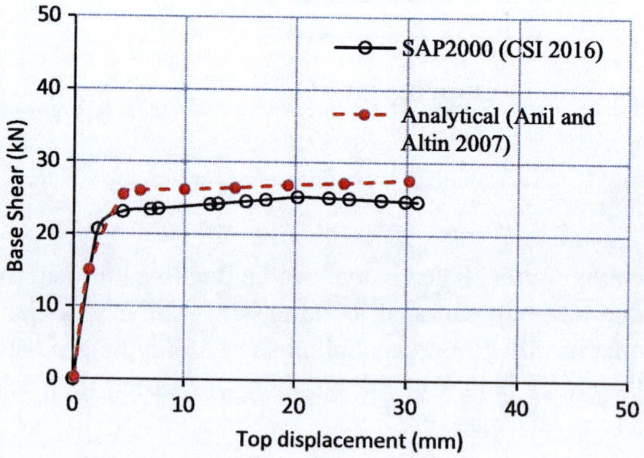

7 Modelling of URM Infill Wall

URM infill was modelled as equivalent diagonal strut considering pivot hysteretic model. It is based on a macro modelling approach, which is reliable and easy to use. Hysteretic pivot law particularization proposed by Cavaleri and Di Trapani [9] was used for modelling infill. According to Cavaleri and Di Trapani [9] and Campione et al. [8], infill wall is mechanically characterised by the parameters,modulus of elasticity, shear modulus of rigidity, and Poisson ratio. The initial stiffness (K_1) and width of the equivalent strut (w) are mentioned (Eqs. (4) and (5)),

$$K_1 = \frac{E_d t w}{d} \tag{4}$$

$$w = \mathrm{d}k \frac{c}{z(\lambda^{*\beta})} \tag{5}$$

$$\lambda^* = \frac{E_d t h}{E_f A_c}\left(\frac{h^2}{l^2} + \frac{1}{4}\frac{A_c}{A_b}\frac{l'}{h'}\right) \tag{6}$$

where E_d is the modulus of elasticity of infill along the diagonal direction, t is the thickness of URM infill wall, d is the strut length, the coefficient k depends on the level of the vertical load acting on a column, z depends on the ratio of length to the height of infill wall, and λ^* is a parametric term, E_f denotes modulus of elasticity of frame element, A_c *and* A_b are the cross-sectional area of column and beam, respectively; l' and h' are the geometrical features of the infill system [9]. Since the URM infill is very weak in tension, the branch corresponding to it coincides with that of the horizontal axis. α is the strength envelope parameter. The modulus of elasticity of masonry infill, which is a function of compressive strength of brick prism (f_m), is given by 550 f_m [6] and is followed for calculation of other parameters.

7.1 *Consideration of Central Opening in URM Infill Wall*

For infill with opening at central locations (doors and windows), suitable reduction factor was computed and multiplied with the equivalent width of infill following the equation proposed by Mondal and Jain [20], for central opening (Eq. (7)),

$$\rho_w = 1 - 2.6\alpha_{co} \tag{7}$$

$$\alpha_{co} = \frac{\text{Area of opening}}{\text{Area of infill}} \tag{8}$$

where, ρ_w is the reduction factor and α_{co} is the area ratio.

8 Results

8.1 Capacity Curve and Seismic Variability Functions

The effect of uncertainty on the building's capacity of different stories and configurations are illustrated in this section. The capacity curves were obtained considering the random parameters for 9 storey SMRF buildings. The curves for bare, infilled and OGS building are shown in Fig. 4.

It can be observed from Fig. 4a, b that the ultimate displacement for the bare frame is nearly twice that of the infilled frame. However, the spectral acceleration is nearly half which is attributed to greater stiffness of the URM infill wall. It was also observed that the variability in the curve at the point of yielding is lesser than that of the ultimate point. Therefore, it is evident to have lesser lognormal variability corresponding to yielding than ultimate. From the analysis, it was observed that for different storied buildings, the lognormal variability in capacity (β_c) ranges between 0.1 and 0.2 for slight damage state and 0.2–0.3 for collapse damage state. In the

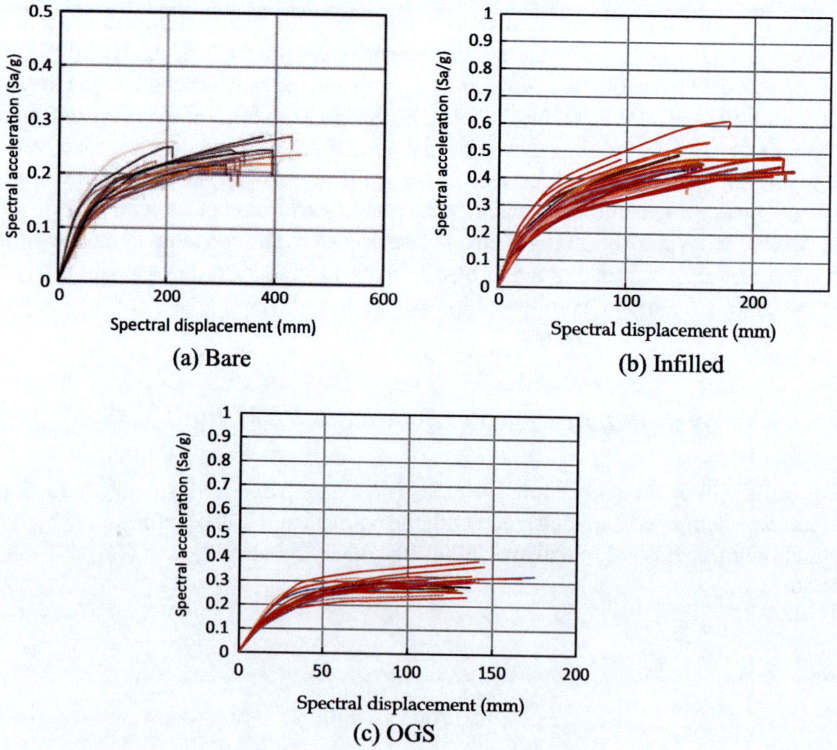

Fig. 4 Capacity curves for 9 storied buildings for different typology

case of OGS (Fig. 4c), owing to the shear failure of open floor columns, the ultimate spectral displacement and yielding displacement reduces significantly but the base shear (or spectral acceleration) shows higher value than the bare framed buildings.

Subsequently, the ground motions developed in the earlier section were utilised to obtain pseudo-acceleration response spectra (Fig. 5a, b) corresponding to far and near field ground motions. Together they are shown in Fig. 5c. The dispersion (β_D) of the spectra was calculated by computing the variability in demand, and the value obtained was 0.435.

The total variability function corresponding to different damage states was computed for 9 storeyed building and illustrated in Table 3. The lognormal variability ranges between 0.73 and 0.74, corresponding to slight and moderate damage

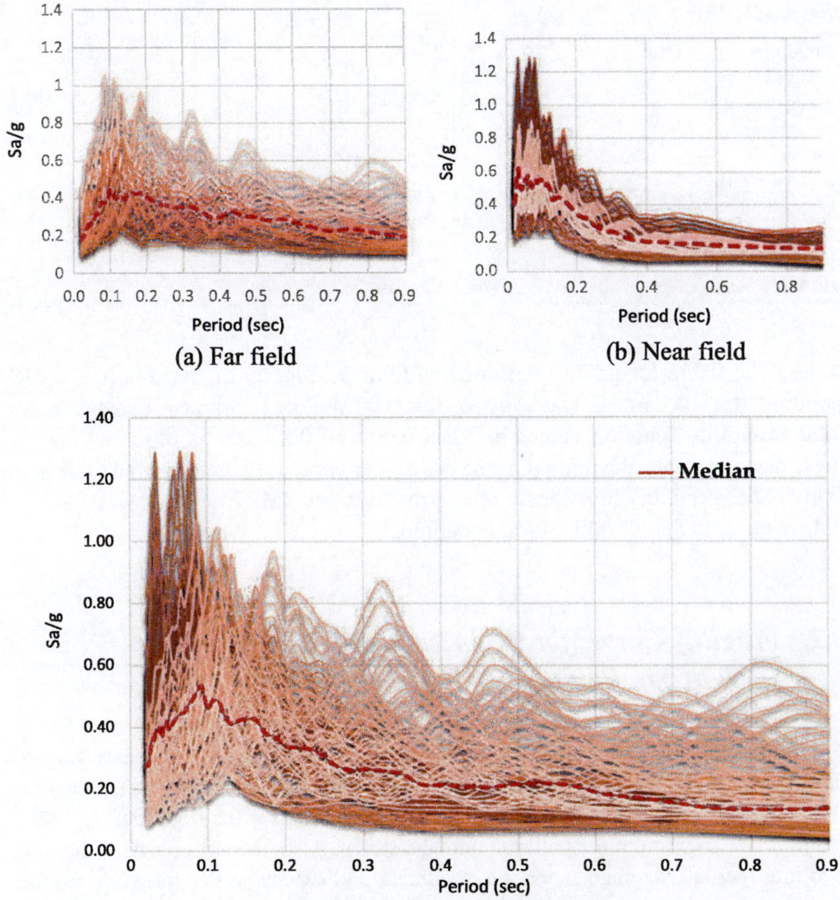

(a) Far field

(b) Near field

(c) Near and far field together

Fig. 5 Response spectra of artificially generated ground motions in Patna

Table 3 Lognormal variability function for buildings of Patna region

Design level	Frame configuration	No of storey	Damage states			
			Slight β_{ds1}	Moderate β_{ds2}	Extensive β_{ds3}	Collapse β_{ds4}
Buildings analysed, designed, and detailed as per earthquake resistant guidelines of [3, 5] for special moment resisting frames (SMRF)	Bare	9	0.73	0.73	0.79	0.81
	Infilled	9	0.73	0.73	0.8	0.81
	Open ground storey (OGS)	9	0.73	0.73	0.8	0.81
Buildings designed for gravity loads as per relevant Indian Standards, BIS [4] without any consideration for earthquake forces	Bare	9	0.74	0.74	0.79	0.81

states. The value for gravity designed building is slightly higher than the SMRF buildings for all damage states. Also, for extensive and complete damage states, total variability function ranges between 0.79 and 0.81. It was observed that for these damage states, the infilled frame has higher variability as compared to the bare frame. The variability function for open-ground storey SMRF building was observed to be similar to that of infilled frame building.

8.2 *Fragility Curves for 9 Storied SMRF and Gravity Designed Building*

The fragility curves for 9 storeyed SMRF and gravity designed (only bare framed) buildings are shown here. It was observed that for all three building's categories (bare, infilled and OGS), variability in the curve was lesser for slight damage states while it was maximum for collapse damage states. Also, the variability in the other two intermediate damage states, i.e. moderate and extensive lies in between slight and complete damage states. The lognormal variability for gravity load designed building was observed to be slightly more than the SMRF buildings.

Mean fragility curve for 9 storied SMRF building corresponding to bare, infilled and OGS are shown in Fig. 6. Figure 7 shows the fragility curve for 9 storeyed (bare framed) gravity load designed buildings with the shear hinge. Since the gravity

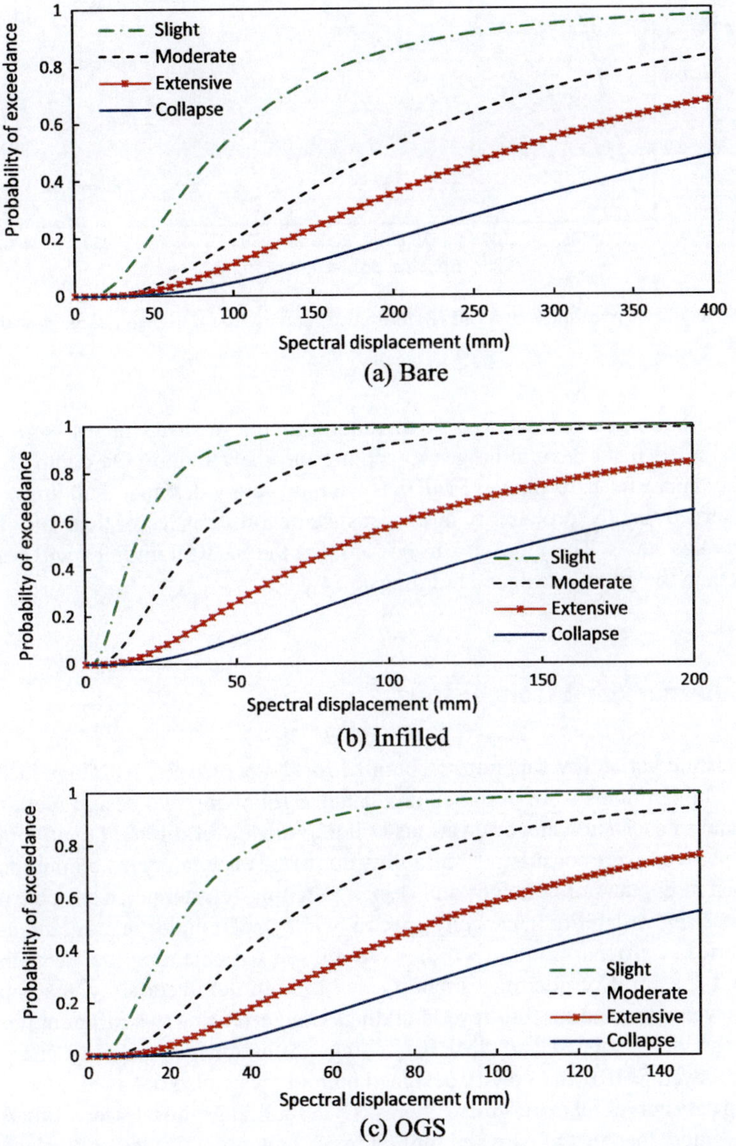

Fig. 6 Average fragility curve for 9 storied SMRF buildings

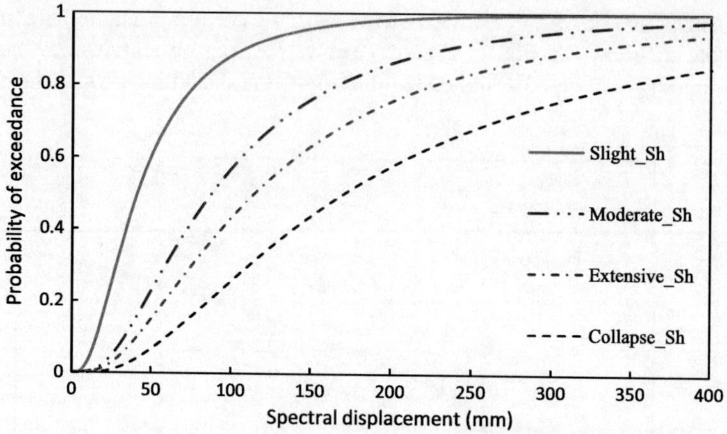

Fig. 7 Fragility curves for 9 storeyed (Bare framed) gravity load designed buildings with shear hinge (_Sh)

designed buildings can undergo shear failure, therefore shear hinges were incorporated apart from flexural hinges to capture the shear failure. On comparing the fragility curve for bare framed SMRF as well as gravity designed buildings, it can be observed that the probability of exceedance of former is lesser than later for all the damage states. Therefore, it can be said that the SMRF buildings will perform better than the gravity designed buildings.

9 Summary and Conclusions

The seismic variability function is obtained for 9 storeyed RC buildings in Patna, India. The buildings were designed and detailed following two design methodologies, and in each sub-category, various building models of different configurations and typologies were considered. Variability due to the uncertainty in column spacing, material strength, storey height and design detailing (epistemic) were considered for obtaining variability in capacity spectra, while uncertainties due to randomness (aleatoric) in ground motion records were utilised for obtaining the variability in demand spectra. Considering variability due to both the uncertainty, the capacity curves were obtained for 9 storeyed buildings designed using two different philosophies: gravity designed and SMRF buildings. Subsequently, fragility curves for nine-storeyed SMRF and gravity designed building were also obtained.

Capacity curves for open-ground storey (OGS) buildings show lesser ultimate roof displacement than that of bare and infilled frame buildings. The lesser displacement is attributed to the brittle shear failure of columns of ground stories. Lognormal variability function corresponding to slight, moderate, extensive and collapse damage states were obtained. The lognormal variability in capacity for slight and moderate damage states corresponding to different categories of buildings ranges between 0.1 and 0.2 depicting lesser variability. However, the value ranges between 0.2 and 0.3 for the extensive and collapse damage states and hence depicting higher variability. The probability of exceedance for gravity load designed building was higher than that of SMRF building. Based on the observations, it is concluded that the SMRF buildings will perform better than the gravity designed building. Also, the lognormal variability corresponding to any damage states for the gravity designed building is more than that of SMRF building.

The current study is limited to the simplified model of representative buildings found in the region. Highly complex buildings were not considered in this study. Also, the buildings were modelled with a fixed base; therefore, accounting soil-structure interaction can be the future scope. An extensive study is further required to determine the variability functions, as well as the fragility curves, for low, and medium-rise buildings of Patna region as well.

References

1. Anbazhagan P, Bajaj K, Patel S (2015) Seismic hazard maps and spectrum for Patna considering region-specific seismotectonic parameters. Nat Hazards 78(2):1163–1195
2. Anil Ö, Altin S (2007) An experimental study on reinforced concrete partially infilled frames. Eng Struct 29(3):449–460
3. BIS (1993) IS 13920: Indian standard ductile detailing of reinforced concrete structures subjected to seismic forces—code of practice. Bureau of Indian Standards, New Delhi
4. BIS (2000) IS 456: Indian standard plain and reinforced concrete—code of practice (fourth revision). Bureau of Indian Standards, New Delhi
5. BIS (2002) IS 1893 (part 1): Criteria for Earthquake resistant design of structures: part 1 general provisions and buildings, 5th rev. Bureau of Indian Standards, New Delhi
6. BIS (2016) Indian standard 1893 (part 1), criteria for earthquake resistant design of structures: part 1 general provisions and buildings, 6th rev. Bureau of Indian Standards, New Delhi
7. BUIDCo (2011) Design, build, operate, manage and maintain water supply system in Patna. Bihar Urban Infrastructure Development Corporation Ltd
8. Campione G, Cavaleri L, Macaluso G, Amato G, Di Trapani F (2015) Evaluation of infilled frames: an updated in-plane-stiffness macro-model considering the effects of vertical loads. Bull Earthq Eng 13(8):2265–2281
9. Cavaleri L, Di Trapani F (2014) Cyclic response of masonry infilled RC frames: experimental results and simplified modeling. Soil Dyn Earthq Eng 65:224–242
10. CSI (2016) SAP2000 user's manual version 16. Computers and Structures Inc., Berkeley
11. Ellingwood BR, Celik OC, Kinali K (2007) Fragility assessment of building structural systems in Mid-America. Earthq Eng Struct Dynam 36(13):1935–1952
12. FEMA (2000) Prestandard and commentary for the seismic rehabilitation of buildings. Report FEMA 356. Federal Emergency Management Agency, Washington, DC, USA
13. FEMA (2005) Improvement of nonlinear static seismic analysis procedures. Federal Emergency Management Agency, Report FEMA 440
14. Halldorsson B, Papageorgiou A (2004) Synthesis of strong ground motions for engineering applications using the specific barrier model. Multidisciplinary Center of Earthquake Engineering Research
15. HAZUS—MH MR5 (2013) Advanced engineering building module (AEBM), Technical and user's manual. FEMA, Washington, DC

16. Hussain MZ (2017) Properties and response of building in Patna for future earthquakes. Master's thesis. Indian Institute of Technology Patna
17. Krawinkler H, Seneviratna GDPK (1998) Pros and cons of a pushover analysis of seismic performance evaluation. Eng Struct 20(4–6):452–464
18. Kwon OS, Elnashai A (2006) The effect of material and ground motion uncertainty on the seismic vulnerability curves of RC structure. Eng Struct 28(2):289–303
19. Lagomarsino S, Giovinazzi S (2006) Macroseismic and mechanical models for the vulnerability and damage assessment of current buildings. Bull Earthq Eng 4(4):415–443
20. Mondal G, Jain SK (2008) Lateral stiffness of masonry infilled reinforced concrete (RC) frames with central opening. Earthq Spectra 24(3):701–723
21. NDMA (2007) Development of probabilistic seismic hazard map of India. Technical report. The National Disaster Management Authority, Govt. of India, New Delhi
22. Polese M, Verderame GM, Mariniello C, Iervolino I, Manfredi G (2008) Vulnerability analysis for gravity load designed RC buildings in Naples-Italy. J Earthquake Eng 12(S2):234–245
23. Roy T, Agarwal P (2015) Comparison of damage index and fragility curve of RC structure using different indian standard codes. In: Advances in structural engineering. Springer, New Delhi, pp 2551–2563
24. Sriwastav RK (2016) Seismic vulnerability assessment of high rise buildings in Patna for future earthquakes. Master's thesis, Indian Institute of Technology Patna
25. Yu X, Lu D, Li B (2016) Estimating uncertainty in limit state capacities for reinforced concrete frame structures through pushover analysis. Earthq Struct 10(1):141–161

Probabilistic Assessment of the Liquefaction Hazard in Central Khartoum, Sudan

Mohammed T. M. Ahmed and Mohamedelamin M. M. Mahmoud

Abstract This paper presents a preliminary evaluation to the risk of earthquake-induced liquefaction in central Khartoum, Sudan. Khartoum is the capital city of the country, it is located in the confluence of the white and blue Niles. The city is very crowded. Even though the area is endured in an intraplate region which is seismically stable, there is a noticeable occurrence of many earthquakes in different regions in the country with the absence of a seismic code or any compliances in the constructions for earthquake loading. Central Khartoum with many high-rise buildings serves as the center for both business and important governmental agencies is characterized by alluvial deposits-locally known as Gezira formation-underlined by Nubian sandstone below at about 25 m depth. This alluvial deposit is consisted of saturated loose to medium dense sands. The study starts with a brief review of the Seismotectonics features of Sudan. Then, sources of earthquakes affecting central Khartoum were identified. The rifts and faults in Abu Dulayq region was found to affect central Khartoum the most. Based on probabilistic analysis, the seismic hazard in central Khartoum was assessed. Using the simplified procedure developed by Seed and Idriss, the study assessed the susceptibility of central Khartoum soils by coupling of seismic hazard and conditional probability of liquefaction. The study explicated that liquefaction risk is low.

Keywords Gezira formation · Liquefaction · Probabilistic analysis · Simplified procedure

1 Introduction

Soil liquefaction has occurred in many earthquakes with very devastating effects, as a result of the instantaneous buildup of pore-water pressure which reduces the shear strength of the soil due to earthquake loading. It usually occurs in cohesionless saturated deposits, pertinently those sandy types of soils [1] and depends on

M. T. M. Ahmed (✉) · M. M. M. Mahmoud
Department of Geotechnical Engineering, Building and Road Research Institute, University of Khartoum, Khartoum, Sudan
e-mail: tawfigciv13@gmail.com

T. G. Sitharam et al. (eds.), *Seismic Hazards and Risk*, Lecture Notes in Civil Engineering 116, https://doi.org/10.1007/978-981-15-9976-7_13

various geological characteristics, and geotechnical properties (e.g., grain size distribution, density of the material). The identification of liquefaction occurrence is mainly observed through surface manifestations like sand boils, ground settlement, and lateral spreading. Nevertheless, the absence of surface manifestation does not substantially deny the incidence of sub-surface liquefaction [2].

Liquefaction is considered an extremely complex phenomenon and hard to predict. Significant efforts have been made to constringe the risk of liquefaction initiation for which the credit goes to Seed and Idriss due to their extensive research and findings.

In the present study, an attempt has been made to assess the hazardous effect of soil liquefaction in central Khartoum, Sudan. Although the region is deemed as seismically stable, a recent seismic activity around Khartoum was reported (e.g., Abu Dulayq earthquakes with a magnitude of 5.3 and 3.5 Mw, respectively, about 149 km from Khartoum). Geological and geotechnical data have shown that an alluvial deposit locally named as "Gezira formation" surrounds the area, comprises of clay, silty sand and loose sand underlain by the Nubian sand stone to about 25 m depth. Concerns were given to central Khartoum due to the extension of the city with more high-rise buildings. In addition to the enormous population of the state which is estimated to be more than 7 million [3]. The study area serves as the center for both business and important governmental agencies and is located at the convergence of the Blue and White Niles. Despite the fact that earthquakes are considered the most subversive natural hazard in the twentieth century [4], they are not taken into account in the design procedure in Sudan, besides the absence of a seismic code which necessitated the need for evaluation.

2 Geological and Geotechnical Description of Central Khartoum

Khartoum is the capital city of the country. It is located in the Northeastern part of Sudan, the study area, central Khartoum known as El-Morgan is located at the conjunction of the Blue Nile and the White Nile shaped in a triangular form (see Fig. 1).

This triangular area, formed by the Blue and White Niles (the main tributaries of the River Nile) consists of deposits ranging from superficial deposits (Quaternary) to pre-tertiary sediments [5]. The superficial soils locally known as "Gezira formation" include windblown sands, the White Nile alluvium and Wadi-deposits (streamer channel) which consists of unconsolidated sand, silt sand, gravel, loosely compacted silt, sandy silt, and clay. The pre-tertiary sediments known as Nubian sandstone was named by [5] which was then used as a description to the Nubian desert in North Africa and made up of conglomerates, grits, sandstone, sandy-mudstone, mudstone, and usually flat-lying sedimentary rock.

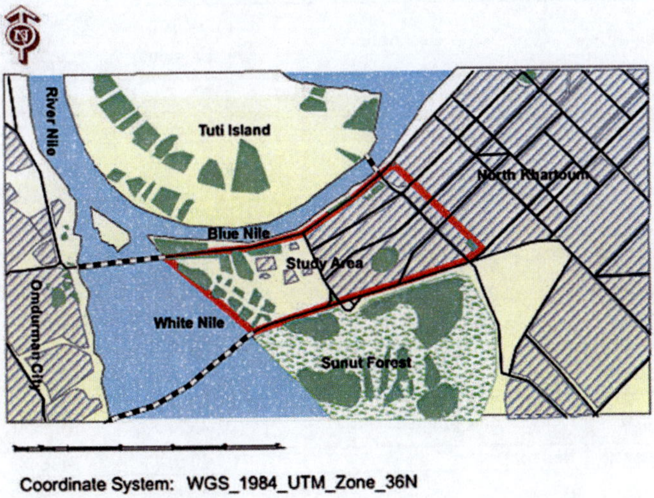

Fig. 1 Study area of central Khartoum

To assess the liquefaction in central Khartoum, standard penetration tests (SPT) were used; although a small portion of the study area was investigated for microzonation by [6, 7]. Sixty-three SPT boring logs in the central Khartoum from consulting firms and research institutions were used in this study. With respect to the subsoil conditions, the area was divided into two zones, the boring logs revealed that the Gezira formation consists of clay and silt soils in the upper 7 m, and various types of sand were found below ranging from about 8 to 25 m depth. Below that, the Nubian Sandstone is found. The sub-surface water varied from 4 m near the Nile to 8 m away from the Nile with an average of 6 m (see Fig. 2).

3 Seismotectonics and Seismicity of the Region

There are three major rift systems in Sudan; (1) the Red sea rifts in Eastern Sudan which are part of the East African Rift system, (2) Central African Rift system in Western Sudan, and (3) Southern Sudan rifts system known as Abu Gabra rifts and its branches including Blue Nile, White Nile, and Atbara River rifts.

Previous studies by [8, 9] have shown that Sudan is seismically stable with infrequent earthquakes of low to moderate magnitude that can give rise to damaging intensity, also they noted that the Southern parts of Sudan are frequently subjected to moderate to high earthquake intensities capable of producing damage (e.g., 1990 earthquake Ms = 7.4 is one of the highest earthquakes in Africa [10]). In central Khartoum, earthquakes were stimulated from many sources (e.g., from rifts and faults in Western State (about 260 km West of Khartoum), rifts and faults in Southern States

Depth(m)	Description	S.P.T (N-value)	Shear Modulus (KN/m2)
0.0	Hard, dark brown, Silty clay of low Plasticity.		6490
2	Hard, light brown, silty clay of High plasticity.		6680
9	Medium dense, poorly graded sand.	14 to 20	79740
11	Loose to medium dense, poorly graded sand.	11 to 17	79740
14	Dense to medium sand and silty Sand	15 to 25	77655
23.0	Dense coarse grained sand	28 to 32	108650
Below 25	Bed rock	30 to 43	3436485

Fig. 2 Average soli profile for central Khartoum

(about 800 km South of Khartoum), faults in Red Sea State (about 500 km North East of Khartoum) and induced earthquakes in Lake Nassir (near Aswan Dam) in Southern Egypt (about 900 km North of Khartoum). Among all of these sources, Abu Dulayq source in western Khartoum (Latitude 15.9° and Longitude 33.84°) is the one that affects central Khartoum area essentially [11], the source of Abu Dulayq has been active recently. An earthquake of 5.3 magnitude on Richter scale was felt in Khartoum in November 2003. The duration of shaking lasted about 40 s causing minor injuries but no substantial damages were reported. Abu Dulayq was also the source of another shock of 3.5 magnitude on Richter scale that was also felt in Khartoum in July, 2010 as reported by [12]. This may validate the elastic rebound theory which states that earthquakes are more likely to occur in any region where small or no seismic activity has been noticed recently [13].

4 Computation of Seismic Hazard

The computation of seismic hazard required estimation of any expected earthquakes and seismic sources through the use of seismic hazard analysis based on either a probabilistic or a deterministic way. On the other hand, seismic performance analysis evaluates the response resistance of the estimated hazard (e.g., soil liquefaction) [14].

4.1 Seismic Hazard Analysis (SHA)

The steps of seismic hazard assessment presented by [14] will be used herein. The past earthquake data was obtained from [15] as well as the historical earthquake data presented by [8] to develop a recurrence model. The equation below was obtained:

$$\lambda(t) = 2.245e^{-0.85M} \quad (1)$$

where; λ = rate of earthquake occurrence per year, M = Richter earthquake magnitude.

Khartoum area is mostly affected by Abu Dulayq source (a distance of 149 km from Khartoum) [11], other sources have unaccountable contribution with an amount of acceleration that could be ignored.

John Douglas in [16] reviewed numerous attenuation relationships published since 1964 around the world, developed from different input parameters. For central Khartoum, the relation proposed by Humbert et al. [17] was used due to the absence of the strong ground motion record, following relation was obtained:

$$\text{Log}(a) = 0.31M - 0.739 \quad (2)$$

where; a = the ground acceleration in (cm/s^2).

Using Eqs. (1) and (2) the relationship between the annual rate of earthquake and the ground acceleration can be obtained. The results are plotted in Fig. 3 for discrete values of ground acceleration.

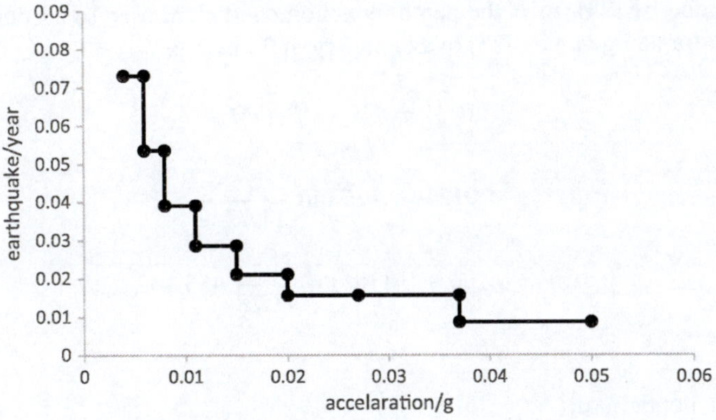

Fig. 3 Relation between acceleration and annual rate of earthquake

4.2 Seismic Performance Analysis (SPA)

The local soil site response against ground shaking (in this case against liquefaction) was evaluated using seed and Idriss simplified method [18], the method is based on empirical correlation of the case history data where the liquefaction has occurred and the adjacent sites where it had not occurred. Whereby analyzing case histories and developing that to a framework using the in-situ indices to identify the resistant. Since the development of the simplified procedure, many correlations were published [19, 20] and were further slightly modified by [21], Idriss and Boulanger in [22–24] to attain for more consistent results.

The dynamic earthquake loading is expressed by the term cyclic stress ratio (CSR), and the resistance to liquefaction (capacity of soil to resist the action) is given by the term cyclic resistance ratio (CRR). The cyclic stress ratio (CSR) is expressed by;

$$\mathrm{CSR} = 0.65\frac{a_{\max}}{g}\frac{\sigma_{v0}}{\sigma'_{v0}}\frac{r_d}{\mathrm{MSF} * K_\sigma} \tag{3}$$

where;

$a_{\max} =$ the maximum peak ground acceleration given by Eq. (2);
$g =$ acceleration due to the gravity;
σ_{v0} and σ'_{v0} are the total and effective overburden pressures.

MSF = magnitude scaling factor used to adjust the relation to different earthquake magnitudes because the relationship above was developed at earthquake (Mw = 7.5).K_σ Is the overburden correction factor used to account for the effective confining pressure. It should be noted that the parameters MSF and K_σ used in Eq. (3) were those defined by [23]. $r_d =$ the depth reduction coefficient used to account for the flexibility for soil given by [23] to a boundary of 34 m depth;

$$\ln(r_d) = \alpha(z) + \beta(z)M \tag{4}$$

$$\alpha(z) = -1.012 - 1.126\sin\left(\frac{z}{11.73} + 5.133\right) \tag{5}$$

$$\beta(z) = 0.106 + 0.118\sin\left(\frac{z}{11.28} + 5.142\right) \tag{6}$$

where;

Z is the depth in m.

CRR can be evaluated from the filed using in-situ test parameters or from the laboratory tests. The filed test is believed to be more reliable [25]. Among all of the in-situ tests, the standard penetration test (SPT-N) is the mostly used for evaluation of cyclic resistance ratio (CRR). Through the use of corrected SPT-N value to 60%

energy and the effect of overburden stress, in addition to other testing procedures' errors correlated to the uniform cyclic loading in CSR.

Liquefaction occurs whenever CSR exceeds the CRR as a function of (SPT-N) value corrected for fine content.

(FC %) for a given earthquake magnitude, therefore at the critical situation the relation yields to;

$$a_{\max} = \left(\frac{g}{0.65 r_d}\right)\left(\frac{\sigma'_{v0}}{\sigma_{v0}}\right)(\text{MSF} * K_\sigma)(N_1)_{60\,CS} \tag{7}$$

The above relation can be expressed by;

$$a_{\max} = \left(\frac{g}{0.65 r_d}\right)\left(\frac{\sigma'_{v0}}{\sigma_{v0}}\right) F(M, N, f) \tag{8}$$

Liquefaction occurs if $a_{\max}$ in Eq. (8) exceeds the acceleration from Eq. (2) (the earthquake acceleration). For any given earthquake magnitude and assuming 15% fine content, the function f can be simplified to;

$$F(M, N, f) = \Phi e^{N\varphi} \tag{9}$$

where Φ and φ are constants determined from the regression,Φ ranging from 0.145 to 0.978 and $\varphi = 0.74$. Considering the uncertainty in (SPT-N) values, a symmetric beta distribution equation presented by [26] is used as given in Eq. (10).

$$F(N) = \frac{\sqrt{N - 0.6N_m} * \sqrt{1.4N_m - N}}{\left(0.25412 N_m^2\right)} \tag{10}$$

where N_m is the mean value of N.

The probability density function (PDF) of acceleration that will cause the liquefaction can be obtained from Eqs. (8)–(10) and the transformation of N can be expressed as below [27];

$$F_{a\max} = F_N(N)\frac{\mathrm{d}N}{\mathrm{d}_{a\max}} \tag{11}$$

Equation (11) gives $F_{a\max}$ for a given earthquake magnitude and given zone (e.g., for zone one and magnitude Mw = 6), $F_{a\max} = 1.05/a_{\max}\sqrt{x_1 x_2}$, where $x_1 = 15.5 \ln \frac{a_{\max}}{1.66}$ and $x_2 = 12.3 - 14.5 \ln a_{\max}/1.66$).

Using the probability density function (PDF) of $a_{\max}$ for a given zone and magnitude, we can get the conditional probability of liquefaction, and with the integration to obtain the cumulative distribution as;

$$\sum_{\text{for all } M} [L|M] = F_{a\max} \tag{12}$$

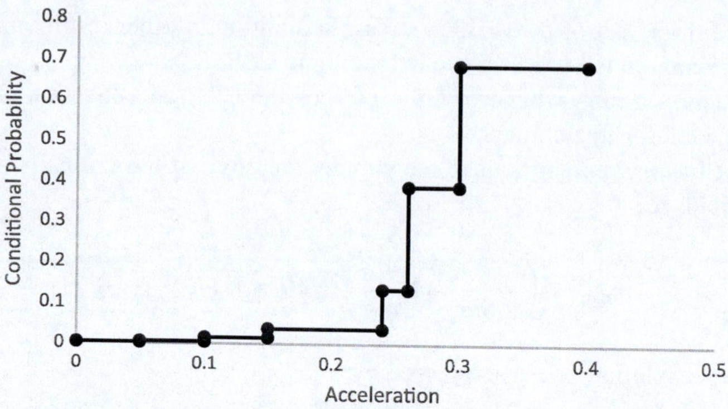

Fig. 4 Relation between conditional probability of liquefaction and acceleration for zone one

Figure 4 shows a typical curve computed using Eq. (12) for zone one. By combining the rate of occurrence for earthquake and the conditional probability of liquefaction, we can obtain the total probability of liquefaction. The number of earthquakes that can cause liquefaction annually is obtained by;

$$N_L = \sum [L|M] * (P(M)) \tag{13}$$

where $P(M)$ is the probability that an actual earthquake of magnitude M will cause liquefaction. Note that the number of earthquakes in $P(M)$ is calculated per year. Considering the liquefaction rate following (n) earthquakes during time interval (0, t) e.g., (during the time of (0) to time (t), (n) of earthquakes per year can cause liquefaction).

$$p(n) = \frac{(N_L * t)^n e^{-N_L t}}{n!} \tag{14}$$

At $n = 0$ (no earthquake will cause liquefaction) is given by;

$$p(0) = e^{-N_L t} \tag{15}$$

Using the complimentary function of Eq. (15) to obtain the probability of at least one event will cause liquefaction given as;

$$p(n \geq 1) = p[L] = 1 - e^{-N_L t} \tag{16}$$

The values of N_L for the two zones were calculated using Eq. (13) and Fig. 3. The results are listed in Table 1 also with using Eq. (16) and Table 1 the hazard of liquefaction for a time (t) for the two zones is presented in Table 2.

As anticipated, the risk of liquefaction increases as the time interval increases. The result showed that zone one is more susceptible to liquefaction compared to

Table 1 Values of N_L for the two zones

Zone	Value of N_L
1	0.00073
2	0.00052

Table 2 Probability of liquefaction (%) for the free field

Time (year)	Zone (1)	Zone (2)
10	0.73	0.52
20	1.45	1.03
30	2.16	1.52
40	2.89	2.05
50	3.54	2.54
100	7.23	5.01
250	17.71	12.03

zone two. This is assigned to the fact that the soil condition in zone one is loose to medium dense sand on average relative to zone two which is medium dense.

5 Conclusion

The following conclusions can be deduced from the present study;

1. The faults in Abu Dulayq area near Khartoum have been active recently and they have a potential of causing earthquakes in the presupposed mild risk area of Khartoum.
2. At central Khartoum the alluvial deposits and subsoil conditions revealed the existence of liquefiable saturated sand layers at various depths.
3. The liquefaction probability at central Khartoum differed between the two zones with respect to subsoil conditions showing minimal liquefaction risk.

References

1. Salgado R, Prezzi M (2014) Penetration rate effects on cone resistance: insights from calibration Chamber and field testing. Soils Rocks 37(3):233–242
2. Juang CH, Li DK (2007) Assessment of liquefaction hazard in Charleston quadrangle, South Carolina. Eng Geol (Amsterdam) 92(1–2):59–72
3. Central Bureau of Statistics (2018) Population indicators in Sudan. https://cbs.gov.sd/index.php/ar/statistics/main/7
4. Walling MY, Mohanty WK (2009) An overview on the seismic zonation and microzonation studies in India. Earth Sci Rev 96:67–91
5. Whiteman AJ (1971) The geology of the Sudan Republic. Clarendon Press, Oxford

6. Mohamedzein YE-A, Abdalla JA, ElSharief AM, Abdelwahab A, Elfati OO (2001) Seismic microzonation of Central Khartoum. In: Proceedings of the forth international conference on recent advances in geotechnical earthquake engineering and soil dynamics, San Diego, CA, USA, 26–31 Mar 2001
7. Mohamedzein YE-A, Abdalla JA, Abdelwahab A (2006) Site response and earthquake design spectra for Central Khartoum, Sudan. Bull Earthq Eng 4:277–293
8. Ambraseys NN, Adams RD (1986) Seismicity of the Sudan. Bull Seismol Soc Am 76:483–493
9. Abdalla JA, Mohamedzein YE, Abdelwahab A (2001) Probabilistic seismic hazard assessment of Sudan and its vicinity. Earthq Spectra 17(3):399–415
10. United States Geological Survey (USGS) (2003) Significant earthquakes of the world, 1990. Compiled by Waverly J. Person. https://neic.usgs.gov/neis/eqlists/sig_1990.html
11. Warage A (2007) Seismotectonics in Central Sudan and local site effect in Western Khartoum. Master thesis. Department of Earth Science, University of Bergen, Norway
12. Ibrahim KY, Babiker NM, Ali MH, Warage AA, Humeda KA (2015) A Unified Mw-based Earthquake catalog of Sudan and adjacent regions for the period 1408–2014. Am J Earth Sci 2(5):150–156
13. Kramer SL (1996) Geotechnical earthquake engineering. Prentice-Hall Inc., Upper Saddle River
14. McGuire RK (2004) Seismic hazard and risk analysis. Inc., Boulder
15. National Geological Research Center, Seismological Research Unite (2018) Earthquake records. 21. Unpublished report
16. Douglas J (2018) Ground motion prediction equations 1964–2018. Technical report. University of Strathclyde, Glasgow, UK
17. Humbert N, Viallet E (2008) An evaluation of epistemic and random uncertainties included in attenuation relationship parameters. In: Proceedings of fourteenth world conference on earthquake engineering, Paper no. 07-0117
18. Seed HB, Idriss IM (1971) Simplified procedure for evaluating soil liquefaction potential. J Soil Mech Found Div 97(9):1249–1273
19. Seed HB et al (1984) The influence of SPT procedures in soil liquefaction resistance evaluations. Report no. UCB/EERC-84/15. Earthquake Engineering Research Center. University of California, Berkeley
20. Seed HB, Tokimatsu K, Harder LF, Chung RM (1985) The influence of SPT procedures in soil liquefaction resistance evaluations. J Geotech Eng 111(12):1425–1445
21. Youd TL, Idriss IM, Andrus RD, Arango I, Castro G, Christian JT, Dobry R, Finn WDL, Harder LF, Hynes ME, Ishihara K, Koester JP, Liao SSC, Marcuson WF (2001) Liquefaction resistance of soils: summary report from the 1996 NCEER and 1998 NCEER/NSF workshops on evaluation of liquefaction resistance of soils. J Geotech Geoenv Eng 127(10):1061–1090
22. Idriss IM, Boulanger RW (2004) Semi-empirical procedures for evaluating liquefaction potential during earthquakes. Soil Dyn Earthq Eng 26:115–130
23. Idriss IM, Boulanger RW (2008) Soil liquefaction during earthquakes. Earthquake Engineering Research Institute, Oakland
24. Idriss IM, Boulanger RW (2012) Examination of SPT-based liquefaction triggering correlations. Earthq Spectra 28(3):989–1018
25. Peck RB (1979) Liquefaction potential: science versus practice. J Geotech Eng Div ASCE 105:393–398
26. Harr ME (1977) Mechanics of particulate media: a probabilistic an approach. McGraw Hill Book Co., New York
27. Elton DJ, Hadj-Hamou T (1990) Liquefaction potential map for Charleston, South Carolina. J Geotech Eng 116(2):244–265

Assessment of Liquefaction Potential Based on SPT Data by Using Machine Learning Approach

Salman Khan, Suvendu Kumar Sasmal, G. Suneel Kumar, and Rabi Narayan Behera

Abstract The evaluation and prediction of the geotechnical properties are critical and cumbersome tasks. This is due to the fact that the estimation of geotechnical properties depends on many factors, and the relationship among these factors is non-linear and complex, but the emerging machine learning techniques have made this task quite easy for geotechnical engineers. In this present paper, an attempt has been made to predict liquefaction based on 68 instances of SPT data with the help of various regression techniques. The regression techniques include Multiple Linear Regression (MLR), Stepwise Linear Regression (SLR), and linear regression with *K*-fold cross-validation. These techniques are modeled and simulated using statistical tool R-Studio. The model outcomes reveal that liquefaction potential can be reasonably predicted using sophisticated regression techniques. These regression models were simulated using seven input parameters such as depth (z), Groundwater Table (GWT), corrected SPT number (N_{60}), bulk unit weight (γ), saturated unit weight (γ_{sat}), fine content (%), and magnitude of the earthquake (M_w). The output parameter was the Factor of Safety (FOS). The analysis of the presented results suggests that the linear regression with *K*-fold cross-validation is the best method for predicting liquefaction, among the considered techniques.

Keywords Liquefaction · Machine learning · Regression analysis · Stepwise linear regression · *K*-fold cross-validation

S. Khan · S. K. Sasmal · G. S. Kumar · R. N. Behera (✉)
Department of Civil Engineering, National Institute of Technology Rourkela, Rourkela, India
e-mail: rnbehera82@gmail.com

S. Khan
e-mail: salmankhan.nitr@gmail.com

S. K. Sasmal
e-mail: suvendukumarsasmal@gmail.com

T. G. Sitharam et al. (eds.), *Seismic Hazards and Risk*, Lecture Notes in Civil Engineering 116, https://doi.org/10.1007/978-981-15-9976-7_14

1 Introduction

The term liquefaction is derived from the word liquefied which was first coined by Allen Hazen in 1918. Since then, it was rarely known, and it comes in big picture when the Niigata earthquake in 1964 (36 dead, 385 injured) and Alaska earthquake in 1964 (131 dead) happened at a disastrous level. During an earthquake, Liquefaction is one of the major causes of damage that occurs when the soil loses its strength and stiffness for a very small time but is enough to cause potential damage. To avoid destruction, the assessment and prediction of liquefaction on sandy soils have attracted considerable attention in recent years [1–5].

The soil liquefaction influenced by the magnitude of earthquake, the distance from the source of the earthquake, relative density, grain size distribution, fines content, plasticity of fines, degree of saturation, site-specific conditions, ground acceleration, type of soil and thickness of the soil deposit, confining pressure, permeability characteristics of soil layer, position and fluctuations of the groundwater table, and reduction of effective stress [6–8]. Liquefaction-induced ground failure is influenced by the thickness of non-liquefied and liquefied soil layers [9].

Predominantly, Standard Penetration Test (SPT) and Cone Penetration Test (CPT) have been widely used to evaluate properties of soil for assessing liquefaction. Other tests which used rarely are Becker's Penetration Test (BPT) and shear wave velocity test.

In this study, an attempt has been made to predict liquefaction based on Standard Penetration Test (SPT) data with the help of various regression techniques. A comparative analysis is presented among three different techniques, namely Multiple Linear Regression (MLR), Stepwise Linear Regression (SLR), and linear regression with cross-validation (K-fold). Based on the statistical study, the best method is suggested. Furthermore, the most important parameters while estimating the liquefaction are derived using the outcomes from the model simulations.

2 Review of Literature

Niigata and Alaska earthquake in 1964, attracted substantial attention of researchers to evaluate liquefaction and settlement caused due to liquefaction. The assessment and prediction of liquefaction are broadly classified into two classes, first being traditional methods, and another is artificial intelligence techniques.

2.1 Traditional Methods

Harry Bolton Seed, father of geotechnical earthquake engineering, was among the first to study liquefaction. Seed and Idrees [1] observed that the most important parameter in happening liquefaction is the relative density of soil. Seed and Idrees [10] developed a simplified procedure to evaluate liquefaction using $(N_1)_{60}$; this method is based on calculating Cyclic Stress Ratio (CSR) and Cyclic Resistance Ratio (CRR). Iwasaki et al. [11] developed Liquefaction Potential Index (LPI) in Japan to estimate the potential of liquefaction to cause foundation damage at a site. It indicates the severity of liquefaction; liquefaction risk is very high if LPI > 15 and low if LPI $\leq$ 5. Tokimatsu and Yoshimi [2] given an empirical chart in terms of dynamic shear stress ratio, SPT value, fine content, and shear strain amplitude to demarcate liquefiable and non-liquefiable conditions. Seed et al. [12] showed the influence of fines content on $(N_1)_{60}$ values and liquefaction strength of the soil. The results showed that the $(N_1)_{60}$ value decreases as the fines content increases. Amini et al. [13] conducted 150 stress-controlled undrained cyclic triaxial tests on stratified silty sand to quantify liquefaction resistance. In this work, the silt content and confining pressure vary from 10–50% and 50–250 kPa, respectively. No significant differences were found in liquefaction resistances in layered and uniform soils. The liquefaction resistance for layered as well as uniform soil conditions decreases with increasing confining pressure but it increases as the silt content increases.

2.2 Artificial Intelligence Techniques

Goh et al. [3] used CPT data in back-propagation neural network algorithm (ANN) for evaluating liquefaction. This model was successfully implemented to predict liquefaction in binary form (0 and 1) with a 94% success rate.

Samui et al. [14] used Relevance Vector Machine (RVM) to predict liquefaction susceptibility of soil based on actual CPT data. The results were compared with predictions done by ANN. It suggests that the normalization of q_c is not required for predicting liquefaction susceptibility. Sen and Akyol [15] developed genetic algorithms for parameter selection and evaluation of liquefaction potential. Based on CPT data, the term Liquefaction Index (LI) was given and summarized that LI value varies from 0 to 1. Liquefaction won't occur if the LI value is lower than 0.5 or vice versa. Bagheripour et al. [16] proposed a new relationship between the probability of Liquefaction (PL), $(N_1)_{60}$ and CSR. The PL and reliability index were calculated by using AFOSM techniques which are related to the genetic algorithm. Samui et al. [17] proposed a new model, Support Vector Machine (SVM) to determine liquefaction susceptibility of soil based on actual CPT. This model obtained an accuracy of 88.99%, and it clearly presented that only two input parameters (q_c and a_{max}) are enough to predict liquefaction.

3 Methodology

3.1 Machine Learning

In today's world, undoubtedly machine learning is one of the most powerful and fastest-growing technologies. More importantly, we are far from seeing its full potential. It is a tool to convert data into knowledge. In recent decades, there is a substantial increase in data generation after carrying out numerous experiments to estimate various properties of soil. These laboratory or in-situ experiments were mainly concerned with targeted properties only but after that, these data become useless until we analyze it and find the valuable hidden patterns within. Machine learning technologies are used to find out these relevant hidden and complex patterns by training the data collected from experiments or survey, and these pieces of information are used to predict the outcome of some future events and perform all kinds of complex decision making. To make it more elaborative and simple, machine learning is further classified into two famous types.

In supervised machine learning, a mathematical model is built by providing past observations to train a model to predict outcomes of some future events. Supervised learning makes use of labeled data. These predictive models can be classified into two groups, regression models and classification models.

Unsupervised machine learning, in contrast, do not build predictive models, instead it makes clusters to separate objects into different groups by identifying patterns and groups of the similar object within a dataset of interest. Principle Component Analysis (PCA) also falls inside unsupervised learning; it basically provides us information about the most important factor contained in the multivariate dataset. It makes use of unlabeled data (Fig. 1).

Fig. 1 Broad classification of machine learning

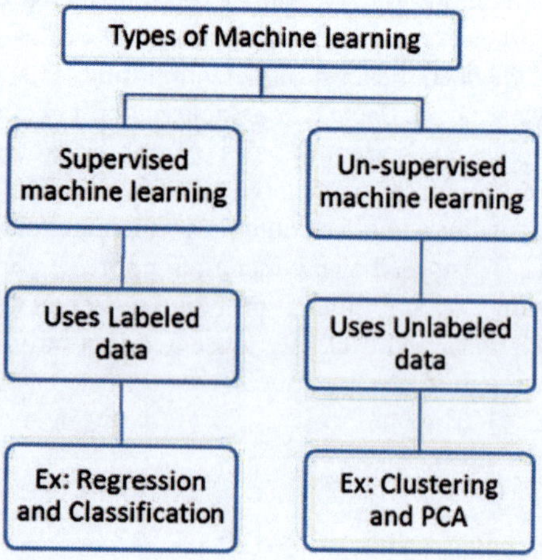

3.2 Regression Analysis

Regression analysis is a subcategory of supervised learning. It is a statistical method to find out the relationship between the quantitative target variable and one or more predictor variables. The target outcome values can be found out by using linear and non-linear strategies. Basically, regression models are used to predict a continuous value. *R*-Studio is a very robust platform, where we can perform a number of regression techniques to predict our desired outcome from the given predictor(s) properties of soil. Regression analysis (or regression model) consists of a set of machine learning methods that allow us to predict a continuous outcome variable (y) based on the value of one or multiple predictor variables (x) James et al. [18]. Briefly, the goal of a regression model is to build a mathematical equation that defines y as a function of the x variables. Regression analysis is further divided into many types; some of the famous regression analysis is simple linear, multi-linear, stepwise linear, polynomial, lasso, ridge, MARS, etc.

3.2.1 Linear Regression

Simple linear regression is the simplest and straightforward regression approach for predicting a quantitative response Y on the basis of a single predictor variable X. It assumes that there is approximately a linear relationship between X and Y, James et al. [19]. Mathematically, we can write this linear relationship as

$$Y \approx \beta_0 + \beta_1 X \tag{1}$$

In Eq. (1), β_0 and β_1 are two unknown coefficients that represent the *intercept* and *slope* terms of the *line of best fit* in a simple linear model. So, before using this Eq. (1) to make predictions, we must use data to estimate the coefficients. Let us suppose, following are n observations

$$(x_1,\ y_1),\ (x_2,\ y_2),\ \ldots,\ (x_n,\ y_n)$$

A line of best fit is a straight line that is the best approximation of the given set of data. On the basis of n observation pairs, the least-square residual approach is commonly applied to plot the line of best fit; the vertical distance between points and the line is termed as residuals or errors (e) which can be seen from Fig. 2. The sum of all the squared error is termed as the sum of squared residuals (SSR).where

$$e_1^2 = (y_1 - (\beta_0 + \beta_1 x_1))^2 \tag{2}$$

$$e_2^2 = (y_2 - (\beta_0 + \beta_1 x_2))^2 \tag{3}$$

similarly,

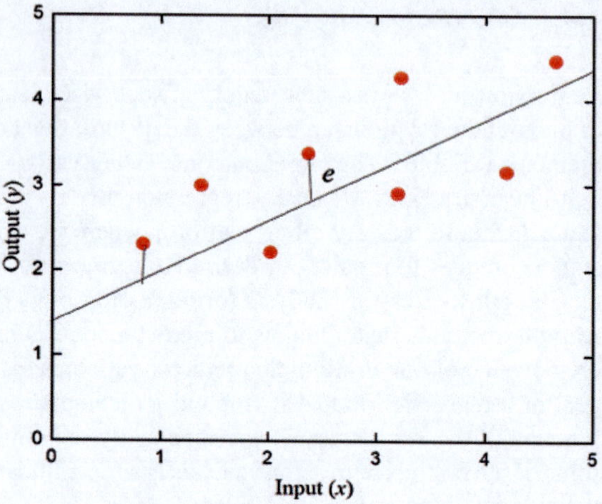

Fig. 2 Scatter plot showing line of best fit and residual error (e)

$$e_n^2 = (y_n - (\beta_0 + \beta_1 x_n))^2 \tag{4}$$

so,

$$\text{SSR} = e_1^2 + e_2^2 + \cdots + e_n^2 \tag{5}$$

To get the least square error, the SSR should be differentiated with respect to β_1 and set the result equal to zero, by differentiating the value of β_0 and β_1 would be as follows,

$$\beta_1 = \frac{N\sum(xy) - \sum x \sum y}{N\Sigma(x^2) - (\Sigma x)^2} = \frac{\text{cov}(x, y)}{\text{var}(x)} \tag{6}$$

and

$$\beta_0 = \frac{\Sigma y - \beta_1 \Sigma x}{N} \tag{7}$$

The equation of the line of best fit could be generated by putting the values of β_0 and β_1 in Eq. (1). Again, to get a measure of how this model fits the data, we go for two most common metrics named coefficient of determination (R^2) and Root mean square error (RMSE) and is defined as;

R^2 represents the squared correlation between the known observed values and the predicted outcome values by the model. The higher the R^2, the better the model.

$$R^2 = 1 - \frac{\text{SSE}}{\text{TSE}} = 1 - \frac{\text{sum of squared errors}}{\text{total sum of squares}} \tag{8}$$

RMSE measures the model prediction error. It corresponds to the average deviation between the known observed values of the outcome and the predicted value by the model. The lower the RMSE, the better the model.

$$\text{RMSE} = \sqrt{\frac{1}{N}\sum_{\text{i}=1}^{\text{N}}\left(\text{mean}(\text{observeds} - \text{predicteds})^2\right)} \tag{9}$$

3.2.2 Multiple Linear Regression (MLR)

Simple linear regression is based on a single predictor variable only. However, in practice, we often have more than one predictor variable; therefore, multiple linear regression comes in the picture as a link between more than one predictor variables and a target variable. The expression for MLR takes the following form,

$$Y \approx \beta_0 + \beta_1 X_1 + \beta_2 X_2 + \cdots + \beta_n X_n \tag{10}$$

3.2.3 Stepwise Linear Regression (SLR)

It is a principal component analysis method that adds or removes predictors in a predictive model in order to achieve the best performing model having the lowest predictive error. The selection of predictors can be done in three ways, which is forward selection, backward selection, and stepwise selection. Forward selection starts with zero predictors and adds the most contributive parameters. The process stops when there is no significant improvement in the model; in contrast, the backward selection model starts with all parameter and iteratively removes the least contributive parameter and stops when a further reduction in parameter results in a significant change in model overall accuracy. Stepwise selection is a combination of both forward and backward.

3.2.4 Linear Regression with Cross Validation

Cross-validation is a term in machine learning which refers to the set of methods for assessing the performance of a given predictive model on a new test dataset. Many approaches are there to quantify the model accuracy such as the validation set approach, Leave One Out Cross-Validation (LOOCV), *K*-fold cross-validation, and repeated *K*-fold cross-validation. R^2, RMSE, and MAE are used to measure the regression model performance during cross-validation.

This approach evaluates the performance of the model on a different subset of the training data and then evaluates the mean prediction error rate. In the following

Table 1 Statistical summary of the parameters

Variables	Depth (m)	SPT $(N)_{60}$	(γ) (KN/m^3)	(γ_{sat}) (KN/m^3)	GWT	F.C. (%)	Mw	FOS
Min	1.50	0.00	14.90	15.17	0.00	35.00	5.00	1.367
Max	15.00	45.00	18.80	20.09	8.00	95.60	6.00	11.432
Mean	5.801	9.147	16.16	16.64	1.988	57.24	5.191	3.827
Standard deviation	3.49	10.56	1.19	1.64	2.86	11.68	0.39	2.28

steps, this cross-validation calculates error; firstly, the dataset would be randomly split into *k* number of subsets (or *k*-folds); secondly, train the model on all subsets except one that would be reserved for testing. In the third step, test the model on the reserved subset and record the prediction error. This process would be repeated until each of the k subsets has served as the test set and finally computes the average errors.

3.2.5 Database Used

The dataset used in this study consists of reported borehole data from field investigations. The method of preparing the dataset along with the numerical values of inputs and the output can be found in Khan [20]. Out of total 68 data, 52 data are used for training and remaining 16 data are used for testing. Factor of safety (FOS) against liquefaction is computed for the present dataset based on the ratio of cyclic resistance ratio (*CRR*) to the cyclic stress ratio (*CSR*). The CRR and CSR are calculated as per Youd and Idriss [21]. The entire analysis is performed using statistical tool R Studio [22].

4 Results and Discussion

4.1 MLR Model

Multiple linear regression could be developed by assigning more than one or all the input parameters. Here, the accuracy obtained by considering all input variable is $R^2 = 0.87$. The R^2 for testing is found to be 0.85.

The MLR model equation is found to be;

$$\text{FOS} = 39.79 - 0.03531 \times d + 0.248 \times (\text{N}_1)_{60} - 0.18 \times \gamma_b + 0.26 \times \gamma_s + 0.817 \times \text{GWT} - 0.07 \times \text{FC} - 7.16 \times \text{Mw} \quad (11)$$

4.2 SLR Model

In a given sequence, the predictors were added to the model; initially, it was only SPT no. $(N_1)_{60}$; and further, it adds Fine Content (FC), GWT, M_w, and depth (z) one by one. The sequence of adding up the predictors was automatically done by *R*-Studio on the basis of the most contributing parameter to the outcome. When the number of variables is four, then this model represents maximum accuracy.

The SLR model equation is;

$$\mathrm{FOS} = 43.314 + 0.28 \times (\mathrm{N}_1)_{60} + 0.98 \times \mathrm{GWT} - 0.126 \times \mathrm{FC} - 7.068 \times \mathrm{Mw} \tag{12}$$

4.3 K-fold Cross-Validation Model

In this method, the total 10 numbers of the fold have been assigned to cross-validate the linear model, and the accuracy of the model increases much and the value $R^2 = 0.89$ for testing data.

This model equation is;

$$\mathrm{FOS} = 37.50 - 0.0214 \times d + 0.245 \times (N_1)_{60} + 0.0064 \times \gamma_b + 0.12 \times \gamma_s + 0.854 \times \mathrm{GWT} - 0.087 \times \mathrm{FC} - 6.67 \times \mathrm{Mw} \tag{13}$$

4.4 Comparative Analysis

A comparative analysis is done to estimate the best model among the considered models. The training and testing outcomes predicted by the models are compared with the actual reported outcomes in Fig. 3. It can be interpreted that all the three methods have reasonable prediction capability as most of the points lie between the ±10% deviation lines. However, in the case of testing data, the coefficient of determination for the *k*-fold method is greater than those of other two methods. Hence, it is clear that the *K*-fold method has good generalization ability.

The RMSE and Mean Absolute Error (MAE) obtained for different methods are illustrated in Fig. 4. The errors are less for the *K*-fold technique. The MLR model shows the highest errors. Therefore, from Figs. 3 and 4, it is inferred that among the models, the linear regression with *K*-fold cross-validation model is the best followed by SLR and MLR.

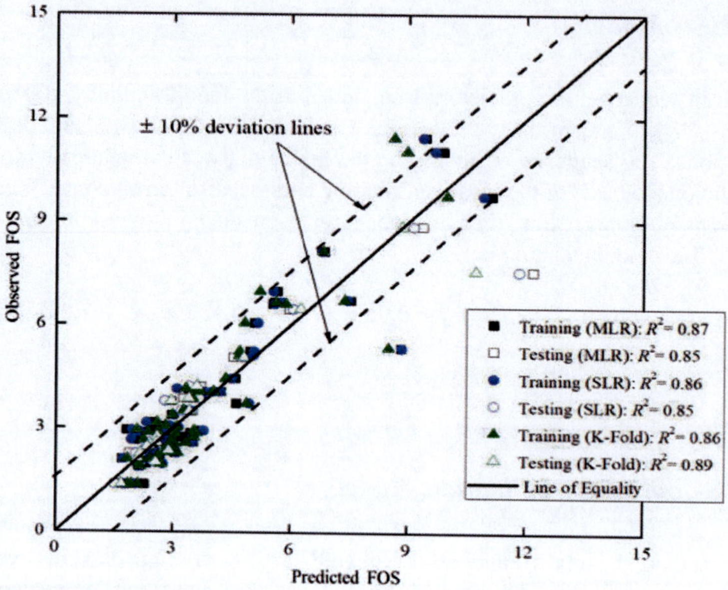

Fig. 3 Comparison between the observed FOS and predicted FOS

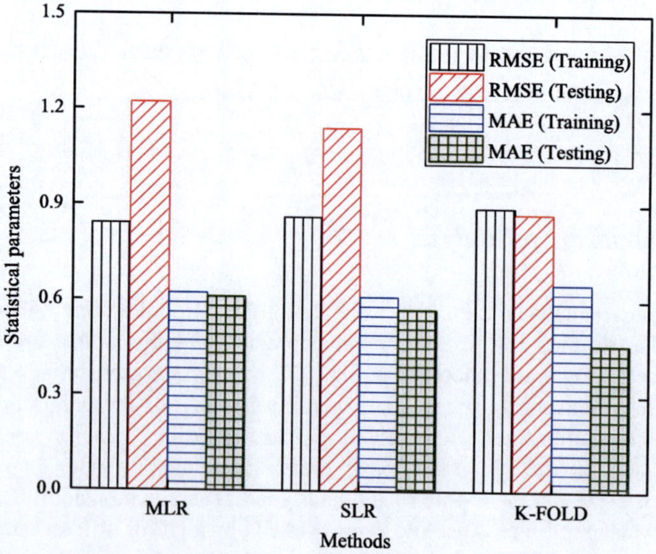

Fig. 4 Statistical comparison among the considered models

5 Conclusions

In this paper, three regression techniques are employed in assessing liquefaction, and the evaluation of these techniques is done based on performance metrics viz. R^2, RMSE, and MAE.

- The K-fold cross-validation model is the best method among the considered regression techniques due to its reasonably good generalization capacity and lower values of RMSE and MAE.
- Three empirical expressions are proposed Eqs. (11), (12), and (13) to estimate the liquefaction, with reasonable accuracy.
- From the analysis of obtained coefficients, it is found that SPT value and fine content are the most contributing parameter in the occurrence of liquefactions.

References

1. Seed HB, Idriss IM (1967) Analysis of soil liquefaction: Niigata earthquake. J Soil Mech Found Div 93:83–108
2. Tokimatsu K, Yoshimi Y (1983) Empirical correlation of soil liquefaction based on SPT N-value and fines content. Soils Found 23:56–74
3. Goh AT (1996) Neural-network modeling of CPT seismic liquefaction data. J Geotech Eng 122:70–73
4. Idriss IM, Boulanger RW (2012) Examination of SPT-based liquefaction triggering correlations. Earthq Spectra 28:989–1018
5. Huang Y, Yu M (2013) Review of soil liquefaction characteristics during major earthquakes of the twenty-first century. Nat Hazards 65:2375–2384
6. Youd TL, Perkins DM (1978) Mapping liquefaction-induced ground failure potential. J Geotech Eng Div 104:433–446
7. Dixit J, Dewaikar DM, Jangid RS (2012) Assessment of liquefaction potential index for Mumbai city. Nat Hazards Earth Syst Sci 12:2759–2768
8. Kumar SS, Krishna AM, Dey A (2013) Parameters influencing dynamic soil properties: a review treatise. In: National conference on recent advances in civil engineering, pp 1–10
9. Ishihara K (1993) Liquefaction and flow failure during earthquakes. Geotechnique 43:351–451
10. Seed HB, Idriss IM (1971) Simplified procedure for evaluating soil liquefaction potential. ASCE J Soil Mech Found Div 97(9):1249–1273
11. Iwasaki T, Tatsuoka F, Tokida K, Yasuda S (1978) A practical method for assessing soil liquefaction potential based on case studies at various sites in Japan. In: 2nd international conference on microzonation. San Francisco, pp 885–896
12. Seed HB, Tokimatsu K, Harder LF, Chung RM (1985) Influence of SPT procedures in soil liquefaction resistance evaluations. J Geotech Eng ASCE 111(12):1425–1445
13. Amini F, Qi GZ (2000) Liquefaction testing of stratified silty sands. J Geotech Geoenviron Eng 126(3):208–217
14. Samui P (2007) Seismic liquefaction potential assessment by using relevance vector machine. Earthq Eng Eng Vib 6:331–336
15. Sen G, Akyol E (2010) A genetic-algorithm approach for assessing the liquefaction potential of sandy soils. Nat Hazards Earth Syst Sci 10:685–698
16. Bagheripour MH, Shooshpasha I, Afzalirad M (2012) A genetic algorithm approach for assessing soil liquefaction potential based on reliability method. J Earth Syst Sci 121(1):45–62

17. Samui P (2013) Liquefaction prediction using support vector machine model based on cone penetration data. Front Struct Civ Eng 7(1):72–82
18. James G, Witten D, Hastie T, Tibshirani R (2017) An introduction to statistical learning: with applications in R. In: Springer texts in statistics, Corrected at 8th printing edn. Springer, New York
19. James G, Witten D, Hastie T, Tibshirani R (2013) An introduction to statistical learning. In: Springer texts in statistics. Springer, New York
20. R studio (version 2.1): computer software. https://rstudio.com/products/rstudio
21. Khan S (2020) Assessment of liquefaction potential based on SPT, CPT and shear wave velocity data using machine learning approach. M. Tech Thesis. National Institute of Technology Rourkela, India
22. Youd TL, Idriss IM (2001) Summary report from the 1996 NCEER and 1998 NCEER/NSF workshop on evaluation of liquefaction resistance of soils. J Geotech Geo-Environ Eng ASCE 127: 297–313. https://doi.org/10.1061/(ASCE)1090-0241(2001)127:4(297)

Comparison of Ballastless and Ballasted Track for High-Speed Train

Naveen Kumar Kedia, Anil Kumar, and Yogendra Singh

Abstract To cope with the demand of an increasing population and to modernize existing railway facilities, Indian Government has started the construction of a high-speed rail between Mumbai and Ahmedabad. Thus, the present paper deals with studying an impact of high-speed train on railway bridges for ballastless and ballasted track. A train has been modeled as the multi-body system, having only vertical degrees of freedom. Both the track system is modeled using finite element method in which rail, slab, and bridge are represented as 2D beam element while rail pads as spring and damper elements. The track irregularity is considered as a major source of vibration and is obtained using the grade six power spectral density curve provided by the Federal Railroad Administration of America (FRA) and is further normalized to keep the irregularity within a permissible limit. A non-linear Hertzian contact theory is used to the study the wheel rail contact mechanism. The dynamic train and track coupled equation are solved using suitable numerical techniques, and since non-linearity is involved, so a further convergence criteria is applied to solve a dynamic equations. It was found that ballasted track is more effective in reducing bridge vibrations compared to ballastless track.

Keywords Finite element model · Wheel rail contact · High-speed rail

N. K. Kedia (✉)
Center for Transportation Systems, IIT Roorkee, Roorkee 247667, India
e-mail: nkedia@ct.iitr.ac.in

A. Kumar
Department of Mechanical and Industrial Engineering, IIT Roorkee, Roorkee 247667, India
e-mail: anil.kumar@me.iitr.ac.in

Y. Singh
Department of Earthquake Engineering, IIT Roorkee, Roorkee 247667, India
e-mail: yogendra.singh@eq.iitr.ac.in

T. G. Sitharam et al. (eds.), *Seismic Hazards and Risk*, Lecture Notes in Civil Engineering 116, https://doi.org/10.1007/978-981-15-9976-7_15

1 Introduction

In India, a rapid growth is taking place in the field of railway engineering, and a prime focus of the government is to upgrade the existing technology and to introduce a new high speed railway corridors. In this regard, national high-speed rail corporation limited (NHSRCL) was set up in 2016 that will be responsible for the construction and operation of high-speed trains in India. The Mumbai–Ahmedabad high-speed rail corridor (MAHSR) has a total length of 508.17 km out of which 460.3 km will be on viaduct, 9.22 km on bridge, 25.87 km will pass through tunnel out of which 7 km stretch will be undersea tunnel passing through reserved area of Thane creek, and 12.9 km will be on embankment/cutting. The maximum operating speed of a trains will be 320 km/h, and rolling stock similar to Japanese Shinkansen trains will be used [1]. Since most of the tracks lie on an elevated structure, so present study is focused on a train–track–bridge interaction to study the effectiveness of ballasted and ballastless track system. Several researchers have studied the train–track–bridge interaction; like Matsuoka et al. [2] had developed a numerical model of train–bridge system to evaluate effect of inclusion of local deck vibration on prediction of maximum acceleration of bridge; Antolin et al. [3] developed a model in which non-linear contact forces are considered for analyzing the dynamic interaction between high-speed trains and bridges; Guo et al. [4] had developed the 3D dynamic model of train–bridge coupled system for a critical assessment of Sesia viaduct located in Italy; Song et al. [5] had proposed 3D dynamic model of train–bridge interaction system in which they used improved non-conforming flat shell (NFS) elements to model deck of a railway bridge; Liu et al. [6] had studied the dynamic analysis of multispan simply supported railway viaducts subjected to moving trains using component mode synthesis reduced order model; Ju and Lin [7] investigated a resonant characteristics of railway bridges under an influence of high speed train; He et al. [8] studied the effect of dynamic bridge–train interaction on a seismic response of Shinkansen system under a influence of moderate earthquakes using a combination of multi-body dynamics and finite element method; Su et al. [9] had used field measurement and analytical simulation to investigate the dynamic interactions between high-speed train and reinforced concrete viaducts. The literature survey highlights that very few studies have been performed to check the effect of track system on bridge vibrations. Therefore, this paper focuses on evaluating the simply supported bridge responses under influence of high-speed train for different types of track system. The paper is arranged as follows, train, track, and bridge modeling which are discussed in Sect. 2; solution technique is explained in Sect. 3; results are discussed in Sect. 4; finally, main outcomes are summarized in conclusion section.

2 Dynamic Model of the Train–Track–Bridge Interaction (TTBI) System

A coupled train–track–bridge dynamic interaction model is used in the present research to study a response of bridge under influence of high-speed train for ballasted and ballastless track system. A train has been modeled as mass spring system, whereas finite element is used for modeling of track and bridge. Interaction among train and track system is achieved through the wheel rail contact force arising at the interface. Analytical functions are used for producing track irregularities responsible for dynamic excitation.

2.1 *Train Model*

A quarter car model of a bullet train, having three degrees of freedom, running on random track irregularity, is used in a present study. A schematic representation of a model is shown in Fig. 1. This vehicle model is a combination of quarter of a car body, half of a bogie and a wheel; primary suspension lies between bogie and wheel and secondary suspension between car body and bogie. The governing equation of motion can be written in matrix form as shown below,

$$[M_v]_{3\times3}\{\ddot{a}_v\}_{3\times1} + [C_v]_{3\times3}\{\dot{a}_v\}_{3\times1} + [K_v]_{3\times3}\{a_v\}_{3\times1} = [P_v]_{3\times1} \tag{1}$$

here, M_v, C_v and K_v are mass, damping, and stiffness matrix, respectively; $\ddot{a}_v$, $\dot{a}_v$, a_v are acceleration, velocity, and displacement vector, respectively; and P_v is a force vector consisting of both gravity components and wheel rail interaction forces $f_{w/r}$. The details of a above matrices can be found in the literature [10]. A non-linear

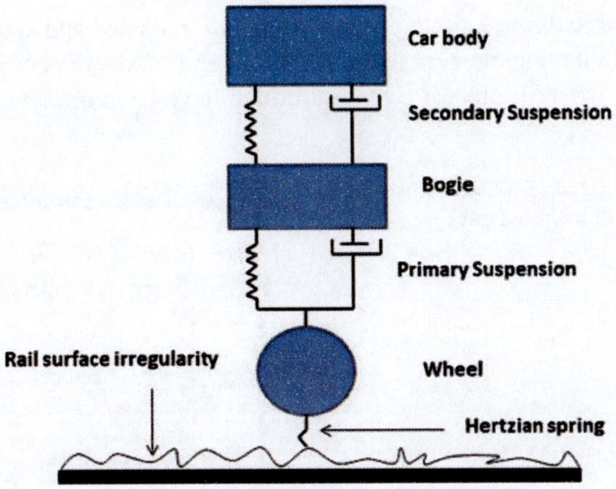

Fig. 1 Quarter car model of bullet train

Table 1 Properties of bullet train [8]

Parameters	Value
Weight of car body	321.6 kN
Weight of bogie	25.9 kN
Weight of wheel	8.8 kN
Primary suspension	443 kN/m
Primary suspension damping	21.6 kN s/m
Secondary suspension	1210 kN/m
Secondary suspension damping	19.6 kN s/m

Hertzian theory is used to describe the interaction forces between wheel and rail and is given by the following equation

$$f_{w/r} = \begin{cases} k_h(a_w - a_r - n_r)^{\frac{3}{2}} \text{ if, } a_w > a_r + n_r \\ 0 \qquad\qquad\qquad \text{else} \end{cases} \tag{2}$$

where k_h, a_w, a_r, n_r are Hertzian spring constant, wheel vertical displacement, rail vertical displacement, and rail surface irregularity, respectively.

The technical specifications of a bullet train are given in Table 1.

2.2 Track—Bridge Subsystem Model

Track and bridge in a present research has been modeled as a single unit using 2D finite element technique in which bridge is considered as the simply supported beam and track as the layered structure. Ballasted track consists of rail, rail pads, sleeper, and ballast placed over bridge structure, whereas ballastless track, i.e., slab track consists of rail, rail pads, precast slab, and cement mortar layer placed over bridge structure. A finite element model of ballasted and slab track is shown in Figs. 2 and 3 having element length of 0.6 m and 12 degrees of freedom (DOFs), respectively. The rail, precast slab, and bridge have been modeled as a Bernoulli–Euler beams,

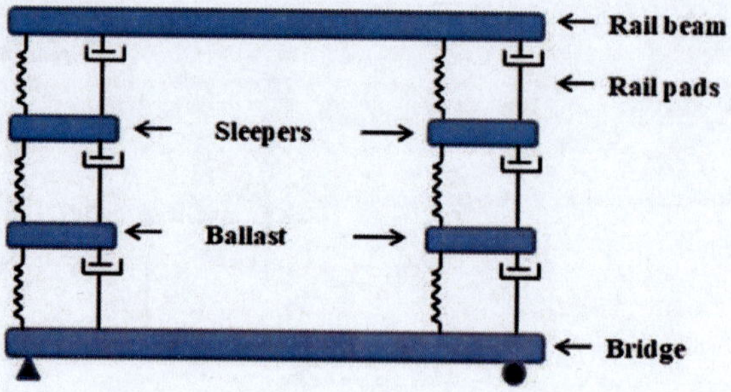

Fig. 2 Finite element model of ballasted track

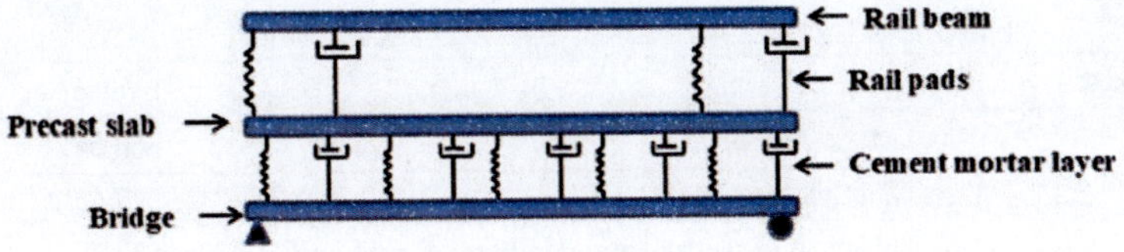

Fig. 3 Finite element model of ballastless track

sleepers, and ballast are considered as a lumped masses placed at a distance of 0.6 m. Rail pads are considered as an equally spaced spring dashpot units, whereas ballast is also considered to be connected to the sleeper and bridge through spring dashpot units. The shear stiffness of a ballast bed is not considered in a present study. Cement mortar layer is represented as a continuous spring damper elements. A track length of 60 m is considered for a present work in which ends are simply supported, and whole track—bridge system is discretized into 100 elements having 606 DOFs.

The properties of a bridge are taken from the literature [6] and are as follows; mass per unit length is 11400 kg/m, and bending stiffness 9.92×10^{10} N m^2. Technical specifications of a track system are given in Table 2. The dynamic equation of track system obtained using finite element assembling rule is shown below,

$$[M_t]_{N\times 1}\{\ddot{a}_t\}_{N\times 1} + [C_t]_{N\times N}\{\dot{a}_t\}_{N\times 1} + [K_t]_{N\times N}\{a_t\}_{N\times 1} = [P_t]_{N\times 1} \tag{3}$$

M_t, C_t, K_t are the global mass, damping, and stiffness matrices of the track-bridge system, respectively; a_t is a displacement vector; P_t is a load vector consisting of gravity force and wheel rail interaction force elements, and N is number of DOFs. The Rayleigh damping model which is a linear combination of mass and stiffness matrices is used to represent damping of rail, precast slab, and bridge. It can be written as follows

$$[C] = \alpha[M] + \beta[K] \tag{4}$$

here, α and β are Rayleigh damping coefficients; a detailed discussion on Rayleigh damping model can be found in the literature [11].

Track irregularity. Track irregularity samples are generated using the power spectral density (PSD) function given by the FRA for vertical profile irregularity and are expressed by the following equation,

$$S_v(\omega) = \frac{kA_v\omega_o^2}{(\omega^2 + \omega_o^2)\omega^2}(\text{cm}^2/\text{rad/m}) \tag{5}$$

In present work, irregularity is obtained for grade 6 using the above function, details of the constants, and solution procedure can be found in the literature [13].

Table 2 Track properties [12]

Parameter		Value
Rail	Mass (kg/m)	60.64
	Section area (cm^2)	77.45
	Moment of inertia around horizontal axis (cm^4)	3217
	Elasticity modulus (MPa)	2.06×10^5
	Distance between ties (m)	0.6
Rail pads	Stiffness coefficient (MN/m)	60
	Damping coefficient (kN s/m)	47.7
Precast slab	Length (mm)	6450
	Width (mm)	2550
	Height (mm)	200
	Density (kg/m^3)	2500
	Elasticity modulus (MPa)	3.9×10^4
Cement mortar	Stiffness coefficient (MN/m)	900
	Damping coefficient (kN s/m)	83
Sleeper	Mass (kg)	340
Ballast	Mass of ballast between two consecutive ties (kg)	2718
	Stiffness of ballast (MN/m)	120
	Damping of ballast (kN s/m)	60

3 Solution Technique

The dynamic equation of train and track–bridge system is solved independently, and interaction among them is achieved through the forces arising at the wheel rail interface. Newmark beta method [14] is used to solve the train and track dynamic Eqs. (1) and (3), respectively. Since non-linearity is involved at wheel rail contact point, convergence check was performed at each time step of numerical integration to optimize the value.

4 Results and Discussion

The results of bridge vibrations for six different train speeds of 40, 50, 60, 80, 100, and 120 m/s are presented and discussed in this section.

Figure 4 compares the midpoint acceleration of bridge at different speeds; it can be seen that at lower speed, not much difference is observed in bridge acceleration for different track system but on the other hand, as the speed increases higher response is observed in the case of ballastless track compared to ballasted track.

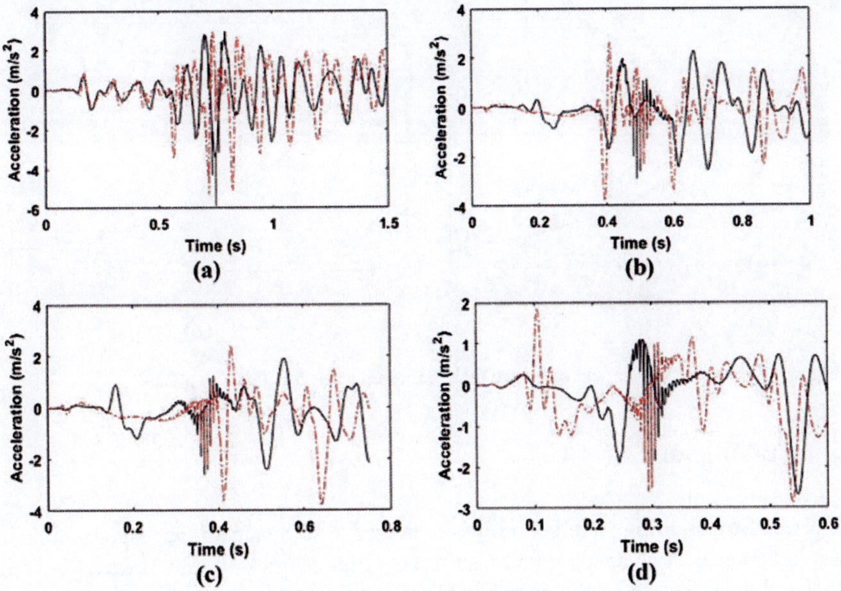

Fig. 4 Midpoint acceleration of bridge for ballasted track (·····) and ballastless track (-----) at different train speeds **a** 40 m/s **b** 60 m/s **c** 80 m/s **d** 100 m/s

Furthermore, a comparison of maximum value of midpoint acceleration of bridge at different train speeds for ballasted and ballastless track is shown in Fig. 5a, b, respectively; a noticeable difference can be seen in the values.

Moreover, a comparison of midpoint displacement of bridge for ballasted and ballastless track is shown in Fig. 6a, b, respectively. Displacements are higher for the ballastless track.

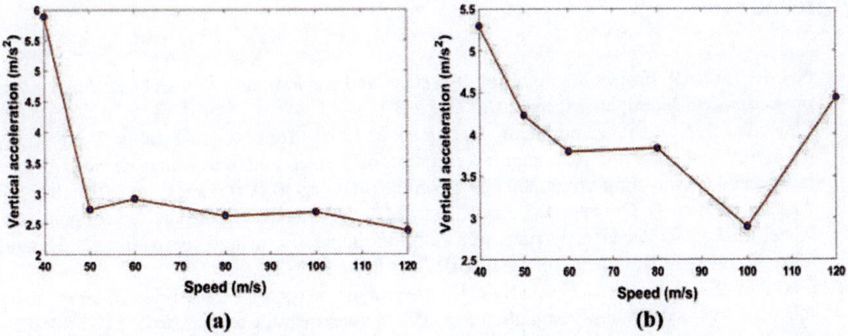

Fig. 5 Maximum value of midpoint acceleration of bridge for **a** ballasted track **b** ballastless track

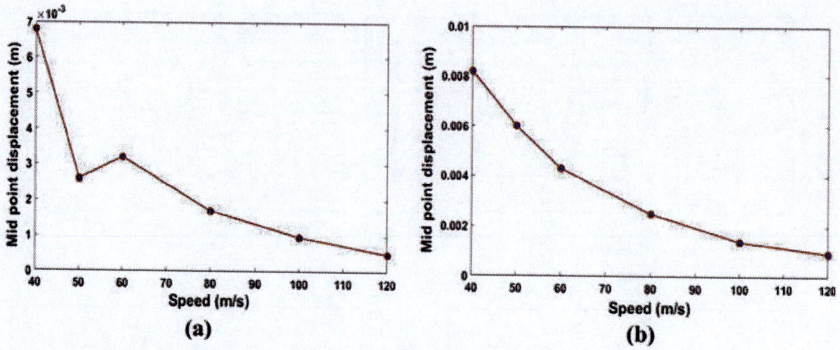

Fig. 6 Midpoint displacement of bridge for **a** ballasted track **b** ballastless track

5 Conclusions

The train–track–bridge interaction system is used to study the bridge responses under the influence of high-speed train for different types of track system. It was observed that rail acceleration in the case of ballastless track is much higher compared to ballasted track, and difference becomes prominent with the increase in speed. Similar trend was observed for the rail displacement. Thus, for high-speed train, use of ballasted track is more effective.

Since the high-speed corridor in India will be mostly elevated, present research is focused on the train–track–bridge interaction but the work can be extended to study the impact of high-speed train running on the ground and the effect of soil properties on train induced ground vibrations. Moreover, to predict more accurate results, actual track irregularity data if available can be used in the present model.

References

1. NHSRCL (2019) Project highlights of high speed rail corridor. https://nhsrcl.in/index.php/en/project/project-highlights. Accessed 29 Dec 2019
2. Matsuoka K, Collina A, Somaschini C, Sogabe M (2019) Influence of local deck vibrations on the evaluation of the maximum acceleration of a steel-concrete composite bridge for a high-speed railway. Eng Struct 200:109736. https://doi.org/10.1016/j.engstruct.2019.109736
3. Antolín P, Zhang N, Goicolea JM, Xia H, Astiz MÁ, Oliva J (2013) Consideration of nonlinear wheel-rail contact forces for dynamic vehicle-bridge interaction in high-speed railways. J Sound Vib 332:1231–1251. https://doi.org/10.1016/j.jsv.2012.10.022
4. Guo WW, Xia H, De Roeck G, Liu K (2012) Integral model for train-track-bridge interaction on the Sesia viaduct: dynamic simulation and critical assessment. Comput Struct 112–113:205–216. https://doi.org/10.1016/j.compstruc.2012.09.001
5. Song MK, Noh HC, Choi CK (2003) A new three-dimensional finite element analysis model of high-speed train-bridge interactions. Eng Struct 25:1611–1626. https://doi.org/10.1016/S0141-0296(03)00133-0

6. Liu K, Lombaert G, De Roeck G (2014) Dynamic analysis of multispan viaducts with weak coupling between adjacent spans. J Bridg Eng 19:83–90. https://doi.org/10.1061/(ASCE)BE.1943-5592.0000476
7. Ju SH, Lin HT (2003) Resonance characteristics of high-speed trains passing simply supported bridges. J Sound Vib 267:1127–1141. https://doi.org/10.1016/S0022-460X(02)01463-3
8. He X, Kawatani M, Hayashikawa T, Matsumoto T (2011) Numerical analysis on seismic response of Shinkansen bridge-train interaction system under moderate earthquakes. Earthq Eng Eng Vib 10:85–97. https://doi.org/10.1007/s11803-011-0049-1
9. Su D, Fujino Y, Nagayama T, Hernandez JY, Seki M (2010) Vibration of reinforced concrete viaducts under high-speed train passage: measurement and prediction including train-viaduct interaction. Struct Infrastruct Eng 6:621–633. https://doi.org/10.1080/15732470903068888
10. El Kacimi A, Woodward PK, Laghrouche O, Medero G (2013) Time domain 3D finite element modelling of train-induced vibration at high speed. Comput Struct 118:66–73. https://doi.org/10.1016/j.compstruc.2012.07.011
11. Yusuf AI, Amin NM (2015) Determination of Rayleigh damping coefficient for natural damping rubber plate using finite element modal analysis. In: Proceedings of international civil infrastructure engineering conference. Springer, Singapur, pp 713–725. https://doi.org/10.1007/978-981-287-290-6
12. Lei X, Zhang B (2011) Analysis of dynamic behavior for slab track of high-speed railway based on vehicle and track elements. J Transp Eng 137:227–240. https://doi.org/10.1061/(ASCE)TE.1943-5436.0000207
13. Yang YB, Yau JD, Wu YS (2004) Vehicle-bridge interaction dynamics. World Sci. https://doi.org/10.1142/5541
14. Bathe K-J (2016) Finite element procedures, 2nd edn. https://web.mit.edu/kjb/www/Books/FEP_2nd_Edition_4th_Printing.pdf

ANN-Based Model to Predict the Screening Efficiency of EPS Geofoam Filled Trench in Reducing High-Speed Train-Induced Vibration

Mainak Majumder and Sayan Bhattacharyya

Abstract Trench or wave barrier is an economical and cost effective way to minimize surface vibrations and screen the vibration of sensitive structures from the unwanted shaking. Previous researches mainly focused either on experimental or on analytical work. Owing to the complexities involved in the mathematical formulation, involvement of vast numbers of parameters, and the requirement of computational time to numerical analysis of an effective vibration screening system. The present study aims to explore the use of an artificial neural network to estimate the vibration screening effectiveness of an EPS geofoam filled trench in reducing train-induced vibration. Identifying different key parameters, a MLR model has been developed to know the influence of key parameters governing the vertical vibration screening of EPS geofoam trench wave barriers. It has been seen that the ANN model can effectively and accurately predict the averaged amplitude reduction ratio of the EPS geofoam infilled trench.

Keywords Vibration screening · EPS geofoam · ANN · Screening efficiency

1 Introduction

High-speed trains are one of the major sources of ground vibrations. The ground vibrations in the form of ground shaking, structural degradation are common in the closely spaced structures. Mitigation of these vibrations has become a critical aspect, especially in densely populated urban and sub-urban areas. Providing base isolation or other damping devices are generally adopted for the buildings with the high seismic activities. On the contrary, trenches or wave barriers are cost-effective solution over the past decades. Open trenches are best solution in this regards but owing to several

M. Majumder (✉)
Department of Civil Engineering, Indian Institute of Technology Kanpur, Kanpur, India
e-mail: mainak91190@gmail.com

S. Bhattacharyya
SMEC India Limited, Gurgaon, New Delhi, India
e-mail: sayan.bhattacharyya93@gmail.com

T. G. Sitharam et al. (eds.), *Seismic Hazards and Risk*, Lecture Notes in Civil Engineering 116, https://doi.org/10.1007/978-981-15-9976-7_16

limitations infilled trenches with different fill materials like concrete, soil-bentonite, soil bags, polyurethane have been used over the years. Expanded polystyrene (EPS) geofoam which is one of the popular products among the geosynthetics family which have been used as vibration screening material [1–5], in reducing different ground vibration sources. EPS geofoam barriers effectiveness depends on the dimensionless geometry and its location from the source of disturbances. The trench width has a minor effect in case of larger depth trench and high-frequency excitations. However, since the problem consists of large number of variables and the above all studies are site specific, it is very difficult to make any conclusive remarks. Moreover, with a single trench at very larger depth is unrealistic in practical condition to achieve the amount of screening efficiency. To resolve this issue, the present study analyze vibration screening system using artificial neural network (ANN). First, 2D FEM analyses have been carried out, and different configurations of EPS geofoam filled trench have been studied varying the main geometric features such as width, depth, distance of trench from the toe of the embankment, shear wave velocity of soil deposit, EPS geofoam density, soil damping ratio, soil Poisson's ratio. The interdependence of different parameters and quantification of the different parameter's influence have been identified through developed ANN-based model.

2 Theoretical Background

A typical ANN model consists of an input layer, an output layer, and one or more hidden layers. The processing nodes or neurons are equal to the number of input and output parameters. The choice of hidden layers used to be depending on the optimization using different trial and error process. The different neurons are connected by different weighted in different nodes. The combined input based on the weighted input signals used to be passed by linear/non-linear transfer function to give output at the node. In this way, complex problem where physical modeling is not possible can be analyzed. The method adopted in the present study is feedforward neural network along with the back propagation algorithm. From the name, it can be understood that feedforward means mean data moves forward from one layer to another layer. The summary of this method can be found out in Fig. 1. Further, the flow chart has been implemented in the MATLAB (R2014b) environment, where log sigmoidal (logsig) function has been used as the transfer function for the hidden layers, whereas pure linear transfer function has been used for the output layer. The weight and bias values have been updated following the *Levenberg–Marquardt* optimization algorithm (LMA).

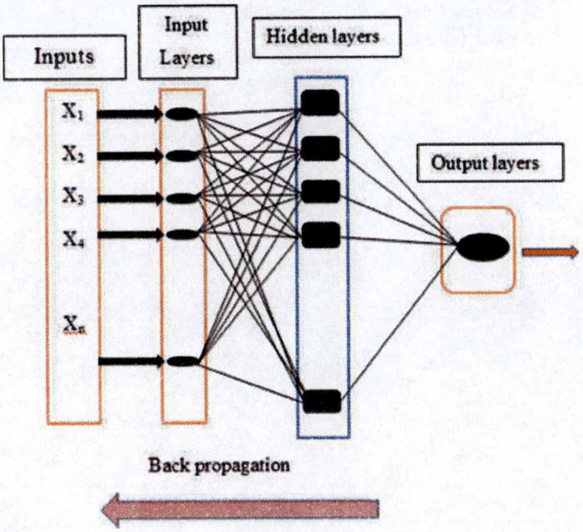

Fig. 1 Typical feedforward neural network along with back propagation algorithm

3 Finite Element Model Development

A 2D finite element model has been developed to carry out an exhaustive parametric study to know the behavior of EPS geofoam infilled trench in high-speed train vibration problem. A moving train with speed V (in kmph) travels on a continuous track resting on embankment soil system as shown in Fig. 2. The screening efficiency of EPS geofoam infilled trench has been evaluated in terms of reduction in vertical velocity amplitude which is quantified as average amplitude reduction ratio (ARR). The ARR over a distance of interest (c) is measured behind the trench which is calculated as per Eq. 1.

$$\text{ARR} = \frac{1}{c} \times \int_0^c \text{VRF} \cdot \text{d}y \tag{1}$$

where VRF is the ratio of the absolute maximum value of vertical velocity at a point with and without trench, and c is the span of the specific region taken as $3\lambda R$. The infinite elements (*CINPE4*) have been provided to model the far-field region. The soil properties have been taken as per the [6]. As reported by Zakeri et al. [6], the soil deposit includes the top1 *m* depth of compacted soil overlain a homogeneous soil deposit up to 10 m as shown in Fig. 2. The EPS geofoam has been modeled according to the low density foam material model (LDF). The PVC sheet has been given as the side cover of the trench to resolve the stability issue. All the FE model has been modeled in the *ABAQUS* (6.14) [7] finite element package. In the present study, the loading pattern has been adopted as per the [8], where the freight train with the speed

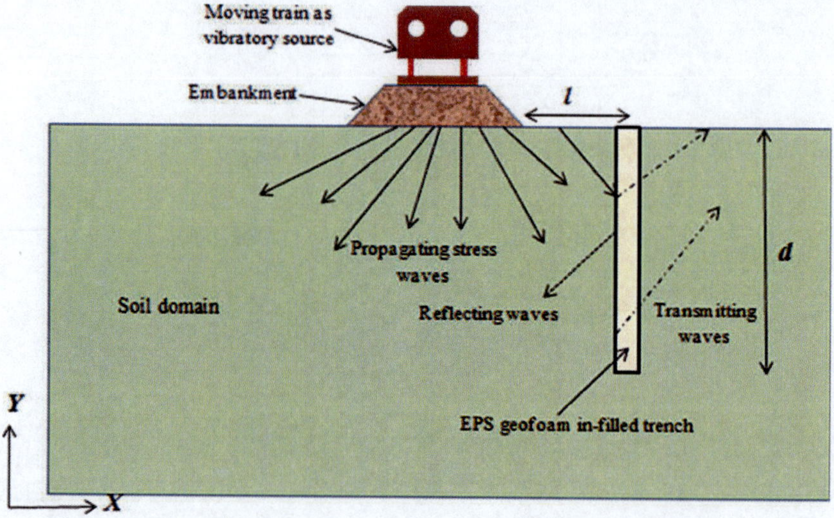

Fig. 2 Schematic diagram of the soil domain along the EPS geofoam filled trench

has been given as the input dynamic excitation considering the bogie distance of 9 m. For different train speed, the loading pattern is calculated based on the pulse time history. In case of the trench, PVC side cover and the soil contacts surface to surface contact elements have been provided to prevent any sliding and slippage. To capture the wave propagation properly, the mesh refinement has been done near the source as shown in Fig. 3. Near the trench also mesh refinement has been adopted. For the static condition, the roller boundary has been provided. The material damping ratio of 5% has been adopted for the soil and 1.2% for the EPS geofoam. The Rayleigh damping coefficients are calculated based on the free vibration analysis. The model

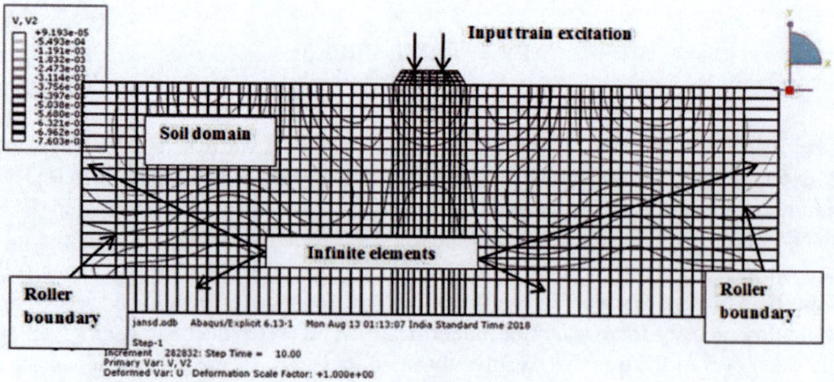

Fig. 3 Discretized domain with finite/infinite elements

has been validated with the existing studies on the EPS geofoam infilled trench as per [1]. Further, model is also validated with the experimental work on the open trench. For both the cases, present FE model predicts very much nearly results as compared to those studies.

4 Parametric Studies from FE Model

The key parameters considered in the present study are: trench depth (d), the distance between the trench and the toe of the embankment (l), the shear wave velocity (V_s), EPS geofoam density, material damping ratio, and train speed (V). All geometric dimensions are normalized with respect to Rayleigh wavelength, λ_R. Before the parametric study without trench analysis has been performed to know the necessity of providing trench in the specific region. To know the influence of different parameters, a nomenclature has been defined as ($L_i W_j D_k V_m$), where L, W, and D are normalized length, width, and depth factor, respectively, and V is the train velocity in kmph. i, j, k, and m are the range of different parameters. The EPS geofoam infilled trench is named as continuously geofoam infilled trench or CF hence after in the text. The location of the CF from the toe of the embankment and the depth of the CF has been varied independently in the present study. The normalized depth is varied from 0.8 to 1.8, and the normalized location is varied from 0.8 to 2.0. The width of the CF has been varied from 0.25 to 0.75 although from practical aspects many researchers have taken it as a constant [1, 9]. In Fig. 4a–d, the optimum values of these geometric parameters of the CF have been obtained by keeping other parameters constant. It can be seen from Fig. 4a that beyond $L = 1.15$, the variation of ARR is in the range of 5–7.2%, which may be considered to be nominal. From Fig. 4b, it can be seen that screening efficiency increases by 14.89% with the increase of W from 0.25 to 0.75. However, keeping the consumption of geofoam as a minimum without significantly sacrificing the screening efficiency, the optimum W is selected as 0.25. Similarly, from Fig. 4c, it can be seen that with the increase in D value screening efficiency increases. But after $D = 1.2$, the amount of increment of screening efficiency is not that much significant. Hence, from conservative point of view, D is assumed to be 1.2. The density of the EPS geofoam is one of the most important properties in this wave propagation problem which controls the amount of reflection, refraction, and mode conversion when stress wave is incident oat the junction between the soil and trench. Five different values of EPS geofoam density starting from EPS12 to ESP28 have been used for the intended purpose. The variation of ARR with the different geofoam density has been shown in Fig. 4d. The density of the geofoam needs to be chosen with utmost care for the field applications. The soil shear wave velocity V_s is also an important property along with the geofoam density which controls the reflection, refraction phenomenon for this wave propagation problem. The effect of V_s on the screening efficiency of CF is explored in the next part. The soil shear wave velocity is varied from 200 to 400 m/s. The soil with the higher value of V_s, i.e., stiffer soil is more effective for the vibration screening. The soil Poisson's ratio (ν)

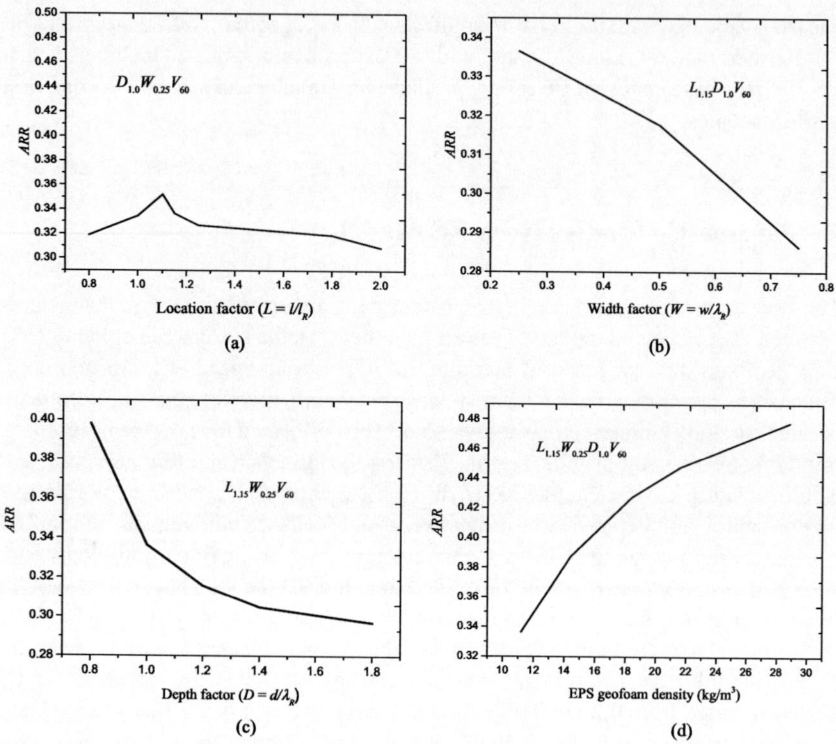

Fig. 4 **a** Variation of ARR with L for CF, **b** variation of ARR with W for CF, **c** variation of ARR with D for CF, **d** variation of ARR with geofoam density for CF

is varied from 0.25 to 0.4 for the optimum values of other parameters. The variation of ν is almost insignificant (Lesser than 5%) with respect to screening efficiency. Further, soil damping ratio (ξ) is varied from 2 to 10%; here, also the changes in the ARR are almost insignificant. Thus, Poisson's ratio and damping ratio of the soil have a very less influence on the screening efficiency. Lastly, a train being the source of vibration, the effect of train speed on the screening efficiency of CF has been studied in terms of Mach number (M_2). Figure 5 shows the variation of ARR with different Mach numbers. It can be observed that the CF is more effective in the super-critical speed region as compared to sub-critical speed. The range of ARR can be found out to be 37–43% for the all optimum values of parameters. The optimum value of ARR has been chosen as 0.45 for the CF to be called effective in the later simulation purpose.

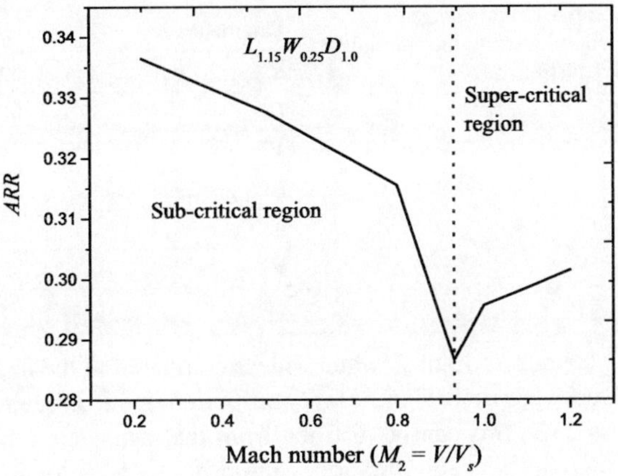

Fig. 5 Variation of ARR with Mach number for CF

5 Neural Network Applications

For the present problem, the neural network have been trained and tested with the data obtained from the FE modeling of the 750 scenarios. The input data chosen for the ANN analysis is as follows: depth of the trench (*d*), distance between the trench and source (*l*), shear wave velocity (V_S), EPS geofoam density, soil damping ratio (ξ), train speed (*V*), while the output data is average amplitude reduction ratio (ARR). The aim of the neural network system is to evaluate the contribution of each input factor. The effectiveness of the CF has been considered based on the following criteria: a trench is effective in reducing the ground vibration if only if for all the parameters the ARR value is lowers than 0.45. The definition of the hidden layers and the nodes with proper weights is important in ANN process. In the LMA algorithm, the error is propagated backward, and the weights are adjusted individually to minimize the error. The ratio of the percentage of data used for the training and simulation has been kept as 75%:25%. In the present analysis, numbers of trial runs have been made with three hidden layers to quantify the performance of different networks. The error term (RMS) which is quantified by mean square error has been computed for the input–output pairs during the whole training process. With increase in the iteration as validation error increased, the training process stopped, and the minimum of the validation error has been saved. RMS error in the training set has been selected in between 0.01 and 0.1, and correlation training set is in between 0.5 and 1. The evaluation of the percentage contribution of each input factor in governing the output factor is major objective. The results obtained from the ANN model can be summarized in Table 1.

By analyzing the simulations, it can be observed that the percentage of contribution to average amplitude reduction ratio of depth (*d*) is close to the 28%. The influence of soil dynamic properties V_S is more significant (25%) as compared to the damping ratio, ξ (3%) which can be justified from the level of strain expected in this wave propagation problem. The influence of distance of CF (*l*) from the dynamic source is

Table 1 Percentage of contribution to the amplitude reduction ratio

Parameters	Percentage of contribution (%)
d	28.00
l	13.50
V_s	10.50
EPS geofoam density	20.00
ξ	3.00
V	25.00

obtained as 13.50% which is less compared to the depth of the trench (d). The train speed has more significant contribution in the screening efficiency which is obtained as 25%; this can be justified from the last section that trench is more effective in high-speed train vibration. Similarly, the EPS geofoam density plays an effective role whose contribution is evaluated as 20%.

6 Conclusions

In the present study, the EPS geofoam infilled trench has been used as vibration screening system in reducing train-induced vibrations in the vibration sensitive site. The outcomes of 2D FE simulations of the present problem have been used for the neural network-based model development. The weights of the different geometric parameters and dynamic properties of the soil have been established from this model. The major conclusions can be summarized as follows:

- The screening effectiveness of EPS geofoam filled trench is mainly influenced by the depth of the trench. The depth factor has been chosen as 1.2 for the present problem. From the ANN analysis, the percentage contribution is obtained as 28%. The width of the trench has very less significant effect in screening efficiency as well as width must not be very large from the practical point of view.
- The static and dynamic properties of soil deposit have important role to play in the screening efficiency. The shear wave velocity of soil has major influence as compared to the damping ratio of the soil. The ANN model evaluated it's percentage contribution as 10.50%. The Poisson's ratio of the soil has very less significant effect on the screening efficiency.
- The train speed has significant role. Especially, the high-speed train (super-critical speed range) vibration is more effectively screened by the geofoam infilled trench as compared to the sub-critical speed train vibration. From the ANN analysis, the percentage contribution is obtained as 25.00% which is very close to the depth of the trench.
- The choice of EPS geofoam density is very crucial in the screening efficiency as well as from the stability of the trench point of view. The contribution of EPS geofoam density in vibration screening efficiency is 20% which is very significant.

References

1. Itoh K, Zeng X, Koda M, Murata O, Kusakabe O (2005) Centrifuge simulation of wave propagation due to vertical vibration on shallow foundations and vibration attenuation countermeasures. J Vib Control11:781–800
2. Murillo C, Thorel L, Caicedo B (2009) Ground vibration isolation with geofoam barriers: centrifuge modeling. Geotext Geomembr 27:423–434
3. Alzawi A, El-Naggar MH (2011) Full scale experimental study on vibration scattering using open and in-filled (GeoFoam) wave barriers. Soil Dyn Earthq Eng 31(3):306–317
4. Majumder M, Ghosh P, Rajesh S (2017) An innovative vibration barrier by intermittent geofoam—a numerical study. Geomech Eng Int J 13(2):269–284
5. Bose T, Choudhury D, Sprengel J, Ziegler M (2018) Efficiency of open and infill trenches in mitigating ground-borne vibrations. J Geotech Geoenviron Eng ASCE 144(8):0018048
6. Zakeri JA, Esmaeili M, Mosayebi S (2012) Effect of vibration in desert area caused by moving train. J Mod Transp 20(1):16–23
7. ABAQUS Inc (2014) ABAQUS version 6.14 user's manual, Providence, Rhode Island, USA
8. Di Mino G, Giunta M, Di Liberto CM (2009) Assessing the open trenches in screening railway ground-borne vibrations by means of artificial neural network. Adv Acoust Vib 2009(i):1–12
9. Ahmad S, Al-Hussaini TM (1991) Simplified design for passive isolation by open and in-filled trenches. J Geotech Eng Div ASCE 117(1):67–68

Time–Frequency Characteristics of Seismic Signal Using Stockwell Transform

Gayatri Devi, P. Sumathi, and Ashvini Kumar

Abstract Ground motions always create great interest for seismologists and engineers worldwide. For these signal, an accurate and precise analysis of non-stationary spectral variation are a longstanding problem aiming at some characteristics of signal like any underlying periodicity. Fourier transform is a conventional tool, used to study the seismic signals. In the last few years, researchers have become attentive to the limitations of the Fourier transform. It decomposes the signal into its constituent frequency components, but does not reveal, where changes in the frequency contents occur. To overcome, it joint time–frequency representations have been introduced which is a representation of both time and frequency. Some conventional method to obtain the desired time–frequency information contained in these signals are short-time Fourier transform (STFT) and Wavelet Transform. These methods show limitation in terms of resolution. The *S*-transform (ST) proposed by Stockwell et al. [10] is fusion of short-time Fourier transform (STFT), and Wavelet Transform. *S*-transform is based on a moving and scalable localizing Gaussian window. It provides frequency-dependent resolution while maintaining a direct relationship with the Fourier spectrum. In this paper, Stockwell transform (ST), STFT, and CWT-based technique for joint time–frequency representation of seismic signal has been used. Effectiveness of ST is evaluated by comparing the result of developed non-stationary synthetic signal and real-time ground motion signal of Uttarkashi earthquake ($M_w = 6.8$, 20 October 1991). Stockwell transform is capable for improving the resolution of non-stationary signals as well as clearly identified the spots of concentration in energy.

Keywords Ground motion · Fourier transform · Time–frequency representation · Stockwell transform

G. Devi (✉)
CSIR-Central Building Research Institute, Roorkee, Uttarakhand 247667, India
e-mail: gayatrikapil@gmail.com

P. Sumathi
Indian Institute of Technology Roorkee, Roorkee, Uttarakhand 247667, India

A. Kumar
H.V.M. (PG) College, Raisi, Uttarakhand, 247671, India

T. G. Sitharam et al. (eds.), *Seismic Hazards and Risk*, Lecture Notes in Civil Engineering 116, https://doi.org/10.1007/978-981-15-9976-7_17

1 Introduction

Earthquake ground motions are inherently non-stationary in nature. For better understanding about the characterization of local site effects subject to earthquake, reliable earthquake signal processing is essential. A time history is a most widely used method for explaining the ground motion. The strong motion duration of an earthquake is the time interval during which most of the energy is contained. Because seismic waves are scattered during propagation, it shows a time evolving frequency composition. The records of these seismic waves exhibit non-stationary characteristics. Seismic signal shows non-stationary characteristics due to attenuation and absorption of seismic energy [1].

Damage of structures depends on earthquake's time duration, their amplitude and frequency content of waves. Fourier-based analysis neglects the time duration information of dominant frequency of strong ground motion signal. Fourier transform can identify the frequencies which are present in signal but it does not reveal, where changes occur in the frequency contents [2].

To overcome this type of problem, joint time–frequency representations have been introduced which is a representation of both time and frequency. For the proper interpretation of seismic data in terms of varying frequency content, there is a need of time–frequency representation techniques jointly [2–4]. A proper analysis of ground motions is very necessary, if building design is to be constructed in seismically active areas, predicting earthquakes, quantification of damage from the recorded motions. To overcome this type of problem, joint time–frequency representations have been introduced which is a representation of both time and frequency. For the proper interpretation of seismic data in terms of varying frequency content, there is a need of time–frequency representation techniques jointly [2].

2 Time–Frequency Distributions: Fundamental Ideas

Time–frequency representation unfolds temporal information and maps a time series into 2D quantity of time and frequency with effective characterization of the time–frequency image. It describes how the spectral content of the signal changes with time. Through the view of mathematical point, this joint distribution will provide fractional energy of signal's total energy at frequency (ω) and time (t) [5].

If the ground motion signal is represented by $x(t)$, then the energy density of signal is represented by $|x(t)|^2$.

The total energy E_x is

$$Ex = \int |x(t)|^2 \mathrm{d}t$$

The frequency domain representation $x(f)$ for the signal $x(t)$ is

$$x(f) = \int_{-\infty}^{\infty} x(t)e^{-j2\pi ft}dt$$

$x(t)$ and $x(f)$ are uniquely related.

The energy or intensity per unit frequency at frequency f is $|x(f)|^2$.

The total energy

$$E_x = \int |x(f)|^2 df$$

This should be equal the total energy of the signal calculated directly from the time waveform

$$E_x = \int |x(t)|^2 dt = \int |x(f)|^2 df$$

Jointly distribution of energy of $x(t)$ over both the time and frequency variables amounts to looking for an energy distribution $P_x(t, f)$ such that

$$E_x = \int_{-\infty}^{\infty}\int_{-\infty}^{\infty} P_x(t, f)dtdf$$

For analysis of non-stationary signal, TFD is an appropriate tool. Various time–frequency representation method has been developed in last two decades for analysis of non-stationary signals. Some early form of JTF representation is STFT and CWT. The *S* transform is also one of JTF representation method [6]. It has some unique advantage as it gives frequency-dependent resolution and also maintain a direct relationship with Fourier spectrum [7].

3 Method

3.1 Stockwell Transform

It is an extension of the ideas of the short-time Fourier transform and is based on a moving and scalable localizing Gaussian window [7] so before to define the Stockwell transform, a general idea about STFT and CWT has been described here.

3.2 Short-Time Fourier Transform

Fourier transform, when made the function of time gives this STFT, i.e. this technique involves the division of signal into narrow time slots using the window template such that the segmented signal is considered to be stationary [8]. The FT of the segmented signal is then taken to get the frequency spectrum of that particular segment. So it can be defined as Fourier transform of the product of the signal and shifted version of window function [9, 10].

The mathematical relation that gives the STFT of a signal $p(t)$ is

$$S(\tau,\omega) = \int_{-\infty}^{+\infty} p(t)w(t-\tau)\mathrm{e}^{-j\omega t}\,\mathrm{d}t$$

where

$p(t)$ is the signal to be transformed.

$w(t)$ is the analysis window.

$S(\tau,\omega)$ is the STFT of the signal.

τ is the centre position of the window.

The width of the window length is fixed, which leads to the disadvantage of fixed time–frequency resolution. Because of fixed window length, there is a difference in number of cycles in that window along frequencies and that makes it inconvenient of having good time resolution comparatively frequency resolution at higher frequency [11].

One another problem through which STFT suffer is leakage due to window effect. Fourier transform of rectangular window used in this transform is a 'Sinc' function that has narrow main lope width and larger side lopes, which result spectral leakage.

Continuous wavelet transform: The basic and main principles of wavelet are review here briefly to understand its application in seismic signal. The continuous wavelet transform is the cross-correlation function, which is calculated using the signal correlated with wavelets. These wavelets are generated through the original mother wavelet by its scaled and translated versions [12–14]. Mother wavelet is termed as the main function, and the modified functions are called wavelets.

Mathematical formulation of the CWT is given by [15]

$$W_{\mathrm{s}}(a,\tau) = \int_{-\infty}^{+\infty} p(t)\frac{1}{\sqrt{|a|}}\psi*\left(\frac{t-\tau}{a}\right)\mathrm{d}t$$

where $p(t)$ is the signal.

$W_{\mathrm{s}}(a,\tau)$ are the CWT coefficients.

$\psi(t)$ is the mother wavelet, dilated by scale a and shifted in time τ.

$\frac{1}{\sqrt{|a|}}$ is known as the multiplication factor, and it ensures the normalization of energy, means the wavelet which has unit energy at all scales.

Like STFT, the wavelet transforms also suffer from uncertainty principle [16]. Both a good time and frequency resolution cannot be achieved simultaneously.

3.3 Stockwell Transform

The Stockwell transform is a mid way between STFT and CWT. This time–frequency representation method has quite similarity to STFT while because of the use of multiresolution tactics, it makes somewhere closer to wavelet transform [11]. A big advantage of *S*-transform is that because of its simple concept, it gives a simple understanding of multiresolution approach as have been introduced in wavelets and hardly require any additional knowledge excepting STFT [11]. In Stockwell transform for window function, Gaussian window is used, which can be time shifted by τ and is inversely proportional to the linear frequency f [10]. The *S*-transform may comparable to the CWT as the Gaussian template is comparable to the mother wavelet with a phase shift.

The mathematical expression for the Gaussian taper is given as

$$w(t) = \mathrm{e}^{\frac{-t^2}{2\sigma^2}},$$

where σ is taper width and inversely proportional to the frequency $\sigma = \frac{k}{|f|}$. The parameter k can be tuned to obtain better frequency localization at the cost of reduced time localization by controlling the width of Gaussian taper [15] here the parameter k is considered 1. Gaussian taper function is normalized to achieve the Gaussian template.

The S transform can be derived from STFT by replacing the window function $w(t)$ with the Gaussian function shown as

$$w(t) = \frac{|f|}{\sqrt{2\pi}}\mathrm{e}^{\frac{-t^2 f^2}{2}}$$

As STFT defined as

$$S(\tau, f) = \int_{-\infty}^{+\infty} p(t)w(t-\tau)\mathrm{e}^{-j2\pi ft}.$$

Then, the S transform is defined as

$$S_{\mathrm{T}}(\tau, f) = \frac{|f|}{\sqrt{2\pi}} p(t)\mathrm{e}^{\frac{-(t-\tau)^2 f^2}{2}}\mathrm{e}^{-j2\pi ft}$$

where $S_{\mathrm{T}}(\tau, f)$ represents the *S*-transform of signal.

$p(t)$ is continuous time signal.

τ parameter controls the position of the Gaussian window on t-axis.

The advantage of S-transform is that it may give multiresolution analysis beside keeping the absolute phase of each frequency [17]. The Gaussian window which localization is inversely proportional to the frequency is an improvement over STFT as fixed width window used in this.

S-transform's phase referenced to time origin which gives useful and accompanying information about the spectra, and this is not obtainable from locally referenced phase information in continuous wavelet transform [10].

4 Result

To demonstrate the performance of ST, one non-stationary synthetic signal and one real-time earthquake signal record are considered.

A. **Synthetic Signal**

A synthetic signal quite similar to synthetic example [18] is generated with three sinusoidal components to illustrate the features of ST. The synthetic signal is a sum of three components which consists of sinusoidal waves. The sampling frequency of the signal is 500 Hz, and the signal-to-noise ratio (SNR) is 10 db. The length of signal is 10 s. Here, η(t) is the Gaussian noise. The details of synthetic signal are given by the following equation and are shown in Fig. 1 with its three components.

$$S(t) = S_1(t) + S_2(t) + S_3(t) + \eta(t)$$

$$S_1(t) = [2 + 0.2\sin(t)] \cdot \sin[2\pi(3t + 0.6\sin(\mathrm{t}))]$$

$$S_2(t) = 0.6\left[1 + 0.3\sin(2t) \cdot \exp\left(-\frac{t}{20}\right) \cdot \cos\left[2\pi\left(5t + 0.6t^{1.8} + 0.3\cos(t)\right)\right]\right]$$

$$S_3(t) = 0.4\sin[2\pi(9t)]$$

The frequency of $S_1(t)$ is the lowest among all the three components but the amplitude of which is highest. The frequency of $S_2(t)$ increases with time, and amplitude becomes smaller when frequency becomes higher. The frequency of $S_3(t)$ is constant, and its amplitude is smallest. The specially designed signal components are so that the higher is the amplitude of a component, the lower the frequency of the component.

Figure 1a shows the synthetic signal and its components considered for the analysis. The TF representations of synthetic signal using STFT, CWT, ST are shown in Fig. 2a–c, respectively. Hanning window with window length (196 in sample) used for STFT and Morse wavelet has been used for CWT; the width factor (K) is 1 for

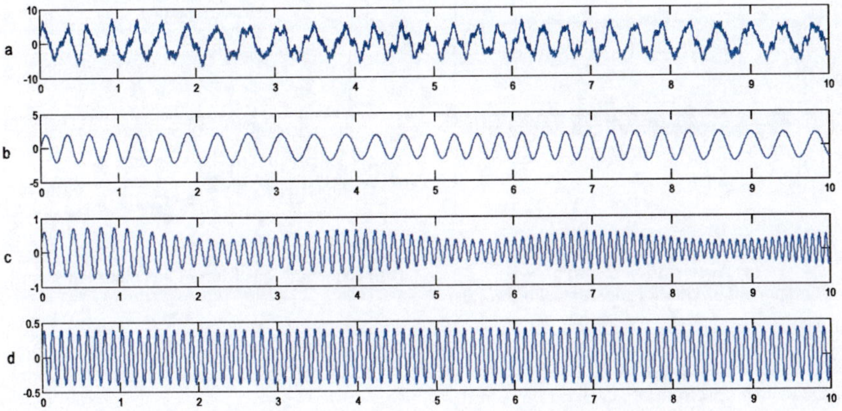

Fig. 1 Developed synthetic signal in (**a**). **b–d** represents the three components $S_1(t)$, $S_2(t)$ and $S_3(t)$, respectively,

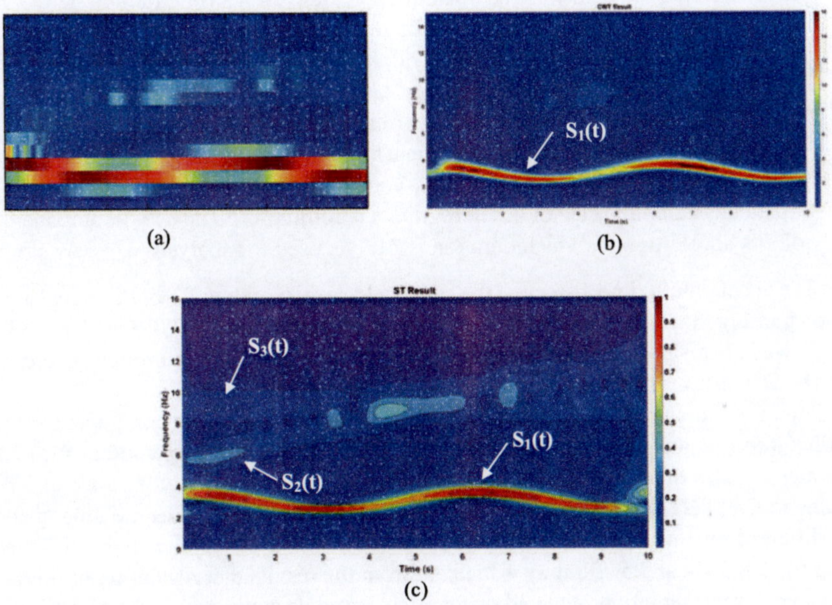

Fig. 2 Time–frequency representations of synthetic signal **a** STFT **b** CWT **c** ST

ST. All three components of $S(t)$, namely $S_1(t)$, $S_2(t)$, $S_3(t)$ are visible in the result of TF spectra of ST while in the result of CWT only one frequency component $S_1(t)$ is clearly visible.

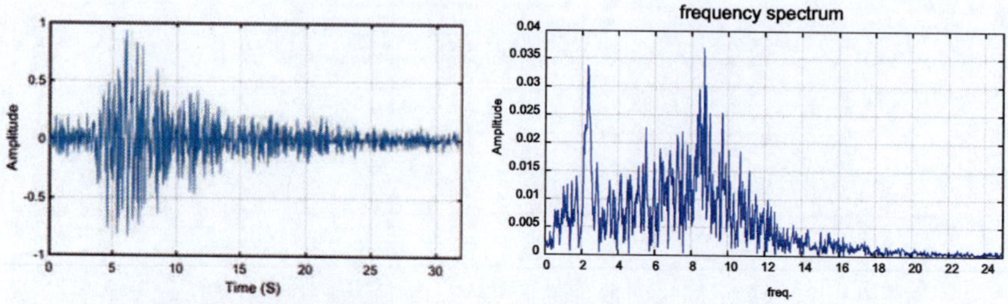

(a) Acceleration record of N-S component of Uttarkashiearthquake recorded at Badkot station

(b) Frequency Spectrum of Acceleration record

Fig. 3 **a** Acceleration record of *N-S* component of Uttarkashi earthquake recorded at Badkot station, **b** frequency spectrum of acceleration record

B. **Real-Time Earthquake Record**

The real-time earthquake record is of Uttarkashi earthquake occurred in October 1991. This earthquake was recorded at 13 stations of a strong motion network installed by Department of Earthquake Engineering, Indian Institute of Technology, Roorkee in Garhwal Himalaya under the research scheme funded by Department of Science and Technology, Govt. of India.

The acceleration record of Uttarkashi earthquake recorded at Barkot station at 20 October 1991 at 02:53 IST with a moment magnitude of 6.8. Signal was recorded through a three channel (North–South, Vertical, East–West) accelerograph. The frequency sampling rate of accelerogram's recording is 50 sps. *N-S* component of this accelerogram, which duration is 31.74 s used for analysis.

The acceleration time history of the signal and corresponding FT of the signal are shown in Fig. 3a, b. *T-F* representation of seismic signal using STFT, CWT and ST is shown in Fig. 4a–c, respectively. The parameter used for analysis through different methods is given in Table 1.

Figure 4a–c shows at least 3–4 frequency components. These components are difficult to find only frequency domain based analysis. These are marked as $f1, f2, f3$ and $f4$. The time localization of $f1, f2, f3, f4$ is very good in the result of ST comparatively STFT and CWT. In the result of ST, the frequencies are annotated at different time as $f1 = 7.5$ Hz at 5.6 s, $f2 = 2.5$ Hz at 6.5 s, $f3 = 9$ Hz at 7.2 s and $f4 = 8.5$ Hz at 8.5 s. It may visualized from the results that resolution of time–frequency representation obtained using ST is better as compared to the STFT and CWT representation.

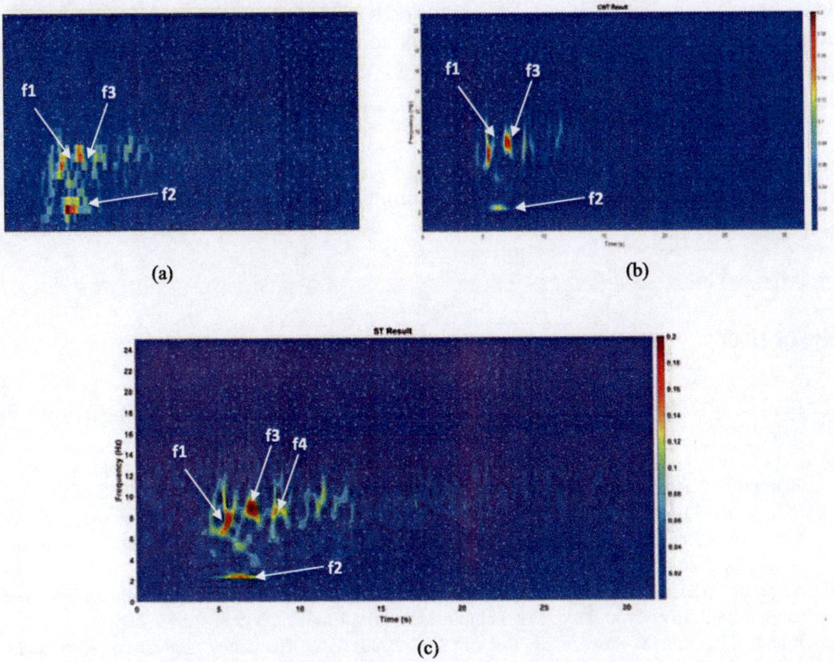

Fig. 4 Time–frequency representations of Uttarkashi station record **(a)** STFT **(b)** CWT **(c)** ST

Table 1 Paramaters used in various T-F Representations

Signal	STFT	CWT	ST
Ground motion acceleration record of Uttarkashi earthquake at Barkot station	Hanning window with window length (196 in sample)	Morse wavelet, cwt uses 10 voices per octave	The width factor (k) = 1

5 Conclusion

The Stockwell transform-based joint time–frequency technique is introduced for seismic signals and is validated for real-time ground motion signal. The time–frequency representations of synthetic signal using STFT, CWT, ST are compared. All three components of $S(t)$, namely $S_1(t)$, $S_2(t)$, $S_3(t)$ are visible in the result of time frequency spectra of Stockwell transform while in the result of CWT only one frequency component $S_1(t)$ is clearly visible. The time–frequency representations of real-time signal using STFT, CWT, ST are also compared. These components are difficult to find only frequency domain-based analysis. These are marked as $f1$, $f2$, $f3$ and $f4$. The time localization of $f1, f2, f3, f4$ is very good in the result of

ST comparatively STFT and CWT. It may visualized from the results that resolution of time–frequency representation obtained using ST is better as compared to the STFT and CWT representation. Stockwell transform is capable for improving the resolution of non-stationary signals as well as clearly identified the spots of concentration in energy. For seismic signals, the *S*-transform provides good time localization. Frequency localization is not so good; it shows frequency smearing. Frequency smearing problem may be minimized through synchrosqueezing of Stockwell transform.

References

1. Bormann P (2009) Seismic signals and noise. New manual of seismological observatory pratice (NMSOP), pp 1–34
2. Devi V, Sharma M (2016) Recent spectral decomposition techniques and its applications in analysis of seismological data: a review. Int J Innov Res Sci Eng Technol 5(1):213–220
3. Kuyuk HS, Kuyuk RT (2010) Investigation on displacement, velocity and acceleration waveforms of 1999 Kocaeli earthquake using wavelet and Stockwell. In: 19th international geophysical congress and exhibition. Ankara, Turkey
4. Nagarajaiah S, Basu B (2009) Output only modal identification and structural damage detection using time frequency and wavelet techniques. Earthq Eng Eng Vib 8(4):583–605
5. Huerta CL, Shin Y, Powers EJ, Roesset JM (2000) Time-frequency analysis of earthquake records. In: Proceedings of 12th WCEE
6. Kuyuk HS (2015) On the use of Stockwell transform in structural dynamic analysis. Sadhana 40(1):295–306
7. Han B (2013) The analysis of signal based on the S-Transform. In: 2nd international conference on measurement, information and control, Harbin, China
8. Allen JB, Rabiner LR (1977) A unified approach to short-time. Fourier analysis and synthesis. Proc IEEE 65(11):1558–1564
9. Black CJ (1998) Dynamic analysis of civil engineering structures using joint time-frequency method. M.A.Sc. thesis, The University of British Columbia
10. Stockwell RG, Mansinha L, Lowe RP (1996) Localization of the complex specturm: the S transform. IEEE Trans Sig Process 44(4):998–1001
11. Ventosa S, Simon C, Schimmel M, Danobeitia JJ, Manuel A (2008) The S-Transform from a wavelet point of view. IEEE Trans Sig Process 56(7):2771–2780
12. Rioul O, Flandrin P (1992) Time-scale energy distributions: a general class extending wavelet transforms. IEEE Trans Signal Process 40(7):1746–1757
13. Daubechies I (1992) Ten lectures on wavelets. CBMS-NSF regional conference series in applied mathematics, vol 61. Society for Industrial and Applied Mathematics (SIAM), Philadelphia, PA
14. Flandrin P (1999) Time-frequency/time-scale analysis. Volume 10 of wavelet analysis and its applications. Academic Press Inc., San Diego, CA
15. Tary JB, Herrera RH, Han J, Baan MVD (2014) Spectral estimation-what is new? What is next? Rev Geophys 52(4):723–749
16. Sinha S, Partha SR, Phil DA, Castagna JP (2005) Spectral decomposition of seismic data with continuous-wavelet transform. Geophysics 70(6):19–25
17. Pinnegar CR, Mansinha L (2003) The S-transform with windows of arbitrary and varying shape. Geophysics 68(1):381–385
18. Huang ZL, Zhang J, Zhao T, Sun Y (2016) Synchrosqueezing S-transform and its application in seismic spectral decomposition. IEEE Trans Geo-sci Remote Sens 54(2):817–825

The Seismotectonic Setting of Indo-Gangetic Plain and Its Importance

Bhukya Naresh, Kumar Venkatesh, and Laxmi Kant Mishra

Abstract This paper gives overview of seismotectonic setting of the Indo-Gangetic plain which is second most seismically active region in the India after Himalayan seismic region. This Indo-Gangetic plain has experienced several great earthquake events like 1934 Nepal Bihar, 2015 Nepal earthquake, where type of soil or medium plays an important role for the devastation of structures in this region. This study is helpful to know the geological feature and properties of the Indo-Gangetic plain, types of faults present in this region (Neotectonic faults, sub-surface structural faults) under the alluvial deposits and ridges (Monghyr-Saharsa and Faizabad Ridges). The thickness of the soil deposits above the crystal basement along the Indo-Gangetic plain which affects diminution of the seismic waves at important places and monumental structures. Many such structures are situated near the fault lines and different types of river basins such as Indus, Ganga River basin and Brahmaputra River system, etc. The seismically active zone of Assam valley along the Brahmaputra river basin and Delhi region in the Indo-Gangetic basin is categorized as seismic Zone-V and seismic Zone-IV (as per IS 1893-part-1). In this region, more than 400 million people are living and also contribute to major economy of India.

Keywords Indo-Gangetic plain · Ganga basin · Punjab basin · Bengal basin · Brahmaputra basin · Fault

1 Introduction

Indian subcontinent is one of the seismotectonic plate which undergone subduction override by the Eurasian plate and formed Himalayas a seismically active belt in the

B. Naresh (✉) · K. Venkatesh · L. K. Mishra
MNNIT Allahabad, Prayagraj, U.P., India
e-mail: bhukya@mnnit.ac.in

K. Venkatesh
e-mail: venkatesh@mnnit.ac.in

L. K. Mishra
e-mail: lkm@mnnit.ac.in

T. G. Sitharam et al. (eds.), *Seismic Hazards and Risk*, Lecture Notes in Civil Engineering 116, https://doi.org/10.1007/978-981-15-9976-7_18

world. The most of the rivers flowing in India are originated in the Himalayas. The important Himalayan rivers are Indus River, Ganga River, Brahmaputra River and Yamuna River, etc. This river system is one of the reason for farming in this region. These river systems make Indo-Gangetic Plain (IGP) as second most seismically active region after Himalayan region in India and formed seismic region of Indo-Gangetic plain.

The seismotectonic study of a region is needed to know the geological features of the region, behavior and the tectonic movements of the plates, subduction of plates, fractures and faults present in the region which causes the earthquake activities in the region. Seismotectonic studies also have good applications in the seismic hazard analysis and to find the earthquake simulating seismic parameters like source, path, site, etc. in the region, for the design of earthquake resistant structures.

The Himalayan foordeep has formed due to the sediments transferred by the Himalayan river systems of Indus and Ganga and created Indo-Gangetic depressions in the solid rock strata. These depressions in the IGP region formed during tertiary age. The sediments carried out by river systems were deposited in this region due to self-weight of sediments. These Himalayan rivers deposited nearly the depth of 15,000–30,000 ft sediments [1].

In this paper, the geological features, seismotectonic setting of region with different types of faults (faults, subsurface faults, neotectonic faults, etc.). These faults are useful for probabilistic seismic hazard analysis of different cities and seismic response of old structures. Further, the populated cities and world heritage structures present in the IGP region are discussed.

2 Geological Features of the Indo-Gangetic Plain (IGP)

The IGP plain has spread over India, Bangladesh, Pakistan and part of Nepal, and the most of the IGP has covered by the India. The IGP basin elevated at average height of 125 m above the mean sea level around the central IGP land surface which is given by Central Pollution Control Board (CPCB). Indo-Gangetic plain is one of the river-based plain, whereas the river Indus, Ganga, Yamuna and Brahmaputra are mainly flowing rivers. This IGP has divided into four basins depending on the flowing river systems are named as (1) Indus Basin (and Punjab basin) (2) Ganga basin (3) Bengal basin and (4) Brahmaputra basin. The IGP starts from the Indus basin from the small portion of the Rajasthan; after that it has spread over Punjab basin which has covered Punjab, Haryana and part of Jammu which is a part of Indus basin. The Ganga basin which contains two main river systems Yamuna and Ganga flow through Delhi, Uttar Pradesh and Bihar. The Bengal basin contains West Bengal, and Bangladesh and Brahmaputra basin have covered with Assam except Shielong plateau as shown in Fig. 1. The crustal thickness of the IGP from the Punjab to Assam varies 35–37 km near the Southern peninsular to 70–72 km at foothills of Himalayas [2].

The IGP has covered with quaternary aged alluvial deposits over granitic bedrock 502 m from the surface due to the flow of the rivers by fluvial process compressing of sand, silt and clay and Pericratonic fill on attenuated continental crust (CPCB). This IGP is monotonous. The basement depth in the U.P. region varies 1000 m near the Allahabad, Varanasi and Ghazipur to 8000 m towards the Himalayas (ex. Gorakhpur). The Indus basin along the Rajastan covered with Mesozoic and Cainozoic sediments.

3 Tectonic Setting of Indo-Gangetic Plain

The IGP has several subsurface faults like Azamgarh-Gorakhpur area E-W trending faults. This Azamgarh-Gorakhpur area is called as Mirganj Graben. Allahabad subsurface fault passing through Allahabad, and Varanasi has located in between the Allahabad subsurface fault and Azamgarh subsurface fault.

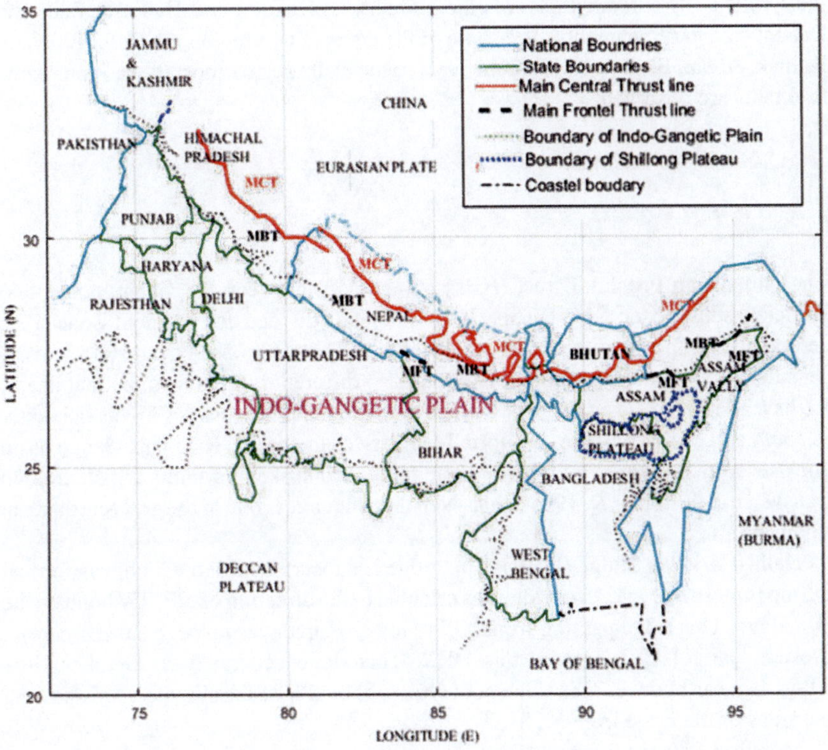

Fig. 1 Indo-Gangetic plain geography in Indian region

3.1 Indus Basin

Complete Indus Basin has classified as three sub-basins (1) Upper Indus basin (2) Lower Indus basin and (3) Panjnad basin (Punjab basin). In the present study, we consider Punjab basin only which is part of the Indian state of Punjab, Haryana, Delhi, Himachal Pradesh, Part of Rajasthan and Part of Uttarakhand. The Punjab basin having the boundaries Suleiman ranges in the north-west and Himalayas in the north-east. This Panjnad plain also filled with alluvial deposits which were transported by the river Indus and its tributaries Jhelum, Chenab, Ravi, Sutlej and Beas [1, 3]. Delhi-Haridwar ridge is covered under alluvial in the western part of U.P., and it is eastern boundary of Indus basin; this ridge behaves like direct thrust sheet [4]. The Punjab basin of Indus having Delhi-Haridwar subsurface fault as boundary in the east. This basin also having in the NW of Chandhigar Ropar fault and in SE of Chandhigar Ropar fault and also one neotectonic fault. North-Delhi fold belt also present in this basin. Mahendragarh and Dehradun subsurface faults are also preset in this basin.

This Punjab basin having some of the neotectonic fault named as Sardar-Sahara fault which has extended towards the Himalayas, Dhaudu and Hathusar fault and Raisingh Nagar lineaments. The some of the cities in this region are Delhi Bhiwani, Rohtak, Alwar, Bharatpur, Mathura. Near to the Delhi region most of the Proterozoic fold belts are observed.

3.2 Ganga Basin

The Himalayan Frontal Thrust (HFT) is boundary between the Ganga plane, and Himalayas has observed [5]. This Ganga plain has divided as Piedmont Zone (PZ), central alluvial plain and marginal alluvial plain. The study based on the GIS, remote sensing and field investigation of Goswami [6] different morphotectonic features of PZ have identified. The Ganga basin having Najibabad blind Fault (NF) as boundary in south and it is parallel to the North boundary of the basin HFT. This Ganga basin consists so many dip slip faults, oblique faults and strike slip faults which crossed transverse to the HFT [7]. 1934 Bihar–Nepal earthquake is one of the great earthquake in this basin.

Delhi-Haridwar Ridge (DHR): This ridge is subsurface ridge which was formed during Proterozoic age. This ridge has oriented in the direction of NE-SW beneath the Himalaya. This ridge passing from IGP to lesser Himalayas through sub-Himalaya. Western part of U.P. covered this DHR. This ridge generated 20 October 1991 Uttarkashi earthquake. This ridge both types of NE-SW strike slip mechanism along the sides of the ridge front [4, 8].

Monghyr-Saharsa Ridge (MSR): The MSR has formed during the upper Archean age. This ridge has extended from Bihar–Nepal border to north eastern part of satpura metamorphics. This ridge also having same mechanism of DHR as strike slip [4, 9]. The Faizabad Ridge: This ridge also extended from north-central U.P. to border of western Nepal. This ridge also having the same mechanism of strike slip like DHR. Najibabad blind Fault (NF), West Patna Fault, East Patna Fault, Lucknow Fault, Allahabad sub-surface fault are other faults in this basin.

3.3 Bengal Basin

The Bengal basin lies north-eastern part of Indian plate and located at the junction of three adjoin plates named as the Indian, Eurasian Plate and Burmese plates. This basin having three geotectonic parts one is the stable shelf which contains Permian–Carboniferous rocks and Precambrian meta-sediments; second tectonic part is the central deep basin which started down warping of plate due to the movement of the Indian plat towards the north, and final one is the Chittagong–Tripura fold belt [10, 11]. This Bengal basin has subjected over northern part of the Ninetyeast Ridge (NER) [12–14]. One of the major delta of the basin is Ganga Brahmaputra River Delta which spreads over West Bengal and Bangladesh and covered with thick sediments; these sediments are carried out from Himalayan and Tibetan plateau and joined in this Bengal fan [15–17]. One of the active sedimentary fan on the earth is Bengal fan which has resulted from the Ganga and Brahmaputra rivers which surpasses deltas, alluvial or fluvial systems. The Bengal basin covered with quaternary alluvium. This basin having the boundaries Indian carton in the west direction and Indo-Burma folds in the east. This basin covers three states of India (West Bengal, part of Assam and Thripura) and Bangladesh. This basin was created due to breaking of the Gondwanaland during the late Mesozoic. The thickness of the sediments in this basin varies from one kilometres to 16 km towards Himalayan fore deep [11]. The researcher divided Bengal basin as active margin and passive margin of setting to divided the active platform as deep basinal and passive margin as the extensional platform, whereas India in this margin as shown in figure [18]. The different types of faults present in this region are shown in Fig. 2. The Farakka and adjoining regions are neotectonically active [19].

Maldah Kishanganj fault: This fault run parallel to the left side of the Ganga River, whereas Rajmahal Fault runs right side of the Ganga. Number of neotectonic faults also presents along this fault as given in Geological Survey of India [20] (Seismotectonic Atlas of India). From 1594 to 1975 seven moderate earthquakes occurred in Farakka and adjoining regions [21]. It is a deep seated fault [19].

Rajmahal fault: This fault has been observed in between the Rajmahal hill and Ganga river in the basin which is parallel to the river. Tista Old Brahmaputra fault: This fault is trending towards to the NW–SE of the basin which has enter into India. It shows fresh or new tectonic activity. Dauki fault is one of the fault at the boundary of Bengal basin and Shilong plateu; this is vertical and deep seated reverse slip fault [22]. This fault experienced three greater earthquakes and ruptured three times, where 1548 and 1897 great Indian earthquake are near about the past earthquakes along this fault [23]. This fault has propagated towards the southern part of Meghalaya

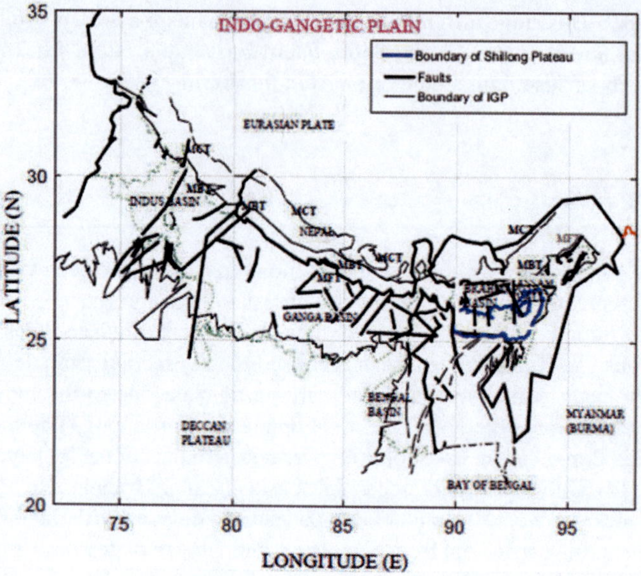

Fig. 2 Seismotectonic setting of Punjab basin which shows geological features

about length of 170 km [24]. Chittagong-Tripura fold belt, Padma lineament, Jamuna depression also present in this basin.

3.4 *Brahmaputra Basin*

One of the unstable region in the North-East India is Brahmaputra basin which is a river basin passes through the Assam. The total area of the Brahmaputra drainage basin is 580,000 km^2, where 33.6%, of basin is lies in India, 50.5% in china, in Bhutan 7.8% and in Bangladesh, it flows 8.1%. Assam valley in this Brahmaputra basin is unstable seismic region [25]. The Brahmaputra valley mostly covered with alluvial soils up to the depth of 200–300 m [20] (geological Survey of India, 2000). The total length of the Brahmaputra River along the Assam valley is 240 km, and average width has increased to 14.03 km in 2005 from 9.74 km in 1915 [26]. The Brahmaputra valley in Assam bounded by the Patkai Naga Arakan belt in the south and Himalayan belt in the North direction. These two belts are tectonically active and most of the active faults also present in this valley region. The major earthquakes occurred in this region are M8.7 in 1897 and 1950 [27, 28]. The seismotectonic setting (faults and Assam valley region) of this region generated one of the massive earthquake which occurs in upper Assam region in 1950 of magnitude *Mw* 8.6.

Himalayan thrust, Naga Patkai thrust: It is outer most boundary fault of Schuppen belt which follows the boundary of the Assam valley, Naga fold, Brahmaputra faults are extended towards the Bengal basin, the Main Boundary Thrust (MBT), Namcha Barwa, Tapu thrusts. Mishmi fault is one of the youngest fault in the region [29].

4 The Important Cities and Monumental

Mirganj Graben: Azamgarh, Gorakhpur and Deoria and Prayagraj (Allahabad). Prayagraj having some monumental structures like Allahabad University, Khushrobhag which are constructed more than 100 years ago.

According to the 2011 data in India, the some of the populated cities in IGP region are Agra, Allahabad, Ghaziabad, Meerut having the population more than 1 million, and cities with population more than 2.7 million are Kanpur and Lucknow in Uttarpradesh. Guwahati nearly 1 million in Assam, Patna in Bihar more than 1.5 million, Chandigarh nearly million, in Delhi more than 16.7 million, Faridabad in Haryana more than 1 million, Amritsar, Ludhiana in Punjab more than 1 million, Haora (Howrah) more than 1 million, Kolkata [Calcutta] nearly 4.5 million (www.citypopulation.de/php/india).

As per Archaeological Survey of India (ASI) India having 32 World Heritage Properties (WHP) in combining both cultural (25) and natural (7) properties. Out of 32 WHP— are present in IGP. The monuments present in IGP are Taj Mahal, Agra Fort, Fatehpur Sikr, Akbar Tomb Sikandra, Mariam Tomb Sikandra, Ram Bagh in Agra, Sheikh Chilli'S Tomb, Suraj Kund, Archaeological Site Museum in Ropar in Chandigarh.

Humayuns Tomb, Qutub Minar, Red Fort, Safdarjung Tomb, Purana Qila, Jantar Mantar, Tughlaqabad Fort, Kotla Firoz Shah, Khan-I-Khana, Sultan Garhi, Purana Qila Museums, Purana Qila Lake in Delhi. Ahom Raja'S Palace, Karenghar of Ahom Kings, Group of Four Maidans, Bishnudol, Ranghar Ruins, Archaeological Site Museum, Sri Suryapahar in Guwahati. Kumbhalgarh Fort, Chittaurgarh Fort, Archaeological Site Museum—Kalibangan in Jodhpur. Hazarduari Palace, Cooch Bihar Palace, Bishnupur Temples, Archaeological Site Museum—Tamluk in Kolkata. Rani Jhansi Fort, Rani Lakshmi Bai Mahal, Site of Sahet—Mahet, Residency Buildings, Kalinjar Fort in Lucknow. Site of Mauryan Palace, Ancient Ruins Vaishali, Sher Shah'S Tomb, Excavated Site, Ancient Site of Vikramshila, Archaeological Site Museum-Nalanda, Archaeological Site Museum-Bodhgaya, Archaeological Site Museum—Vaishali, Archaeological Site Museum—Vikramshila in Patna (ASI) [30].

5 Conclusion

This paper gives the overview about the seismotectonic setting of IGP and the monumental structures in this region. The seismotectonic setting includes different types of faults like subsurface faults and neotectonic faults. The study of these faults are useful to find the maximum magnitude of earthquake in the probabilistic seismic hazard analysis of different cities and locations in the region. The monumental structures are helpful to study the seismic performance of buildings.

References

1. Burrard S (1915) On the origin of the Indo-Gangetic trough, commonly called the Himalayan foredeep. J Proc R Soc Lond A 91:220–238. https://doi.org/10.1098/rspa.1915.0014
2. Choudhury SK (1975) Gravity and crustal thickness in the Indo-Gangetic plains and Himalayan region, India. Geophys J R Astr Soc 40:441–452. https://doi.org/10.1111/j.1365-246x.1975.tb04141.x
3. Bajracharya SR, Maharjan SB, Shrestha F (2015) Glaciers in the Indus Basin. International Centre for Integrated Mountain Development, Kathmandu, Nepal
4. Sati D, Nautiyal SP (1994) Possible role of Delhi-Haridwar subsurface ridge in generation of Uttarkashi earthquake, Garhwal Himalaya, India. Cur Sci 67:1
5. Goswami PK (2012) Geomorphic evidences of active faulting in the northwestern Ganga Plain, India: implications for the impact of basement structures. Geosci J 16(3):289–299. https://doi.org/10.1007/s12303-012-0030-7
6. Goswami PK, Yhokha A (2010) Geomorphic evolution of the Piedmont Zone of the Ganga Plain, India: a study based on remote sensing, GIS and field investigation. Int J Remote Sens 31(20):5349–5364. https://doi.org/10.1080/01431160903303013
7. Goswami PK (2018) Using DTMs to delineate active faults of the proximal part of the ganga plain, Uttarakhand, India. In: Proceedings Asian conference on remote sensing (conference)
8. Kanaujia J, Mitra S, Kumar A, Sharma ML, Gupta SC (2018) Crustal seismic anisotropy in the Garhwal Lesser Himalaya. In: 16th symposium on earthquake engineering, IIT Roorkee, India
9. Chapter-III, Geology (https://shodhganga.inflibnet.ac.in/bitstream/)
10. Alam M, Alam MM, Curray CMLR, Gani MR (2003) An overview of the sedimentary geology of the Bengal Basin in relation to the regional tectonic framework and basin-fill history. Sed Geol 155:179–208
11. Mukherjee A, Fryar AE, Thomas WA (2009) Geologic, geomorphic and hydrologic framework and evolution of the Bengal basin, India and Bangladesh. J Asian Earth Sci 34:227–244. https://doi.org/10.1016/j.jseaes.2008.05.011
12. Curray JR, Emmel FJ, Moore DG, Raitt RW (1982) Structure, tectonics and geological history of the northeastern Indian Ocean. In: Nairn AEM, Stehli FG (eds) The Ocean basins and margins. The Indian Ocean, vol 6, pp 399–450
13. Rao DG, Krishna KS, Sar D (1997) Crustal evolution and sedimentation history of the Bay of Bengal since the Cretaceous. J Geophy Res Solid Earth 102(B8):17747–17768. https://doi.org/10.1029/96jb01339
14. Michael L, Krishna KS (2011) Dating of the 85 °E ridge (northeastern Indian Ocean) using marine magnetic anomalies. Curr Sci 100(9):1314–1322
15. Palamenghi L, Schwenk T, Spiess V, Kudrass HR (2011) Seismostratigraphic analysis with centennial to decadal time resolution of the sediment sink in the Ganges-Brahmaputra subaqueous delta. Cont Shelf Res 31:712–730
16. Sclater JG, Fisher RL (1974) Evolution of the East: Central Indian Ocean, with emphasis on the tectonic setting of the ninetyeast ridge. Geol Soc Am Bull 85:683–702. https://doi.org/10.1130/0016-7606(1974)85%3c683:eoteci%3e2.0.co;2
17. Ingersoll RV, Dickinson WR, Graham SA (2003) Remnant-Ocean submarine fans: largest sedimentary systems on earth. Geol Soc Am 370:191–208
18. Mukherjee S Tectonics and structural geology: Indian context

19. Singh CL, Singh VP (1989) Tectonic stability of the Farakka dam and its adjoining regions in East India. Phys Earth Plan Int 58:87–93. https://doi.org/10.1016/0031-9201(89)90098-8
20. Sujit D, Narula PL, Acharyya SK, Banerjee J (2000) Seismotectonic atlas of India and its environs. Geological Survey of India
21. Sinha R, Ghosh S (2012) Understanding dynamics of large rivers aided by satellite remote sensing: a case study from Lower Ganga plains, India. Geocarto Int 27(3):207–219. https://doi.org/10.1080/10106049.2011.620180
22. Talukdar SC, Murthy MVN (1971) The Sylhet Traps, their tectonic history, and their bearing on problems of indian flood basalt provinces. Geological Survey of India, Shillong
23. Morino M, Kamal ASMM, Akhter SH, Rahman MZ, Ali RME, Talukder A et al (2014) A paleo-seismological study of the Dauki fault at Jaflong, Sylhet, Bangladesh: historical seismic events and an attempted rupture segmentation model. J Asian Earth Sci. 91:218–226. https://doi.org/10.1016/j.jseaes.2014.06.002
24. Srinivasan V (2005) The Dauki fault in Northeast India: through Remote Sensing. J Geol Soc 66(4)
25. Goswami DC, Das PJ (2002) Hydrological impact of earthquakes on the Brahmaputra River regime, Assam: a study in exploring some evidences. My Green Earth 3(2)
26. Lahiri SK, Sinha R (2012) Tectonic controls on the morphodynamics of the Brahmaputra River system in the upper Assam valley, India. Geomorphology 169–170:74–85. https://doi.org/10.1016/j.geomorph.2012.04.012
27. Valdia (1987) Environmental geology: Indian context. Tata McGarw Hill Pub. Co. New Delhi, pp 217
28. Gupta ID (2006) Delineation of probable seismic sources in India and neighbourhood by a comprehensive analysis of seismotectonic characteristics of the region. Soil Dyn Earthq Eng 26(8):766–790. https://doi.org/10.1016/j.soildyn.2005.12.007
29. Singh VP, Nayan Sharma N, Shekhar P, Ojha CSP (2004) The Brahmaputra basin water resources. Water Sci Technol Library 47
30. Archeological Survey of India. https://asi.payumoney.com/

Stochastic Simulation of Near-Fault Earthquakes in Delhi Region Using Long-Period Velocity Pulse

Hemant Shrivastava, G. V. Ramana, and A. K. Nagpal

Abstract Near-fault ground motions are characterized by long-period velocity pulse. This pulse type ground motion is produced due to forward directivity effect, where fault rupture propagates towards the site with velocity close to shear wave velocity. Such impulsive ground motion consists high-input seismic energy at the beginning of the record. In this study, specific barrier model is calibrated to simulate ground motion at bedrock level in Delhi region. The Delhi earthquake (M_w 4.1, 2007) is used to estimate the seismic parameters (global and local stress drop) of specific barrier model. The estimated seismic parameters and long-period velocity pulse are then used to simulate the near-fault earthquake (M_w 5, 5.5 and 6) at bedrock level in Delhi region. The response spectra, *Sa*, 5% damping for Delhi region which falls in zone IV of the seismic zoning map of India is not conservative for ground motion with long-period velocity pulse.

Keywords Near-fault earthquakes · Long-period velocity pulse · Delhi region

1 Introduction

Seismic activity in Delhi region is attributed to the northward movement of Indian plate [40]. Delhi region is surrounded by seismogenic sources such as Delhi-Haridwar ridge, Delhi-Lahore ridge, Aravalli-Delhi fold, Sohna fault, Mathura fault, Rajasthan great boundary fault and Moradabad fault [33]. In the past years, Delhi region experienced some moderate earthquakes (M_w 6.5, 1720; M_w 6.0, 1745; M_w 6.7, 1956;

H. Shrivastava (✉)
Civil Engineering Department, Bundelkhand Institute of Engineering & Technology, Jhansi, Uttar Pradesh 284128, India
e-mail: hemant.ce@bietjhs.ac.in

G. V. Ramana · A. K. Nagpal
Department of Civil Engineering, IIT Delhi, Hauz Khas, New Delhi, Delhi 110016, India
e-mail: gvramanaiitdelhi@gmail.com

A. K. Nagpal
e-mail: aknagpal_iitd@yahoo.co.in

T. G. Sitharam et al. (eds.), *Seismic Hazards and Risk*, Lecture Notes in Civil Engineering 116, https://doi.org/10.1007/978-981-15-9976-7_19

M_w 6.0, 1960; M_w 5.8, 1966) from near source that caused damage, injury and loss of lives [38]. The 1960 earthquake that occurred between Delhi cantonment and Gurgaon was reassessed by Singh et al. [32], and the magnitude of earthquake was estimated to be M_w 4.8 which is less than M_w 6.0 as reported earlier in the literature. Future major earthquake in Delhi region get generated due to accumulation of seismic strain energy in the northern part of the Aravalli and adjoining Gangetic plains [40]. Shukla et al. [34] studied the fault plane solution of small magnitude earthquakes (M_w 3.0–4.2) and showed that seismicity is oriented in NNE-SSW and NW–SE directions. Prakash and Srivastava [27] reviewed the seismicity of Delhi region and identified three main zones of seismic activity located to west of Delhi due to activation of several faults and lineament of the DFB. The Delhi-Haridwar ridge is significant in the nucleation of seismicity in this region. Sharma et al. [33] estimated seismic hazard parameters for Delhi region from near sources and provided the variation of return period with magnitude of earthquake. The return periods of M_w 6.0 earthquakes are seen to vary from 93 to 291 years from several near seismogenic sources of Delhi-NCR region. It was estimated that the earthquake magnitudes corresponding to 20% exceedence value in 50 years from near source vary between 5.85 and 6.6. Bansal et al. [3] used Gutenberg-Richter relation for Delhi region developed by Shukla et al. [34] and estimated recurrence period about to be 130 years for $M_w \geq 5.0$ earthquakes.

The ground motion with long-period velocity pulse is generated by forward directivity effect which occurs when the fault rupture propagates towards the site at a velocity nearly equal to the propagation velocity of the shear waves, and the direction of fault slip is aligned with the site and causes the wave front to arrive as a single large pulse [35]. The ground motion with long-period velocity pulse can cause very large inelastic deformation demand on a structure and can be critical in the design of some structures [4]. In the near source region, large peak ground velocities are often associated with this pulse, particularly within 10–20 km of the rupture surface [14, 22]. The effect of long-period ground motion on the performance of flexible structures has been extensively reported in the literature [2, 11, 17–19, 35, 37] and indicates that amplitude and period of the long-period velocity pulse in the near-fault earthquakes are the key parameters that control the performance of structures [2, 4, 11, 29, 41].

Several researcher efforts were made in simulation of earthquakes from near sources in Delhi region [3, 16, 21, 28]. Different methodologies have been used to simulate ground motion: composite source model [16], empirical green function [3, 21] and specific barrier model [28]. However, these reported studies do not take into account the long-period velocity pulse which is important characteristic of near-fault ground motion [36].

In the current work, the hybrid methodology proposed by Halldorsson et al. [13] has been used for simulating near-fault earthquakes (M_w 5.0, 5.5 and 6.0) at bedrock level in Delhi region. The global and local stress drop parameters of specific barrier model were estimated by the observed ground motions of Delhi earthquake (M_w 4.1, 2007).

2 Tectonic Setting in Delhi Region

Delhi region lies on an extension of the peninsular shield and comprises of Indo-Gangetic alluvial plains and Aravalli hill ranges. The terrain of Delhi is flat except for a low NNE-SSW trending ridge. This low-ridge portion is an extension of the aravalli ranges which get buried under Yamuna alluvium in the north and eastern part of Delhi. Sharma et al. [33] studied the seismogenic sources around Delhi region are: the Delhi-Haridwar ridge, the Delhi-Lahore ridge, the Aravalli-Delhi fold, the Sohna fault, the Mathura fault, the Rajasthan great boundary fault and the Moradabad fault and concluded that the southern and western parts of Delhi region are more affected by the Mathura fault and Sohna fault, respectively. A thorough discussion of Srivastava and Somayajulu [38] is presented by geological structures and seismicity of the Delhi. Their study indicated that Sonepat-Delhi-Sohna dislocation is responsible for frequent earthquakes in and around Delhi city. Verma et al. [40] have studied several events of magnitude >3 around Delhi region and observed the north-easterly trend of seismicity along Delhi-Haridwar ridge upto main boundary fault. The NW–SE trending DSR (Delhi-Sargodha Ridge/Thrust) runs sub-parallel to Himalayan thrust system and typically exhibits a flexural bulge on the Himalayan foreland basin and is a potential zone of seismic importance [9]. Earthquakes of magnitude 5–6 are found to have occurred in this area, prominent among which is the Gurgaon earthquake (1960) of magnitude ~ 6 that caused damage in Delhi cantonment and adjoining Gurgaon areas.

3 Strong Ground Motion Database

The data consists of 20 horizontal records of Delhi earthquake (M_w 4.1, 2007) obtained at 10 stations (from strong motion network of 20 stations) deployed in the Delhi region by Department of Earthquake Engineering, IIT Roorkee [20]. The locations of these 10 strong motion stations as well as the epicentre of Delhi earthquake are shown in Fig. 1.

4 Simulation of Strong Ground Motion in Delhi Region

The specific barrier model proposed by Papageorgiou and Aki [23, 24] is free from the disadvantages of point source model [12, 26]. The specific barrier model is characterized by few physical input parameters and scaling of these input parameters with earthquake magnitude has been well established [23, 24]. It allows consistent ground motion simulation over the entire frequency range and for all distances of engineering interest.

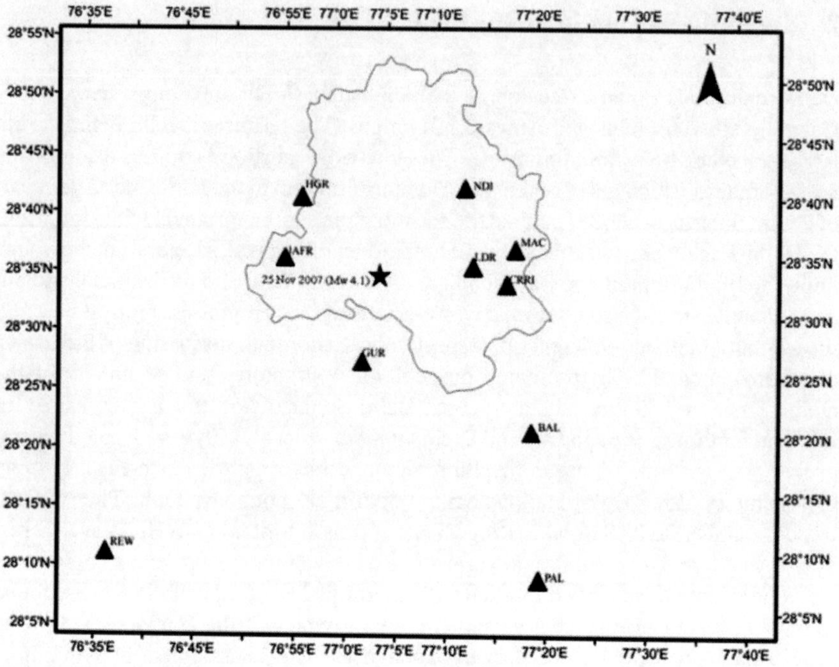

Fig. 1 Location of the November 25 2007 Delhi earthquake and the station which recorded the earthquake

4.1 *Simulation of High-Frequency Ground Motion*

The Fourier amplitude spectrum of shear waves of horizontal strong ground motion, $Y(M_0, r, f)$, at distance R from the source, can be expressed as

$$Y(M_0, R, f) = E(M_0, f)G(R)A_n(f)D(f) \tag{1}$$

$$E(M_0, f) = c * S(M_0, f) \tag{2}$$

where c is a constant; $S(M_0, f)$ is source spectrum; $G(R)$ represents path effects; $A_n(f)$ is path attenuation term; $P(f)$ is site term; M_0 is the seismic moment; f is the frequency; and R represents distance in km.

In specific barrier model, the heterogeneous seismic fault plane is assumed as rectangular fault plane of length L and width W and is divided into circular cracks, of equal radius ρ_0, representing each subevent as shown in Fig. 2. As rupture front sweeps over the fault plane, rupture nucleates at the centre and spreads radially with constant rupture velocity and stops abruptly at a distance equal to radius of each crack. The radiation of elastic waves emitted from each crack as its breaks

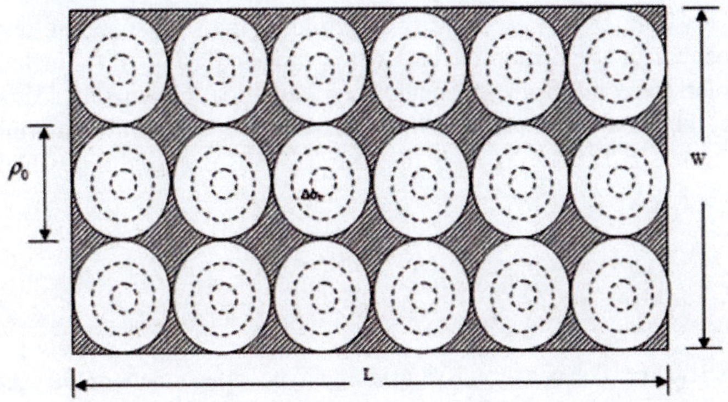

Fig. 2 Schematic view of specific barrier model [23]

is based on a physical description of source processes using kinematic dislocation theory [1]. Global stress drop represents the size of entire source, and it controls the lower frequencies of generated ground motion. Local stress drop is stress drop of each subevent and controls the higher frequencies. The subevents are assumed to break randomly and independently. The acceleration source spectrum [25], the sum of high-frequency ground motion from each subevent, is expressed as follows

$$S(M_0, f) = (2\pi f)^2 \left\{ N \left[1 + (N-1) \left(\frac{\sin(\pi f T)}{(\pi f T)} \right)^2 \right] \right\}^{\frac{1}{2}} \overline{M}_{0i}(f) \quad (3)$$

where $M_{0i}(f)$ is the acceleration source spectrum of the individual subevent; $N = (\pi/4)^3(\Delta\sigma_L/\Delta\sigma_G)^2$ is the total number of subevents that compose the rupture; $\Delta\sigma_G$ is global stress drop of the main event; and $\Delta\sigma_L$ is local stress drop of the subevent. The high frequency radiation from the each subevent of the specific barrier model sums up incoherently and their spectrum as given by Eq. (1). $M_{0i} = (16/7)\Delta\sigma_L\rho_0^3$ is the seismic moment released by the crack, and the patch corner frequency f_2 is corresponding to the crack radius, ρ_0, by $f_2 = C_s\beta/2\pi\rho_0$, where β denotes the source region S-wave velocity, C_s is an increasing function of the υ/β ($1.72 \leq C_s \leq 1.85$ for $0.7 \leq \upsilon/\beta \leq 0.9$) [1]; and υ indicates the propagating rupture velocity inside the circular cracks. Time of rupture, T, is related to the corner frequency f_1 by $T = C/f_1$, where in C is taken as 0.47 [25]. The size of subfault, ρ_o, is estimated based on empirical relationship [5, 12] with moment of target event (M_w)

$$\log 2\rho_0 = -2.58 + 0.5M_w \quad (4)$$

The source term $E(M_0, f)$ is computed as a product of $S(M_0, f)$ by a frequency independent scaling factor of $c = FSR_{\varphi\theta}V/4\pi\rho\beta^3$, where $R_{\varphi\theta}$ accounts for the average S-wave radiation pattern; FS denotes free surface amplification; V indicates partitions of the total S-wave energy into two horizontal components. The ρ and β represent the mass density and the S-wave velocity surrounding the source region, respectively. The anelastic path attenuation factor $A_n(f) = \exp(-\pi f R/\beta Q(f))$ [7] includes all the losses which have not been accounted by geometrical attenuation factor. $Q(f)$ is the quality factor, which includes both anelastic absorption and scattering. $Q(f)$ is assumed to be $253f^{0.8}$ [3]. $G(R)$ is the geometrical spreading term which may be taken as [31]

$$G(R) = \begin{cases} R^{-1} \text{ for } R \leq R_x \\ (RR_x)^{-1/2} \text{ for } R > R_x \end{cases} \tag{5}$$

with R_x as 100 km. The site term is considered to be independent of the source to site travel distance. The factor $D(f)$ [6] accounts for the path-independent loss of high frequency in the ground motion and is given by

$$D(f) = \left[1 + (f / f_{\max})^8\right]^{-1/2} \tag{6}$$

where $f_{\max}$ is the maximum cut off frequency.

4.2 Simulation of Long-Period Velocity Pulse

In near-fault ground motion, most of the elastic energy arrives coherently in a single, intense, relatively long-period velocity pulse at the beginning of the record, representing cumulative effect of almost all the seismic radiation from the fault [36]. The phenomenon is even more pronounced when the direction of slip on the fault plane points towards the site as well. Mavroeidis and Papageorgiou [22] presented an analytical expression for long-period velocity pulse, given as

$$v(t) = \begin{cases} \dfrac{A}{2}\left[1 + \cos\left(\dfrac{2\pi f_p}{\gamma}(t - t_0)\right)\cos[2\pi f_p(t - t_0) + \nu]\right], \\ t_0 - \dfrac{\gamma}{2f_p} \leq t \leq t_0 + \dfrac{\gamma}{2f_p} \text{ with } \gamma > 1 \\ 0, \text{ otherwise} \end{cases} \tag{7}$$

Here, parameter A controls the amplitude of signal; f_p is prevailing frequency; υ is phase of amplitude modulated harmonic; γ is a parameter that defines the oscillatory character of the signal; and t_0 specifies the epoch of the envelope. The pulse duration, pulse amplitude, as well as the number and phase of half cycles are the key parameters that define the waveform characteristics of near-fault velocity pulses. The prevailing frequency ($f_p = 1/T_p$) of the signal is the inverse of duration of pulse. An empirical

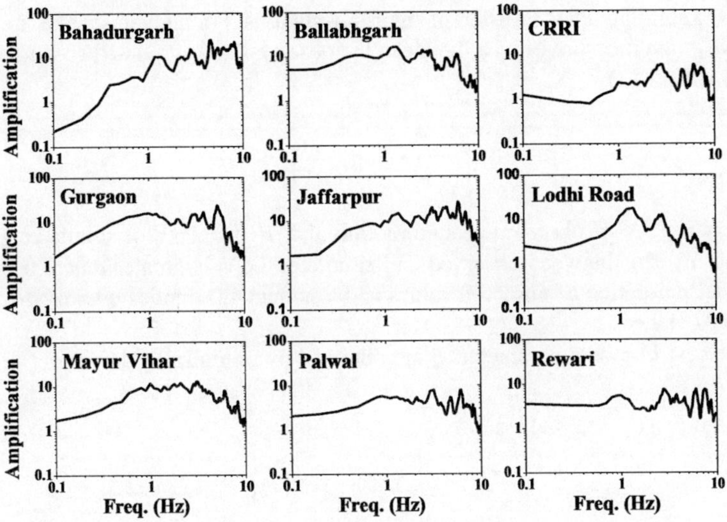

Fig. 3 Site amplification at different sites during Delhi earthquake

relationship between pulse period and the moment magnitude of the earthquake are given by [22].

$$\log T_{\mathrm{p}} = -2.9 + 0.5M_{\mathrm{w}} \tag{8}$$

5 Site Amplification

Site amplification at each station has been estimated using the standard spectral ratio technique [8]. The site effects are commonly estimated by dividing the spectrum obtained at target site by that obtained at a nearby reference site which is preferably on the bedrock. The site amplification of strong motion recorded at different stations during Delhi earthquake is shown in Fig. 3 with ridge observatory station as reference station.

6 Calibration of Specific Barrier Model with Delhi Earthquake

The specific barrier model, most important parameters are $\Delta\sigma_{\mathrm{G}}$ and $\Delta\sigma_{\mathrm{L}}$ that control the Fourier spectra at lower and higher frequencies [12]. These model parameters are calibrated through a comparison of observed and simulated PSV. $\Delta\sigma_{\mathrm{G}}$ and $\Delta\sigma_{\mathrm{L}}$ are varied to simulate PSV at different recording stations. For a given set of θ, i.e. $\Delta\sigma_{\mathrm{G}}$ and $\Delta\sigma_{\mathrm{L}}$, the measure of error between the observed and simulated ground motion is defined as [12]

$$e(\theta) = \log\left(\frac{y}{\mu(\theta)}\right) \tag{9}$$

where y = PSV of observed ground motion and $\mu(\theta)$ = PSV of simulated ground motion. If error between observed and simulated PSV is greater than 1.0 implies an under prediction of observed values and less than 1.0 implies over prediction of observed values.

The best fit model parameters θ are obtained by minimizing the norm (Γ) given by

$$\Gamma = \sqrt{\sum_{i}^{N_e}\sum_{j}^{n_i \cdot n_j}\left[e_{ij}(\theta)\right]^2} \tag{10}$$

where N_e is the number of earthquake event considered in the analysis; n_i represents the number of observation stations; and n_j represents the number of discrete oscillator frequencies (f).

For minimizing the above norms, genetic algorithm optimization technique [10] has been used. In this study, $n_j = 12$ has been chosen (f = 0.5, 0.7, 0.9, 1.1, 1.5, 2.0, 2.6, 3.4, 4.4, 5.8, 7.6 and 10.0 Hz) [12]. The norm (Eq. 10) used as fitness function.

Necessary seismological parameters for simulation of ground motion time histories during Delhi earthquake using specific barrier model, other than $\Delta\sigma_{\mathrm{G}}$ and $\Delta\sigma_{\mathrm{L}}$ obtained from Singh et al. [30] are given in Table 1. From calibration, the optimum evaluated $\Delta\sigma_{\mathrm{G}}$ and $\Delta\sigma_{\mathrm{L}}$ obtained to be 99 bars and 375 bars, respectively.

The results of simulated ground motion are compared with the observed horizontal ground motion in terms of PGA and their corresponding *Sa* response spectra. A comparison of 5% damped *Sa* of observed ground motion with the simulated spectra at various stations for Delhi earthquake is shown in Fig. 4. It is observed that simulated *Sa*-period trend is similar to the observed *Sa*-period trend. At stations Bhadurgarh,

Table 1 Seismological parameters used in the calibration for $\Delta\sigma_{\mathrm{G}}$ and $\Delta\sigma_{\mathrm{L}}$ parameters of specific barrier model

Parameters	Values
$F, V, R_{\theta\phi}$	2, 0.71, 0.55
Crustal density (kN/m^3)	3.2
Rupture velocity (km/s)	4.2
Geometrical spreading	$G(R) = R^{-1}$ for $R \leq R_x$ $G(R) = (RR_x)^{-1/2}$ for $R > R$
$Q(f) = Q_0 f^n$	$253 f^{0.8}$
$f_{\max}$ (Hz)	35

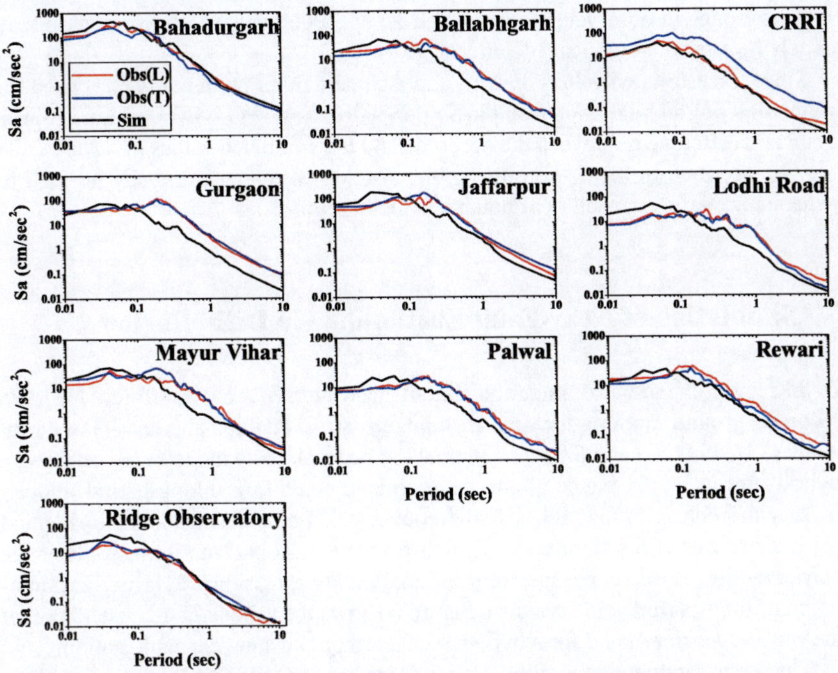

Fig. 4 Comparison of simulated and observed *Sa* (5% damping) response spectra at different stations during Delhi earthquake

CRRI, Mayur Vihar, Rewari and RO stations simulated ground motion matches well with observed ground motion. In addition, a comparison of simulated and observed PGA values with hypocentral distance is shown in Fig. 5. The trend of decay of simulated PGA values with hypocentral distance is similar to observed PGA values.

Fig. 5 Comparison between observed and simulated PGA with hypocentral distance

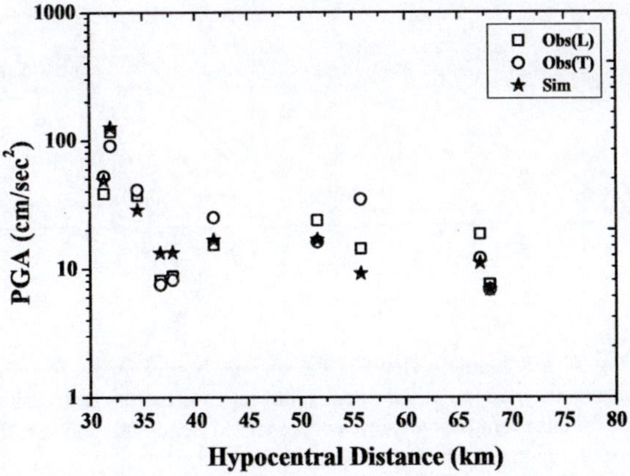

The difference in observed and simulated *Sa* and PGA values at some station may largely be attributed to local site effects.

Since the estimated values of $\Delta\sigma_G$ and $\Delta\sigma_L$ and other input parameters used for simulation of ground motion result in *Sa*-period trend and PGA-hypocentral distance trend and differences between the observed and the simulated values of *Sa* and PGA may be ascribed to local site effects, therefore, these parameters may be used to generate scenario near-fault earthquake in Delhi region.

7 Simulation of Near-Fault Earthquake in Delhi Region

Delhi region experienced an earthquake of maximum M_w 6.0 as of now [33]. No recorded ground motions for such earthquakes are available. The near-fault earthquakes (M_w 5.0, 5.5 and 6.0) are generated at bedrock level in Delhi region using a hybrid approach [13] that combines deterministic modelling at long-period velocity pulse with stochastic modelling at high frequencies. The stochastic modelling method [6] has been used for simulation of high-frequency ground motion using specific barrier model [25] as source spectrum and analytical expression [22] is used for simulation of long-period velocity pulse. Figure 6 shows the schematic representation of the specific barrier model for a hypothetical earthquake and near-fault station.

The input parameters for simulating the long-period velocity pulse are as below:

1. **Amplitude**

In present study, attenuation relationship among the peak ground velocity (PGV), moment magnitude (M_w) and closest distance of the site to the fault (R) has been used for estimate amplitude of long-period velocity pulse [37]. This relationship cannot use for distance less than 3 km.

$$\log_{10} \text{PGV} = -1.0 + 0.5M_w - 0.5\log_{10} R \tag{11}$$

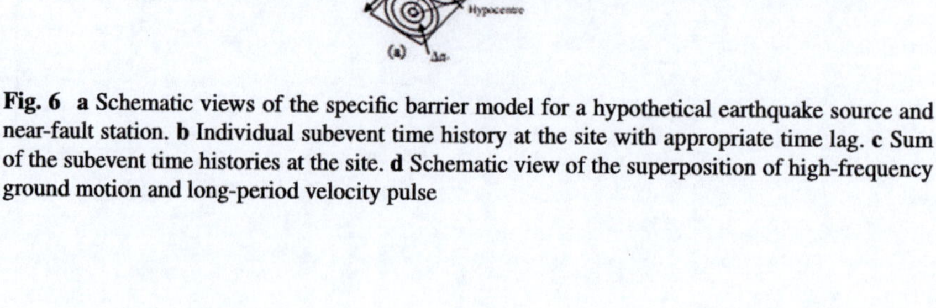

Fig. 6 **a** Schematic views of the specific barrier model for a hypothetical earthquake source and near-fault station. **b** Individual subevent time history at the site with appropriate time lag. **c** Sum of the subevent time histories at the site. **d** Schematic view of the superposition of high-frequency ground motion and long-period velocity pulse

2. **Number of pulses**

No models are currently available for predicting the number of significant pulses in the velocity time history. For most, number of pulses will vary between 1 and 3, with number of pulse equal to 2 being a good general value to use in seismic evaluation [39]. In this study, number of significant pulse is assumed to be equal to two (i.e. one full cycle of pulse type ground motion).

3. **Phase angle and t_0**

There are no clear guidelines available in the literature regarding selection of phase angle and t_0 for simulating the long-period velocity pulse. To examine the influence of phase angle and t_0 a, hence, sensitivity analysis is carried out while keeping amplitude, pulse period and number of pulses being kept constant. The phase angle and t_0 are taken as 150°, 235° and 0.7 s, 1.5 s, respectively. Fifteen random time histories are generated for each phase angle and t_0, i.e. a total of 60 sets are generated, and response spectrum is computed. The mean of response spectrum computed for each phase angle and t_0 is shown in Fig. 7 and observed that the variation in response spectra for different t_0 and phase angle is negligible and indicating that t_0 and phase angle do not affect the simulation of ground motion with long-period velocity pulse.

To generate near-fault ground motion, the seismological parameters for simulation of high-frequency ground motion and parameters for simulation of long-period velocity pulse are given in Tables 1 and 2, respectively. The fault location is assumed within the Delhi region. Scenario earthquakes are simulated at the bedrock level. The acceleration time histories of near-fault earthquakes of M_w 5.0, 5.5 and 6.0 at bedrock level for two epicentral distances 10 km and 20 km are shown in Figs. 8 and 9, respectively. Response spectra (with and without long-period pulse) are also shown in Figs. 8 and 9, and the effect of long-period pulse is clearly observed at

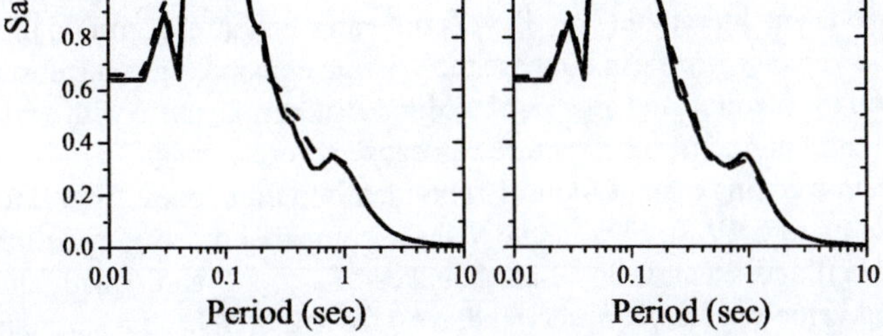

Fig. 7 Comparison of response spectra for **a** epoch of envelope peak (t_0) and **b** phase angle (υ)

Table 2 Input parameters for long-period velocity pulse

Parameters	Value					
Magnitude	5.0	5.5	6.0			
Distance (km)	10	20	10	20	10	20
Amplitude (cm/s)	10	7.07	17.78	12.57	31.62	22.36
ϒ	1	1	1	1	1	1
γ	220	20	120	10	150	145
t_0	0.3	0.6	0.42	0.8	0.8	1

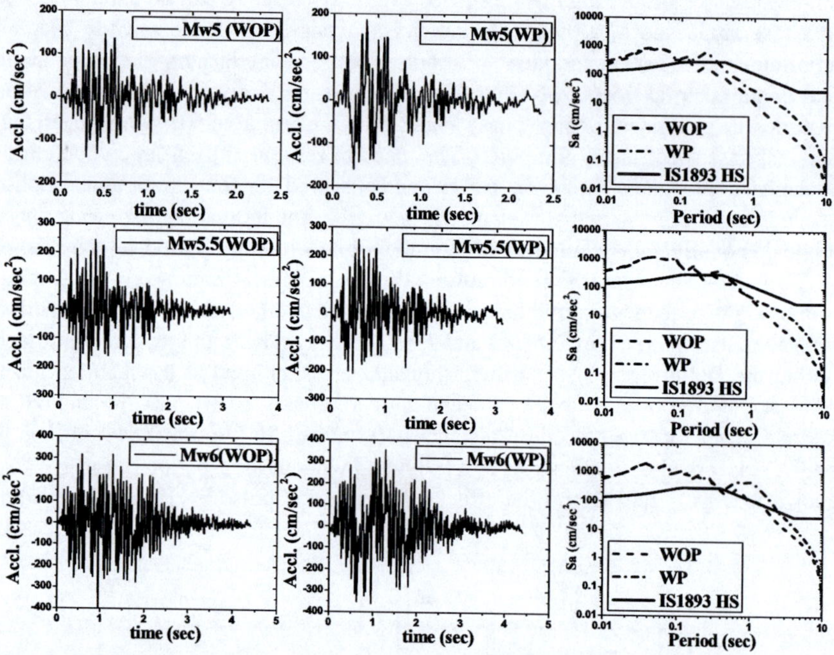

Fig. 8 Acceleration time history and response spectra for M_w 5, 5.5 and 6 earthquakes (wop—without pulse; wp—with pulse) for $R = 10$ km in Delhi region at bedrock level

higher period. The literature [2, 4, 11, 17] indicates that strong ground motion with long-period velocity pulse can increase the seismic demand (base shear, storey drift, ductility) on the medium and high rise building. In addition, pulse with a width larger than the natural period of the structure can cause severe damage.

The seismic zoning map of India [15] divided into four zones, from II to V. Delhi region falls in zone IV. In Figs. 8 and 9 show comparison between *Sa* for zone IV [IS1893 (HS)] and scenario local earthquakes M_w 5, 5.5 and 6 with and without long-period velocity pulse. It observed that *Sa* for M_w 5, 5.5 and 6 earthquakes with

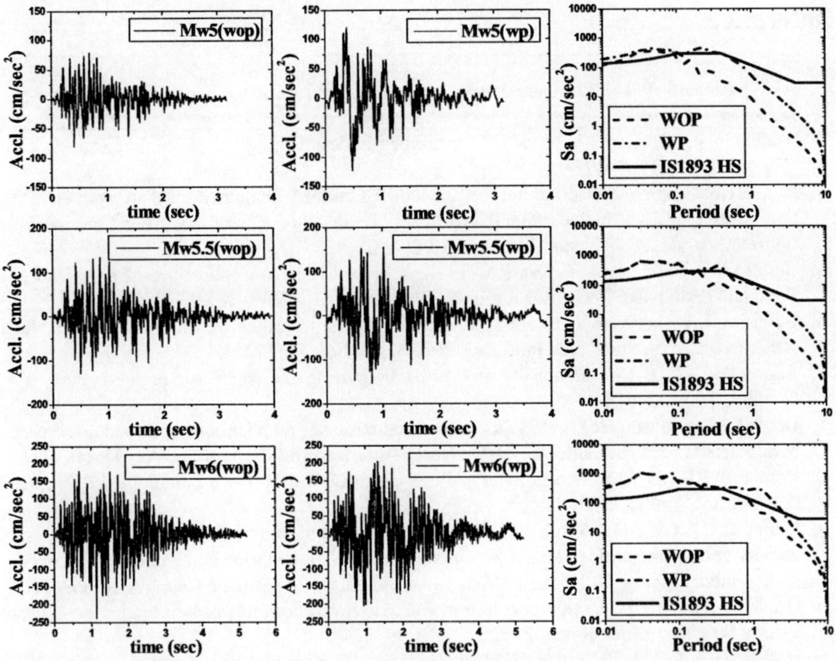

Fig. 9 Acceleration time history and response spectra for M_w 5, 5.5 and 6 earthquakes (wop—without pulse; wp—with pulse) for $R = 20$ km in Delhi region at bedrock level

and without long-period pulse exceeded *Sa* for IS1893 (HS) at shorter period but *Sa* for M_w 6 earthquake with long-period pulse exceed IS1893 (HS) at higher period.

8 Conclusions

There is no recorded ground motion of near-fault earthquakes having characteristics of long-period velocity pulse is available in Delhi region. In this study, methodology available in the literature has been used for simulation of near-fault earthquakes with long-period velocity pulse at bedrock level in Delhi region. The specific barrier model has been used for simulation of high-frequency ground motion. The parameters of specific barrier model are estimated by comparing the simulated and observed ground motions of Delhi earthquake. The specific barrier model with estimated parameters and a long-period velocity pulse has been used to simulate the near-fault earthquakes (M_w 5.0, 5.5 and 6.0) in Delhi region. The *Sa* response spectra [IS1893 (HS)] for Delhi region are not conservative for M_w 6.0 ground motion with long-period velocity pulse.

References

1. Aki K, Richards PG (1980) Quantitative seismology theory and methods. W. H. Freeman, San Francisco
2. Anderson JC, Bertero VV (1987) Uncertainties in establishing design earthquakes. J Struct Eng ASCE 113:1709–1724
3. Bansal BK, Singh SK, Dharmaraju R, Pacheco JF, Ordaz M, Dattatrayam RS, Suresh G (2009) Source study of two small earthquakes of Delhi, India, and estimation of ground motion from future moderate, local events. J Seismol 13:89–105
4. Bertero VV, Mahin SA, Herrera RA (1978) A seismic design implications of near fault San Fernando earthquake Records. Earthquake Eng Struct Dyn 6:31–42
5. Beresnev, I.A., Atkinson, G.M.: Source parameters of earthquakes in eastern and western North America based on finite-fault modeling. Bull Seism Soc Am (2002)
6. Boore DM (2003) Simulation of ground motion using the stochastic method. Pure Appl Geophys 160:635–676
7. Boore DM, Atkinson GM (1987) Stochastic prediction of ground motion and spectral response parameters at hard rock sites in Eastern North America. Bull Seismol Soc Am 77:440–467
8. Borcherdt RD (1970) Effects of local geology on ground motion near San Francisco Bay. Bull Seismol Soc Am 60:29–61
9. Dubey CS, Shukla DP, Singh RP, Sharma M, Ningthoujam PS, Bhola AM (2012) Present activity and seismogenic potential of Himalayan sub-parallel thrust faults in Delhi: inferences from remote sensing, GPR, gravity data and seismicity. Near Surface Geophys 10:369–380
10. Goldberg DE (1989) Genetic algorithms in search optimization and machine learning. Addison-Wisely Publishing Inc., Reading, MA
11. Hall JF, Heaton TH, Halling MW, Wald DJ (1995) Near-source ground motion and its effects on flexible buildings. Earthquake Spectra 11:569–605
12. Halldorsson B, Papageorgiou AS (2005) Calibration of the specific barrier model to earthquakes of different tectonic regions. Bull Seismol Soc Am 95:1276–1300
13. Halldorsson B, Mavroeidis GP, Papageorgiou AS (2011) Near fault and far field strong ground motion simulation for earthquake engineering applications using the specific barrier model. J Struct Eng ASCE 137:433–444
14. Heaton TH, Hall JF, Wald DJ, Halling MW (1995) Response of high rise and base isolated buildings to a hypothetical M_{w} 7.0 blind thrust earthquake. Science 267:206–211
15. IS 1893 (2016) (Part I): Criteria for earthquake resistant design of structures—Part 1: general provisions and buildings. Bureau of Indian Standards, New Delhi
16. Kumar D, Sarkar I, Sriram V, Teotia SS (2012) Evaluating the seismic hazard to the National capital (Delhi) region, India from moderate earthquakes using simulated accelerograms. Nat Hazards 61:481–500
17. Liao WI, Loh CH, Wan S (2001) Earthquake responses of RC moment frames subjected to near fault ground motions. Struct Des Tall Build 10:219–229
18. MacRae GA, Morrow DV, Roeder CW (2001) Near fault ground motion effects on simple structures. J Struct Eng ASCE 127:996–1004
19. Malhotra PK (1999) Response of buildings to near field pulse like ground motion. Earthquake Eng Struct Dyn 28:1309–1326
20. Mittal H, Kumar A, Ramhmachhuani R (2012) Indian national strong motion instrumentation network and site characterization of its stations. Int J Geosci 3:1151–1167
21. Mittal H, Kumar A, Kamal (2012) Ground motion estimation in Delhi from postulated regional and local earthquakes. J Seismolog. https://doi.org/10.1007/s10950-012-9340-5
22. Mavroeidis GP, Papageorgiou AS (2003) A mathematical representation of near fault ground motions. Bull Seismol Soc Am 93:1099–1131
23. Papageorgiou AS, Aki K (1983a) A specific barrier model for the quantitative description on inhomogeneous faulting and the prediction of strong ground motion. I Description of the model. Bull Seismol Soc Am 73:693–722

24. Papageorgiou AS, Aki K (1983b) A specific Barrier model for the quantitative description of inhomogeneous faulting and the prediction of strong ground motion. Part II. Appl. Model Bull. Seismol. Soc. Am. 73:953–978
25. Papageorgiou AS (1988) On two characteristics frequencies of acceleration spectra: patch corner frequency and f_{max}. Bull Seismol Soc Am 78:509–529
26. Papageorgiou A (2003) The barrier model and strong ground motion. Pure Appl Geophys 160:603–634
27. Prakash R, Shrivastava JP (2012) A review of the seismicity and seismotectonics of Delhi and adjoining areas. J Geol Soc India 79:603–617
28. Rao HCh (2006) Ground response analyses and liquefaction studies for soils of Delhi. Ph.D. thesis, Indian Institute of Technology Delhi, New Delhi
29. Sasani M, Bertero VV (2000) Importance of severe pulse-type ground motion in performance based engineering: Historical and critical review. In: 12th World conference earthquake engineering, Paper No. 1302
30. Singh SK, Kumar A, Suresh G, Ordaz M, Pacheo JF, Sharma ML, Bansal BK, Dattatrayam RS, Reinoso E (2010) Delhi earthquake of 25 November 2007 (M_w 4.1). Curr Sci 7:939–947
31. Singh SK, Mohanty WK, Bansal BK, Roonwal GS (2002) Ground motion in Delhi from future large/great earthquakes in the central seismic gap of the Himalayan arc. Bull Seismol Soc Am 92:555–569
32. Singh SK, Suresh G, Dattatrayam RS, Shukla HP, Martin S, Havskov J, Perez-Campos X, Iglesias A (2013) The Delhi 1960 earthquake: epicenter, depth and magnitude. Curr Sci 105:1155–1165
33. Sharma ML, Wason HR, Dimri R (2003) Seismic microzonation of the Delhi region for bedrock ground motion. Pure Appl Geophys 160:2381–2398
34. Shukla AK, Prakash R, Singh RK, Mishra PS, Bhatnagar AK (2007) Seismotectonics implicationsof Delhi region through fault plane solutions of some recent earthquakes. Curr Sci 93:1848–1853
35. Somerville PG, Smith NF, Graves RW, Abrahmson NA (1997) Modification of empirical strong ground motion attenuation relations to include the amplitude and duration effects of rupture directivity. Seismol Res Lett 68:199–222
36. Somerville PG (2005) Engineering characterization of near fault ground motions. In: New Zealand society of earthquake engineering conference, New Zealand
37. Somerville PG (1998) Development of an improved representation of near fault ground motions. In: Proceedings of the SMIP98 seminar on utilization of strong ground motion data, 1–20, Oakland CA
38. Srivastava LS, Somayajulu JG (1966) The seismicity of the area around Delhi. In: Third symposium on earthquake engineering. University of Roorkee, Roorkee
39. Stewart JP, Chio SJ, Bray JD, Graves RW, Somerville PG, Abrahnmson NA (2001) Ground motion evaluation procedures for performance based design. PEER report 2001/09. University of California, Berkeley
40. Verma RK, Roonwal GS, Kamble VP, Mohanty WK, Dutta U, Gupta Y, Chatterjee D, Kumar N, Chauhan PKS (1995) Seismicity of Delhi and its surrounding region. J Himalayan Geol 6:75–82
41. Zhang Y, Iwan WD (2002) Active interaction control of tall buildings subjected to near-field ground motions. J Struct Eng ASCE 128:69–79

Variability in the Period of Velocity Pulse Based on Nucleation Position on the Fault

K. S. K. Karthik Reddy and Surendra Nadh Somala

Abstract Spatial variability is observed in ground motions depending on various factors like path, site, and source. Based on board classification concerning the path characteristics, the ground motions can be classified as near field and far field depending on the distance of the station from the source. The elementary difference between the near-field and far-field ground motions lies in the reflection of pulse in the velocity field of a near field which is not apparent in far-field ground motions. Less emphasis is placed on variation in pulse period of these ground motions with the change in source kinematics. In order to explore the content to a deeper extent, one can rely on dynamic simulation. The objective is to run dynamic rupture simulations using an open-source code SPECFEM3D to study the variability of pulse period by modifying the source parameters like the positioning of the nucleation region to explore the variation in the amplitude and time period content of pulse exhibited in near-field stations. The alignment of stations is equidistant with a similar rupture distance from the fault. Two scenarios were considered by placing the nucleation region at the center of the fault in scenario-1 and away from the center from the axis of fault symmetry in scenario-2. Baker's method was employed to quantify pulse period for chosen race track configuration of stations. The results depict that change of nucleation center transforms the non-pulse-type ground motions to pulse-like ground motions for similar (Mw = 6.5) magnitude earthquake, and for stations in front and behind the fault, there is considerable difference of 6% in pulse period for scenario-2.

Keywords Pulse period · Nucleation asperity · Dynamic rupture · SPECFEM3D

1 Introduction

Amid all-natural calamities, earthquakes are causing most of the socio-economic loss around the globe due to their unpredictable nature of occurrence and propagation.

K. S. K. Karthik Reddy (✉) · S. N. Somala
Department of Civil Engineering, Indian Institute of Technology, Hyderabad, India
e-mail: ce17resch01003@iith.ac.in

T. G. Sitharam et al. (eds.), *Seismic Hazards and Risk*, Lecture Notes in Civil Engineering 116, https://doi.org/10.1007/978-981-15-9976-7_20

Though the earthquakes are unavoidable, the destruction caused can be minimized by exploring their characteristics such as faulting style, magnitude, the zone of influence, and soil characteristics. Most of the earthquakes are of tectonic origin and takes place in the faults.

The motion of the fault occurs when the shear stress on the fault plane overcomes the strength of the material or the friction which holds the ruptured surface together. The total energy is dissipated into the surrounding medium in the form of elastic waves, called seismic energy [1]. The response of civil infrastructure is regulated in terms of response spectrum, where the ground motion is converted to response spectra and an equivalent force and is applied to the structure. Typically, near-field stations experience pulse-type ground motions [2] that result in significant damage to structures.

Traditional analysis methods do not employ the dynamics of fault rupture hence are inadequate to capture the full effects of these pulse-type ground motions. Computational seismology overcomes this limitation and plays an important role to simulate dynamic earthquake ruptures [3]. The subsequent instant step trailed once evaluating the velocity field is assessing the structural response to these ground motions. A reliable way of gaging the performance is by executing an incremental dynamic analysis by means of near-field excitations and estimating the seismic capacity.

2 Features of Near-Field Ground Motions with Pulse Content

The ground motions are broadly distinguished based on their distance from the fault rupture; near-fault ground motions are clearly recognizable from far field ones [2]. The near-field velocity time histories are categorized by high amplitude and longer duration which can be reflected in their velocity content. The pulse content can vary based on the orientation of the site; further, the type of fault styling also modifies the pulse content.

There is a need to understand the concept of directivity while trying to understand the velocity pulses in near-field stations. Directivity is a consequence of a fault rupturing, wherein earthquake ground motion in the direction of rupture propagation is more severe than that in other directions from the earthquake source (USGS earthquake grocery). If the rupture propagates toward a site at a velocity that is as large as the shear wave velocity, then it causes most of the seismic energy to arrive in a single pulse which represents the cumulative effort of all the seismic radiation from the fault. These variations become significant at a time period of 0.6 s and generally in size with an increasing time period [4].

It is also recommended to understand the manifestation of backward directivity which is inconsistent with the phenomena of forward directivity, wherein the velocity pulses are less significant though the backward directivity station is located at an identical distance like forward directivity station [5]. A simple parametric study proposed by Bray and Rodriguez-Marek [6] based on amplitude, pulse period, and the amount of notable pulses to capture the characteristics of these pulses. Empirical relations were also prescribed to estimate the peak ground velocity and period of the velocity pulse, adding to the fact that such attenuation relationships do not account for kinematics in the fault. The near-field ground motions which are rich in pulse content vary while we alter the kinematics of source model in dynamic rupture, in a dynamic rupture the inputs of background stress, the positioning of nucleation region, and friction coefficients are fed into the simulation model. The influence of change in the positioning of nucleation region and its influence on the pulse content of the ground motion are emphasized. In the case of nucleation region placed at the extreme end of the fault face, the conflicting behavior of backward and forward directivity ground can be sensed, whereas in the case of asperity in the center of the fault, both stations represent similar field vectors. The reason for the conflicting behavior of field vectors at stations in front and behind the fault for case 2 is a result of allowing a lengthier path for rupture propagation only in one direction and arresting the rupture propagation in the opposite direction. The velocity pulses resulting from rupture propagation superimpose and give rise to a single-velocity pulse of high amplitude and time period. In case of case 1, the position of asperity is located at the axis of symmetry of the fault giving liberty for the rupture to propagate in both the direction with limited path length affecting the stations in front and behind the faults in the same manner.

3 Fracture Mechanics Approach to Earthquake Simulations

The ground motion of earthquakes is forecasted by physics-based ground motion simulations using numerical methods by assimilating the physics of the earthquake source and the outcoming propagation of seismic waves. Stochastic simulation methods combine the physics and kinematics of the earthquake source employing indirect approaches. These approaches signify the physics of earthquakes but do not essentially solve the accepted mathematical concepts that describe the physics of source dynamics and wave propagation [7]. The finite difference (FD) was first introduced by Alterman and Karal [8] in seismology. These methods were also used to describe the propagation of waves in stratified media and sedimentary basins to assess ground motion amplification [9]. The initial 3D simulations were performed by Frankel and Vidale [10] who simulated and validated Mw 4.4 aftershock of the 1989 Loma Prieta, California, earthquake using a point source model; with an increase in the computational facility, vigorous models were created and simulated in the spectrum of obligatory frequencies [11, 12].

4 Methodology

4.1 Elasto-Dynamic Equation

Earthquake ground motion simulation involves obtaining the solution of the linear momentum equation, which can be written in Cartesian coordinates and indicial notation as

$$\frac{\partial \tau_{ij}}{\partial x_j} + F_i = \rho \frac{\partial^2 u_i}{\partial t^2} \tag{1}$$

τ_{ij} represents the Cauchy stress tensor, ρ is the mass density, and F_i and u_i are the body forces and displacements in the ith direction within a bounded domain. In earthquake simulations, our objective is to calculate ground displacements varying with time due to conditions on the fault plane using 3D spectral element method [3]. Rupture models are based on assumptions; ground motions can either be generated by means of kinematic ruptures or dynamic ruptures; kinematic ruptures are not physics-based simulations; we consider rupture as moving slip discontinuity pulse (predefined) traveling at a velocity. Earthquake rupture is assumed as dynamically running shear crack on a frictional surface, embedded in an elastic medium. We give stress field across fault plane as input to simulation, based on fracture mechanics principles. The correlation between stress drop and slip is related by using linear slip weakening friction laws [13].

5 Modeling and Simulation

The physics-based simulation follows a workflow, initially, a region of interest for simulation is selected. The region contains topographical features and fault positioning which is followed by the selection of source rupture parameters that highlights the friction parameters, stress drop in asperity as well as background stresses on the fault. The material properties of the domain like the profile of p, s wave velocities, and densities are defined for the domain. These parameters together are called as modeling parameters. Meshing schemes are applied to the domain; the size of the mesh typically depends on the maximum resolving frequency of interest; one can obtain higher resolving frequencies with finer mesh; the mesh size also depends on minimum velocity of the wave; the lower the velocity the finer the mesh should be for particular resolving frequency in the model. The simulation is performed with the help of high-performance computing (HPC).

A three-dimensional vertical strike-slip fault is modeled in a half space (Fig. 1). Earth's surface with a size of 75 km by 150 km lies in the XY plane. Fault plane lies in the XZ plane. Figure 2a, b shows the geometry of the fault plane and earth domain considered for simulations. Rupture is allowed to propagate and is limited to

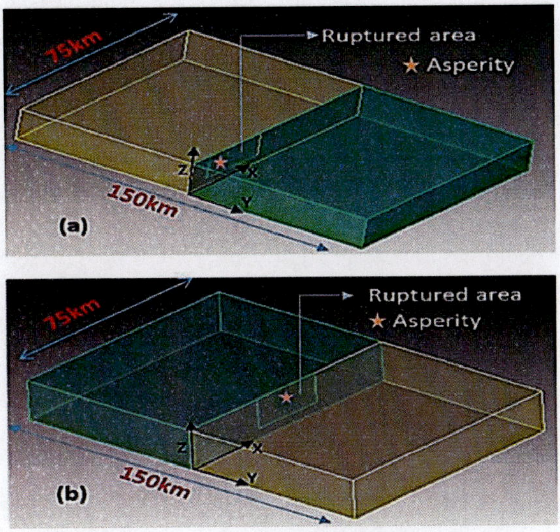

Fig. 1 Positioning of the fault along with asperity to simulate. **a** Case 1 (unilateral rupture), **b** Case 2 (bilateral rupture)

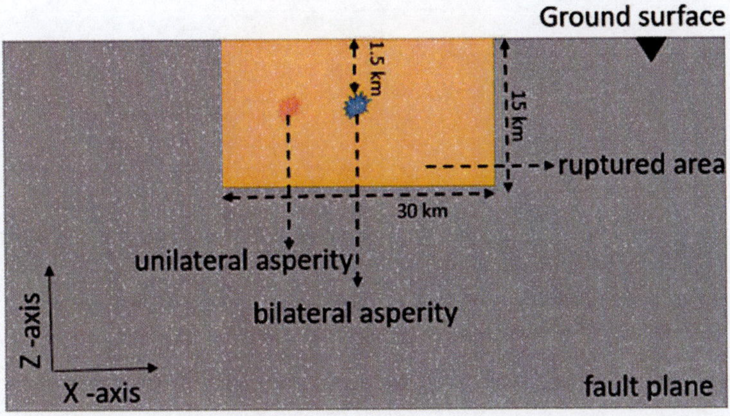

Fig. 2 Positioning of the fault along with the nucleation patch to simulate. **a** Case 1 and **b** Case 2

the fault of size 30 km × 15 km. Rupture initiates in a 3 km × 3 km patch (asperity). Figure 3 depicts Case 1 (unilateral rupture), where rupture initiation is at the end and Case 2 (bilateral) where rupture initiates at the center of the fault. In this nucleation patch, we give high shear stresses to begin fracture in fault plane.

Both the simulations are done by using alike dimensions and orientation of fault plane and earth domain. Similar material properties and friction law parameters (critical slip weakening distance, dynamic, and static friction coefficients) tabularized in Table 1 is engaged in both cases. The stations considered are aligned around the fault in a race track faction, such configuration of stations helps us to understand the variation in pulse content based on their orientation around the fault. To learn the

Fig. 3 Positioning of stations around the fault

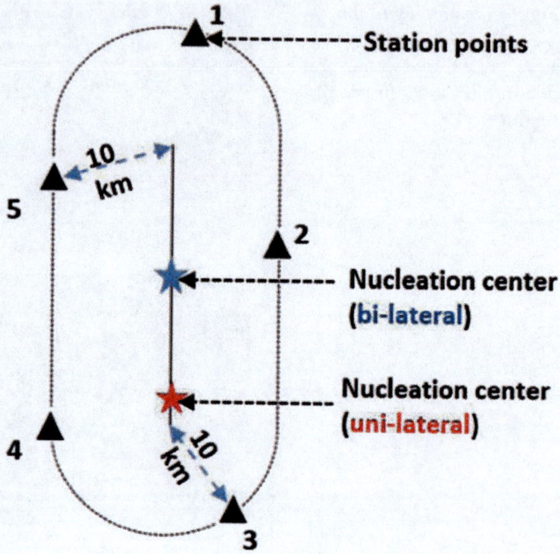

Table 1 Simulation parameters of the fault in both cases were tabulated

Fault geometry	
Length of fault	30 km
Width of fault	15 km
Spectral element size	400 m
Density (ρ)	2670 kg/m^3
Shear wave velocity (c_s)	3464 m/s
Friction parameters in the rupture zone	
Slip weakening distance (D_c)	0.4 m
Dynamic frictional stress (σ_k)	63 MPa
Static frictional stress (σ_s)	81.24 MPa
Nucleation patch size	3 km
Static frictional stress ($\sigma_{s,nuc}$)	81.24 MPa
Initial shear stress ($\sigma_{in,nuc}$)	81.6 MPa

directivity effects for both the cases, station 1 is considered, and one can rely on station 3 to understand backward directivity effect; stations 2, 4, and 5 lies perpendicular and at random orientation to the faults. Figure 4 depicts the orientation of the stations.

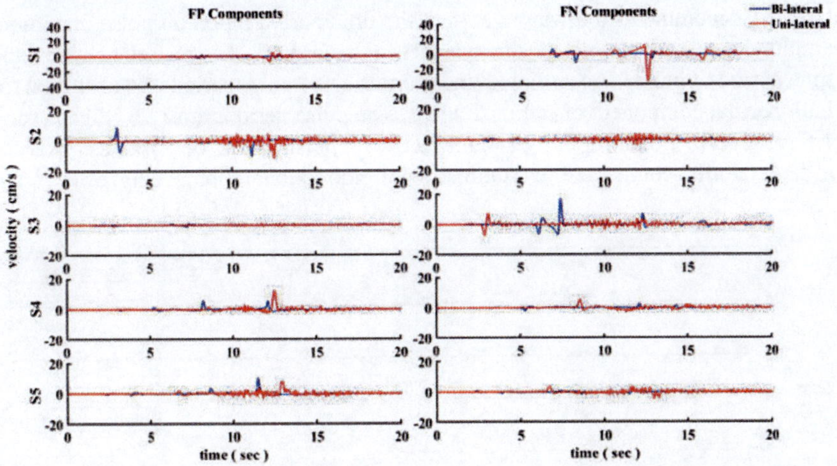

Fig. 4 A comparison of velocity–time histories bilateral ruptures (blue), unilateral ruptures (red)

6 Results and Discussion

The plots show a comparison in velocity–time histories for fault parallel and normal components for the scenario ruptures (unidirectional and bidirectional ruptures). The plots on the left and right are a comparison of fault parallel and normal components, respectively. Station 1 clearly depicts the case of forward directivity as fault normal dominates the fault parallel components. A shift of phase and increase in amplitude can be observed for the unilateral case. The amplification is about 100% with a change in the nucleation region. Fault parallel components dominate at station 2; this fact can be attributed to the location of station 2, where fault parallel component radiate from the nucleation region, and this being a mutually exclusive event, fault normal component can be barely visible. Station 3 is oriented similarly to station 1 but located behind the fault. Fault normal components clearly dominant in both the scenarios, though the station is located close to the fault, it acts as a backward directivity station for unilateral rupture, hence contains a lower peak when compared to the bilateral scenario. Station 4 and station 5 are randomly orientated and contain significant peaks in parallel components for both scenarios but the lack of pulses. Such stations can be considered for placement of piers in long-span bridges, where fault crossing is inevitable.

6.1 Extraction of Pulses Based on Baker's Wavelet Analysis

Near-field ground motions contains pulses, we need to focus on extracting the pulse period of these velocity pulses and relate them to the time period of the building

[14, 15], recommended a way to extract the time period based on pulse maximum amplitude and arrival time. Mother wavelets of the order 1–4 were used to extract the time period of pulses. According to these criteria, the pulse period was identified for fault normal component of station 1 and 3. The pulse periods are 1.76 (Fig. 5a) and 1.6 for unidirectional rupture case, respectively; pulse period of 1.14 was observed in fault parallel component for station 2 with bidirectional rupture (Fig. 5b).

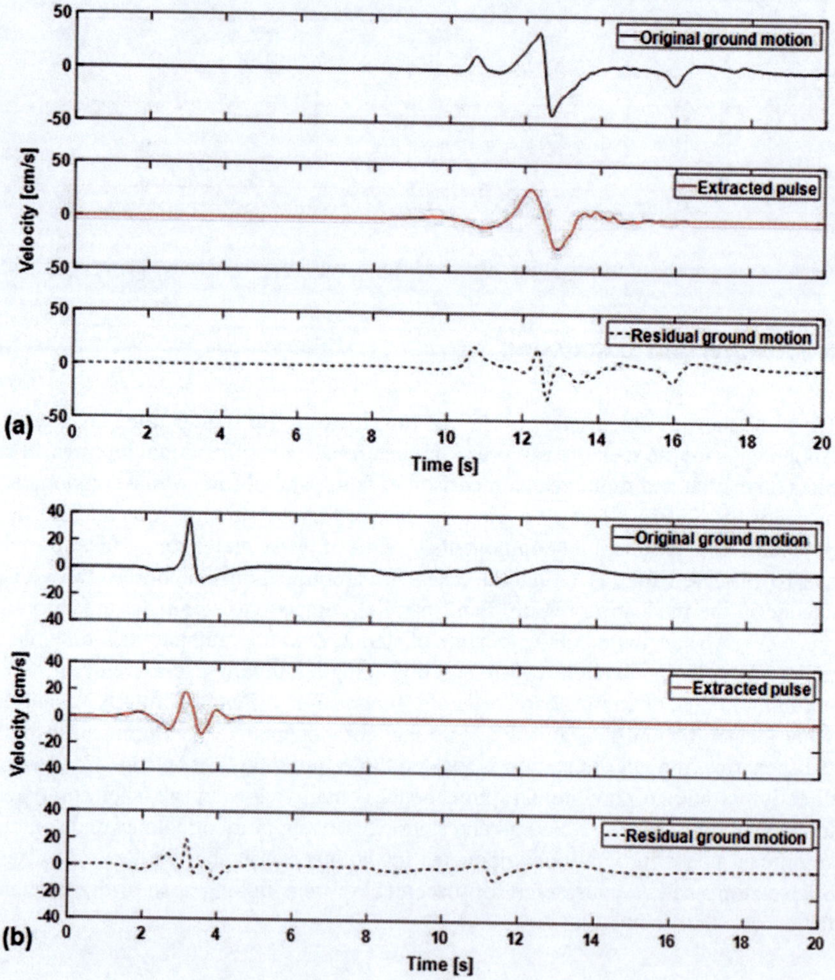

Fig. 5 Plot shows the extracted pulse based on baker's wavelet analysis. **a** Normal component for station 1 for unidirectional rupture, **b** Parallel component for station 1 for bidirectional rupture

7 Conclusions

- Though differential peaks of velocity amplitude were observed at all stations, only few velocity fields were classified as pulse-like ground motions. Station 1 is clearly differentiated to contain velocity pulses and has also been termed as forward directivity station; the directivity pulse is detected for a unilateral rupture but not for a bilateral rupture for fault normal component.
- A significant velocity pulse was observed for station 2 for fault parallel component, though the period is less when compared to directivity ground motions yet it poses pulse period.
- Station 3 can be labeled as back work directivity station for unilateral rupture; this station is also classified as pulse-like ground motion; the only contradiction observed is though the peak amplitude is less when compared to station 1, yet returned similar pulse period.
- At station 4 and 5, though the peak velocity amplitudes lie in par with station 2 yet were unable to return any pulse period, as they do not process a full pledged pulse profile; station 5 can be considered as pier points in order to establish bridges crossing faults as they are less vulnerable to velocity pulses as well as high ground motion intensities.

Acknowledgements We thank Indian institute of tropical metrology (IITM) for giving access to HPC Supercomputing resources to run SPECFEM3D. Funding from MoES/P.O (Seismo)/1(304)/2016 is greatly acknowledged.

References

1. Udías A, Vallina AU, Madariaga R, Buforn E (2014) Source mechanisms of earthquakes: theory and practice. Cambridge University Press
2. Kohrangi M, Vamvatsikos D, Bazzurro P (2019) Pulse-like versus non-pulse-like ground motion records: Spectral shape comparisons and record selection strategies. Earthquake Eng Struct Dynam 48(1):46–64
3. Galvez P, Ampuero JP, Dalguer LA, Somala SN, Nissen-Meyer T (2014) Dynamic earthquake rupture modelled with an unstructured 3-D spectral element method applied to the 2011 M 9 Tohoku earthquake. Geophys J Int 198(2):1222–1240
4. Somerville PG, Smith NF, Graves RW, Abrahamson NA (1997) Modification of empirical strong ground motion attenuation relations to include the amplitude and duration effects of rupture directivity. Seismol Res Lett 68(1):199–222
5. Poiata N, Miyake H, Koketsu K (2013) Generation mechanism of near-fault ground motion pulses for dip-slip faulting. 15 WCEE (2012) 8. In: Shi Z, Day SM (2013) Rupture dynamics and ground motion from 3-D rough-fault simulations. J Geophys Res: Solid Earth 118(3):1122–1141
6. Bray JD, Rodriguez-Marek A (2004) Characterization of forward-directivity ground motions in the near-fault region. Soil Dyn Earthquake Eng 24(11):815–828
7. Shi Z, Day SM (2013) Rupture dynamics and ground motion from 3-D rough-fault simulations. J Geophys Res: Solid Earth 118(3):1122–1141

8. Alterman Z, Karal FC Jr (1968) Propagation of elastic waves in layered media by finite difference methods. Bull Seismol Soc Am 58(1):367–398
9. Smith WD (1975) The application of finite element analysis to body wave propagation problems. Geophys J Int 42(2):747–768
10. Frankel A, Vidale J (1992) A three-dimensional simulation of seismic waves in the Santa Clara Valley, California, from a Loma Prieta aftershock. Bull Seismol Soc Am 82(5):2045–2074
11. Taborda R, Bielak J (2013) Ground-motion simulation and validation of the 2008 Chino Hills, California, earthquake. Bull Seismol Soc Am 103(1):131–156
12. Taborda R, Roten D (2014) Physics-based ground-motion simulation. Encycl Earthquake Eng 1–33
13. Andrews DJ (1976) Rupture velocity of plane strain shear cracks. J Geophys Res 81(32):5679–5687
14. Baker JW (2007) Quantitative classification of near-fault ground motions using wavelet analysis. Bull Seismol Soc Am 97(5):1486–1501
15. Karthik Reddy KSK, Somala SN (2021) Fracture mechanics based unilateral and bilateral earthquake simulations: application to cable-stayed bridge response. In: Saha SK, Mukherjee M (eds) Recent advances in computational mechanics and simulations. Lecture notes in civil engineering, vol 103. Springer, Singapore. https://doi.org/10.1007/978-981-15-8138-0_27

Establishing Seismic Site Class for five Recording Stations in Delhi Based on Theoretical Horizontal to Vertical Spectral Ratio

N. H. Harinarayan and Abhishek Kumar

Abstract PESMOS is an extensive database of strong-motion records in India having accelerograms from more than 300 strong-motion recording stations (RSs), located across the country. PESMOS classifies the RSs based on the physical description of local geology. Field studies on some of the RSs reported in previous studies indicate the ambiguity regarding the site class (SC) of RSs assigned by PESMOS. Accurate assessment of local soil conditions of RSs is essential for the proper utilization of accelerograms in seismic hazard and ground response studies. Numerous studies have shown that the predominant frequency (f_{peak}) obtained from Horizontal to the Vertical Spectral Ratio (HVSR) method can be effectively used to establish SC of a RS. In the current work, theoretical HVSR curves are developed for five RSs in the Delhi region using a MATLAB tool called Model HVSR to obtain the f_{peak}. Theoretical HVSR curves are modeled as the ratio between transfer functions relative to *S* waves and *P* waves. In addition, HVSR curves based on strong motions records are also estimated for each RSs. The value of f_{peak} obtained from both the above methods are found to be closely matching. Based on the values of f_{peak} estimated in the present work, SC of RSs is assigned. Present analyses show a clear difference in the SC assigned by PESMOS and the one obtained from the present study. The SC proposed in the present work are also matching with the results of existing field studies.

Keywords HVSR · Model HVSR · Site classification · PESMOS

1 Introduction

Modification of amplitude, duration, and frequency content of the incoming seismic waves from bedrock by the subsequent soil layers is termed as local site effect (LSE) and can cause amplification or de-amplification in bedrock motion at surface. Studies on damage patterns during 1985 Mexican EQ and 1989 Loma Prieta EQ have shown

N. H. Harinarayan (✉) · A. Kumar
Indian Institute of Technology, Guwahati 781039, India
e-mail: n.harinarayan@iitg.ac.in

T. G. Sitharam et al. (eds.), *Seismic Hazards and Risk*, Lecture Notes in Civil Engineering 116, https://doi.org/10.1007/978-981-15-9976-7_21

that LSE is the main reason for severe damages to buildings built on soft soils, even though they were located far away from the epicenter. Similar observations were reported during 2001 Bhuj EQ, which caused extensive damages to buildings on soft alluvial soil in Ahmedabad city, located 300 km from the epicenter [14]. LSE is quantified by site amplification function and is one of the crucial factors for seismic microzonation of a region. There are several empirical methods proposed by various researchers based on spectral analysis of EQ waveform to evaluate LSE (e.g., standard spectral ratio method [5], generalized inversion method [1], HVSR [21]). Comparative studies between the various empirical methods of evaluating site parameters reported by Field and Jacob [6] and Harinarayan and Kumar [8] highlighted that HVSR can accurately estimate f_{peak} of RSs. HVSR is a modification of the method of original proposed by Nakamura [16] (for estimating f_{peak}, based on ambient noise measurement using horizontal to vertical ratio of Fourier spectrum of ambient noise at the site of interest). Nakamura [16] proposed that the site effect is retained only in the horizontal component, while the source and path factors are retained in horizontal as well as vertical component of ambient noise. Therefore, the ratio of the horizontal component to the vertical component of ambient noise gives the site factor. Zhao et al. [21], based on the assumptions of Nakamura [16], determined f_{peak} of RSs considering the ratio of horizontal and vertical spectral acceleration of ground motion records considering 5% damping. The method developed by Zhao et al. [21] (denoted by HVSR) has been used widely across the globe for determining the site factor in terms of f_{peak} values.

India has endured four devastating EQs in the past 120 years, namely the 1897 Shillong EQ, 1905 Kangra EQ, 1934 Bihar-Nepal EQ, and 1950 Assam EQ. The Bureau of Indian Standards [2] classify most of the north and northeastern parts of India as Zone IV and V. For the purpose of tracking the continuing seismicity of the region, ministry of earth science under the Government of India had installed state of the art RSs in various seismically active regions of the country. Ground motions from these RSs are maintained by program for excellence in strong-motion studies (PESMOS). The instrumentation of PESMOS-based RS consists of an internal AC-63 GeoSIG triaxial force balanced accelerometers capable of recording EQ at a sampling rate of 200 samples per sec in trigger mode [13].

Along with ground motion records, PESMOS gives SC details of each of the RSs. The SC gives by PESMOS which is not based on in-situ field tests, but is based on the physical description of local geology obtained from SEISAT [19] and geological maps of India [15]. Based on the physical description of local geology, the value of average shear wave velocity for 30 m (V_{s30}) is assumed to each PESMOS recording station, and SC is assigned using the Borcherdt [5] classification scheme [15]. Field studies on some of the RSs in Delhi and Uttarakhand reported by Pandey et al. [17, 18] have highlighted the inaccuracy of SC given by PESMOS. With ground motion records of EQ events since 2004 and from more than 300 RSs across the country, PESMOS is an important ground motion database for regional seismic hazard studies and development of regional ground motion prediction relationships. Hence, accurate SC assessment of PESMOS-based RSs is necessary to utilize the database for EQ engineering studies effectively.

The present study aims at analyzing EQ records based on HVSR studies to provide site characterization of five RSs located in the Delhi region. Based on the analyses, values of f_{peak} obtained from average HVSR curves from all the five RSs are estimated. In addition, the estimated average HVSR curves from the present study are compared with the theoretical HVSR curves developed for the selected five RSs in this work. The theoretical HVSR curves are generated using Model HVSR tool developed by Herak [9].

2 Study Area

Delhi, the capital city spreading across an area of 1484 km^2 with a population of approximately 20 million (according to the 2012 census) is one of the most densely populated regions in India. The region has been affected by EQs of local origin as well as from the Himalayan origin. Important seismotectonic features of the region include Delhi–Hardwar ridge, Delhi–Lahore ridge, the Aravalli–Delhi fold, the Sohna fault, the Mathura fault, and the Moradabad fault [10]. The seismicity of the region is mainly attributed to the Delhi–Hardwar ridge, trending northeast-southwest [20]. The region is located 200 km from the Himalayan plate boundary systems including the main central thrust (MCT) and the main boundary thrust (MBT) and has experienced ground shaking during the EQs of Himalayan origin including 1991 Chamoli EQ and 1999 Uttarkashi EQ [11]. The terrain of Delhi is flat and is mostly covered with quaternary alluvium and pre-Cambrian metasediments (GSI [7]).

3 Dataset

The five RSs selected for analysis in this present work are located over an area between 77°E and 82°E longitude and 28°N and 32°N latitude as shown in Fig. 1. The location details of each of the five recording stations are given in Table 1. The database used for the HVSR study comprises 16 ground motion records from four EQ events of magnitude ranging from 3.5 to 4.6. The details of EQ events considered in the present work are summarized in Table 2.

4 HVSR Calculation

HVSR is computed for each of the five RS as the ratio of 5% damped response spectra of horizontal and vertical components smoothened using a Konno and Ohmachi [12] window of parameter $b = 20$. The horizontal component is computed by the geometric mean of east–west and north–south components.

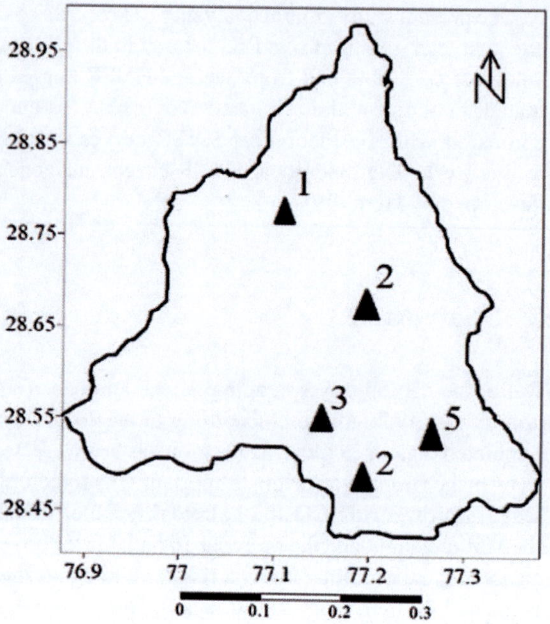

Fig. 1 Study area showing five recording stations considered for the present work (1-ZAKI; 2-DJB; 3-JNU; 4-IGN; 5-IIT)

5 Model HVSR

Model HVSR is a MATLAB code used to analyze and interpret ambient noise measurements. The tool is developed by Herak [9] based on the assumption that the ambient vibrations constitute body waves such that the vertical ground motion components are controlled by *P*, and horizontal ground motion components are controlled by *S* wave amplitudes. Thus, if *P* and *S* waves have the same amplitude at the bedrock, then HVSR at the surface reflects the amplifications produced based on characteristics of local soil above the bedrock. This HVSR curve is calculated as the ratio between the transfer functions to *S* waves (F_S) and *P* waves (F_P) as given below;

$$\text{HVSR} = \frac{F_S}{F_P} \tag{1}$$

The input parameters for the Model HVSR include shear wave velocity (V_S), layer thickness (h), density (ρ), and quality factor for *S* and *P* waves (Q_S and Q_P). The V_S profile for the recording stations is obtained from Pandey et al. [17, 18] based on the geophysical investigation. The value of Q_P and Q_S for the region is obtained from Banerjee and Kumar [3]. Using the above-mentioned input parameters, theoretical HVSR curves for the recording stations are developed.

Table 1 Detail of strong-motion recording stations

Station	Lat. (°) (N)	Long. (°) (E)	HVSR		Model HVSR		PESMOS classification scheme		
			f_{peak}	A_{peak}	f_{peak}	A_{peak}	SC given in PESMOS	SC based on f_{peak}	SC as per NEHRP
ZAKI	28.6	77.2	3.9	3.1	3.9	1.5	C	B	C
DJB	28.7	77.2	14	4.5	18	2.1	A	A	A
JNU	28.5	77.2	14.5	2.7	16	1.0	A	A	A
IGN	28.5	77.2	4.5	3.9	4.3	1.5	A	B	C
IIT	28.6	77.3	4.5	6	4.2	4.1	C	B	C

Table 2 Details of earthquakes considered for analysis

Event no.	dd/mm/yyyy	Lat.	Long	Depth	Magnitude
1	14-12-2005	30.9	79.3	25.7	5.2
2	07-05-2006	28.7	76.6	20.2	4.1
3	29-11-2006	27.6	76.7	13.0	3.9
4	10-12-2006	31.5	76.7	33.0	3.5

6 Results and Interpretation

6.1 Comparison of Observed HVSR with Theoretical Spectra

HVSR curves computed from EQ records are shown in Fig. 2 along with the theoretical HVSR computed by Model HVSR MATLAB routine. The frequency corresponding to the maximum value of HVSR (denoted by A_{peak}) is identified as f_{peak}. It can be observed from Fig. 2 that the f_{peak} values obtained from theoretical HVSR and from HVSR based on EQ data are found to be matching for all the RSs. The values of f_{peak} range from 3.9 to 16 Hz. Values of f_{peak} and A_{peak} obtained from both the methods are tabulated in Table 1. Even though the theoretical HVSR and HVSR from same EQ data are giving identical f_{peak} values, a significant disparity in the value of A_{peak} obtained using the two methods is seen for all RSs. A_{peak} values are found to be lower for Model HVSR curves compared to HVSR curves for same EQ data. It must be mentioned here that HVSR from EQ data is corresponding to a particular magnitude, whereas theoretical HVSR is corresponding to very low strain versus This might be the reason for the difference in A_{peak} values based on two approaches. Further, it has to be mentioned here that the purpose of this study is to estimate SC of RSs based on f_{peak} value. Hence, differences in A_{peak} values will not change the outcomes from this study.

6.2 Grouping of HVSR Based on f_{peak} Value

Based on the value of f_{peak} obtained in the present study, RS is classified first based on the classification scheme used by PESMOS and then based on NEHRP classification scheme as discussed below.

The site classification scheme used by PESMOS was modified after Borcherdt [4, 5] There are three SCs, namely SC A, SC B, and SC C corresponding to hard rock, soft rock, and soil sites (Table 3). The frequency range corresponding to V_{s30} for each SC is calculated using Eq. 2.

$$f_{\text{peak}} = V_z/4H \tag{2}$$

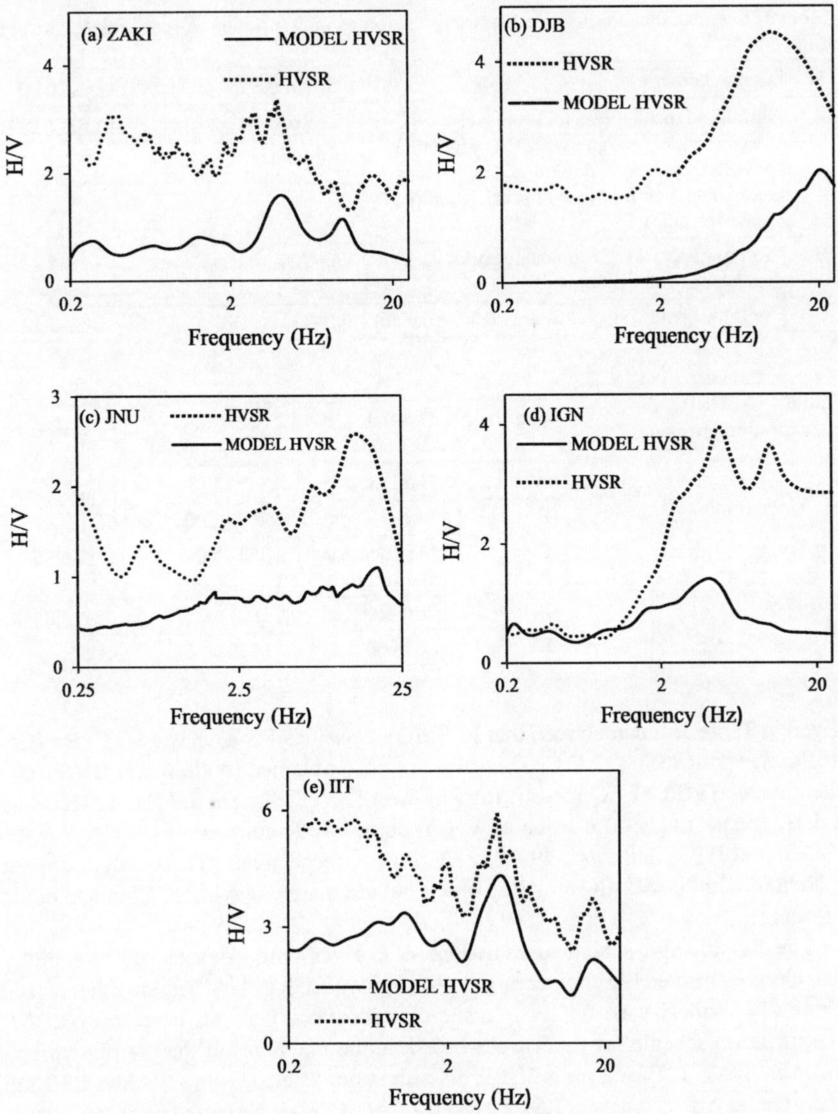

Fig. 2 HVSR and Model HVSR curves for ZAKI, DJB, JNU, IGN, and IIT recording stations

In Eq. 2, V_z represents the shear wave velocity for layer thickness H (considered as 30 m). SCs of five RSs considered in the present study are assigned by comparing the value of f_{peak} obtained using HVSR studies with the range of f_{peak} calculated for each SC. Summary of PESMOS-based SC for the five RSs is presented in Table 1. Based on the present study, RSs ZAKI, IGN, and IIT are classified as SC B, whereas JNU and DJB are classified as SC A. The original SC given by PESMOS is also

Table 3 Site classification scheme for strong-motion recording stations as per PESMOS (as per Borcherdt [4, 5])

SC	General description	Shear wave velocity (m/s)	f_{peak} (Hz)
A	Firm/hard rocks (Fresh and compact metamorphic, e.g., gneiss, schist, migmatites, phyllites, quartzites, dolomites, and igneous rocks, e.g., granites, granodiorites, granitoid, basic volcanic)	700–1620	5.83–13.33
B	Soft to firm rocks (Sedimentary rocks, e.g., sandstone, shale, limestone)	375–700	3.12–5.83
C	Soils (Alluvium, slope wash material, Aeolian)	200–375	1.66–3.12

Table 4 NEHRP site classification provisions

Site class	General description	Shear wave velocity (m/s)	f_{peak} (Hz)
A	Hard rock	>1500	>12.70
B	Rock	760–1500	6.35–12.70
C	Very dense soil and soft rock	360–760	3.05–6.35
D	Stiff soil	180–360	1.52–3.05
E	Soft soil	<180	<1.52

given in Table 1. It can be seen that PESMOS classifies RS ZAKI as SC C, RS IGN as SC A, and RS IIT as SC C. The values of f_{peak} obtained (both Model HVSR and theoretical HVSR of EQ record) for RSs ZAKI, IGN, IIT are 3.9 Hz, 4.5 Hz, and 4.5 Hz, respectively. The value of V_{s30} reported by Pandey et al. [17, 18] for RS ZAKI, and IIT is 388 m/s^2, 493 m/s^2, 392 m/s^2, respectively. These values are not matching with the SC given by PESMOS, but are matching with SC obtained in the present study.

The classification scheme used by PESMOS gives a broader sense of the geology that has very limited EQ engineering applications. NEHRP classification scheme (see Table 4) is widely used for seismic site classification. The SCs based on NEHRP classification scheme for the five RS based on the outcomes of the present work is given in Table 1. Based on NEHRP classification scheme, RSs JNU and DJB are classified as SC A, whereas RSs ZAKI, IIT, and IGN are classified as SC C.

7 Conclusion

The present study highlights the inaccuracy of SC given by PESMOS which is based on surface geology alone. Without accurate information regarding the SC of RSs, ground motion records from these RSs cannot be used for any EQ engineering applications. In this study, SC for five RSs located in the Delhi region is estimated based on f_{peak} values obtained from HVSR analyses. HVSR curves based on the theoretical model and from EQ data are developed. A comparison between the theoretical HVSR curves and that obtained from EQ data for each recording station was found matching in terms of f_{peak} value. Based on the f_{peak} values, SC for each of the recording stations is determined and compared with the SC given by PESMOS. Mismatch in the SC given by PESMOS and the present study is observed for RSs ZAKI, IGN, and IIT. The SC proposed for these RSs in the present study is matching with the existing studies for the region. In the absence of in-situ test data, HVSR provides an alternative to estimate the SC of recording stations.

References

1. Andrews DJ (1986) Objective determination of source parameters and similarity of earthquakes of different size. Earthq Source Mech 259–267
2. BIS (2002) IS 1893 (Part 1): general provisions and buildings: criteria for earthquake resistant design of structures. Bureau of Indian Standards, New Delhi, India
3. Banerjee S, Kumar A (2017) Determination of seismic wave attenuation for the Garhwal Himalayas, India. Geosci. Res. 2(2):105–126
4. Borcherdt RD (1970) Effects of local geology on ground motion near San Francisco Bay. Bull Seismol Soc Am 60(1):29–61
5. Borcherdt RD (1994) Estimates of site-dependent response spectra for design (methodology and justification). Earthq Spectra 10:617–653
6. Field EH, Jacob KH (1995) A comparison and test of various site-response estimation techniques, including three that are not reference-site dependent. Bull Seismol Soc Am 85(4):1127–1143
7. GEOLOGICAL SURVEY OF INDIA (2000) Seismotectonic atlas of India and its environs, published by Geological Survey of India
8. Harinarayan NH, Kumar A (2017) Seismic site classification of recording stations in Tarai Region of Uttarakhand, from multiple approaches. Geotech Geol Eng 36(3):1–16
9. Herak M (2008) ModelHVSR—A Matlab® tool to model horizontal-to-vertical spectral ratio of ambient noise. Comput Geosci 34(11):1514–1526
10. Iyengar RN, Ghosh S (2004) Microzonation of earthquake hazard in greater Delhi area. Curr Sci 87(9):1193–1202
11. Iyengar RN (2000) Seismic status of Delhi megacity. Curr Sci 78(5):568–574
12. Konno K, Ohmachi T (1998) Ground motion characteristics estimated from spectral ratio between horizontal and vertical components of microtremor. Bull Seismol Soc Am 88(1):228–241
13. Kumar A, Mittal H, Sachdeva R, Kumar A (2012) Indian strong motion instrumentation network. Seismol Res Lett 83(1):59–66
14. Mahajan AK, Kumar SK (2004) Macroseismic field observations of January 26th, 2001 Kachchh earthquake and its seismotectonics. J Asian Earth Sci 23:17–23
15. Mittal H, Kumar A, Ramhmachhuani R (2012) Indian national strong motion instrumentation network and site characterization of its stations. Int J Geosci 3(06):1151
16. Nakamura Y (1989) A method for dynamic characteristics estimation of subsurface using microtremor on the ground surface, vol 30, issue 1. Railway Technical Research Institute, Quarterly Reports
17. Pandey B, Jakka RS, Kumar A (2016) Influence of local site conditions on strong ground motion characteristics at Tarai region of Uttarakhand, India. Nat Hazards 81(2):1073–1089

18. Pandey B, Jakka RS, Kumar A, Mittal H (2016b) Site characterization of strong-motion recording stations of Delhi using joint inversion of phase velocity dispersion and H/V Curve. Bull Seismol Soc Am 106(3):1254–1266
19. SEISAT (2000) Seismotectonic Atlas of India and its environs. Published by Geological Survey of India
20. Sharma M, Wason H, Dimri R (2003) Seismic zonation of the Delhi region for bedrock ground motion. Pure Appl Geophys 160:2381–2398
21. Zhao JX, Irikura K, Zhang J, Fukushima Y, Somerville PG, Asano A, Ohno Y, Oouchi T, Takahashi T, Ogawa H (2006) An empirical site-classification method for strong-motion stations in Japan using H/V response spectral ratio. Bull Seismol Soc Am 96:914–925

Love Wave Dispersion in an Inhomogeneous Earth's Crust Lying Over a Pre-stressed Inhomogeneous Mantle

Sumit Kumar Vishwakarma, Rupinderjit Kaur, and Tapas Ranjan Panigrahi

Abstract Study has been made on the propagation of Love wave in a non-homogeneous anisotropic substratum over an initially stressed non-homogeneous isotropic half space. The rigidities and density vary exponentially for the upper layer, whereas the rigidity, density and initial stress vary as a product of two exponential function for the lower half space. The displacements for both the media have been calculated analytically, and the closed form dispersion equation has been derived using suitable boundary conditions. Numerical simulations have been performed, and graphs have been plotted to classify the impact of inhomogeneity parameters on phase velocity. The study may help in different areas of science and engineering for better understanding of the wave phenomenon, especially in refining our knowledge of crustal and sub-crustal region near the earth surface.

Keywords Initial stress · Inhomogeneity · Love wave · Phase velocity · Dispersion equation · Half space

1 Introduction

Generally, it has been found that in theory and practical applications, earth has been assumed to be isotropic, or at most, to be composed of isotropic layers. However, Shearer [1] in his investigations has indicated the presence of anisotropy. This investigation has confirmed the directional dependence of such anisotropy on elastic waves, although there are other anomalous observations which suggest indirectly a departure from isotropy. Many seismologists showed that if the anisotropy is neglected, then it may end up with large errors in depth determinations that occur during seismic exploration. It was Anderson [2] who has explicitly demonstrated an argument that explains about the existence of anisotropy in angular material. As the dynamic behaviour of

S. K. Vishwakarma (✉) · R. Kaur · T. R. Panigrahi
Department of Mathematics, BITS-Pilani, Hyderabad Campus, Hyderabad 500078, India
e-mail: sumitkumar@hyderabad.bits-pilani.ac.in

T. G. Sitharam et al. (eds.), *Seismic Hazards and Risk*, Lecture Notes in Civil Engineering 116, https://doi.org/10.1007/978-981-15-9976-7_22

a matter generally associated with its physical and chemical properties, the investigations of the nature of seismic wave propagation within earth's interior are an important approach to this end.

Love [3] was the first to discover the horizontally polarized surface waves, named after him as Love waves. It is a result of the interference of many shear waves (S–waves) guided by an elastic layer, which is welded to an elastic half space on one side while bordering a vacuum on the other side. In elastodynamics, Love waves are also known as Q waves that cause horizontal shifting of the earth during an earthquake. In particular, it is the resultant surface waves that trap up in the earth's crustal layer during the reflection and refraction of SH-waves. Chattopadhyay and Bandyopadhyay [4] have demonstrated the behavioural impact of SH-wave in an infinite monoclinic crystal plate. With due course of time, Chattopadhyay and Choudhury [5] have explained reflection and transmission of SH-waves in self-reinforced medium that clearly shows the trapping up of Love wave and its propagation behaviour. References can be given to Mistri et al. [6] and Vishwakarma et al. [7] for their works to refine the knowledge of SH-wave propagation in different geo-media.

Love wave frequency equation under a specific physical conditions which are likely to exist in the interior of the earth has been studied in the present problem. Again, it was the immense observation of Love [8], who has first observed that there exists a high initial stress inside the earth. Due to gravity vibration, creep, atmospheric pressure, creep, difference in temperature, large initial stresses have been found inside the earth. These stresses that exist in an elastic body even though external forces are absent and are termed as initial stresses, and the body is said to be initially stressed. These stresses eventually exert significant influence on the elastic waves produced during earthquakes and explorations. Therefore, it is important to consider the presence of initial stress while studying the propagation behaviour of surface seismic waves. Biot [9] in his elaborative studies pointed out the influences of initial stress and its impact on waves to a great extent. Love waves in the fibre-reinforced layer over a gravitating porous half space has been investigated by Chattaraj and Sapan [10], whereas Nam et al. [11] studied the surface wave propagation in a layered half space under the impact of initial stress. Also, a similar work has been performed by Shams [12] on Love wave, where he considered the wave propagation at the boundary between a layer and a half space under the effect of initial stress.

The crust region of our planet is composed of various inhomogeneous layers with distinct geological parameters. It is very clearly explained out by Bullen [13] that the earth's density varies at different rates with different layers within the earth. He has also approximated the density law inside the earth as a quadratic polynomial in depth parameter for 413–984 km depth. For depth from 984 km to the central core, Bullen has approximated the density as a linear function of depth parameter. He has also investigated in his study that these variations in rigidity and density as well as initial stress can be a function of depth, such as exponential, hyperbolic, polynomial with degree $n \in R$. Earth being highly anisotropic and inhomogeneous, there are several books on the geology of earth and the theory of wave mechanics [14–18] which claims that different part of earth has different compositions in terms of material properties with variations in rigidity, density and initial stress as function of depth. This has motivated many mathematicians and geophysicists worldwide to consider various permutations and combinations of layers of various property one beneath the other. Such is the work of Sari and Salk [19] who have assumed the density of the layers with depth as a hyperbolic function. It is Wang et al. [20] who has explained the propagation behaviour of Love waves in an inhomogeneous orthotropic medium obeying the exponential and generalized power law models. He, on the basis of the theory of elasticity, has generated the analytical solutions of the Love wave dispersion equation for the inhomogeneous orthotropic medium, for which the Young's modulus $\left(E_x, E_y, E_z\right)$, shear modulus $\left(G_{xy}, G_{yz}, G_{xz}\right)$, and medium density (ρ) are obeying the exponential law model $\left(E_x e^{\alpha z}, E_y e^{\alpha z}, E_z e^{\alpha z}, G_{xy} e^{\alpha z}, G_{yz} e^{\alpha z}, G_{xz} e^{\alpha z}, \text{and } \rho_x e^{\alpha z}\right)$ and the generalized power model $\left[E_x(a+bz)^m, E_y(a+bz)^m, E_z(a+bz)^m, G_{xy}(a+bz)^m, G_{yz}(a+bz)^m, G_{xz}(a+bz)^m, \text{and } \rho(a+bz)^m\right]$. Similarly, there are many authors who have taken variations which are functions of depth and are being same for shear modulus and density. Many phenomenal results have been drawn out of such research work which has received appreciations and applauds all over the world. Kielczynski et al. [21] have given the idea of the effect on Love wave propagation when the medium is filled with viscous liquid. In the year 2013, Gupta et al. [22] and Gupta et al. [23] have given two important remark on the very possibility of Love wave existence under the presence of varying rigidity, density and initial stress. References can be given to the work of Singh [24–28] for their commendable investigations in this regard.

In the present paper, the crustal layer has been taken overlying a semi-infinite space. The layer exhibits an inhomogeneity, where directional rigidities and density vary as the function of depth in the exponential of a constant, whereas in the half space, the variation has been taken as product of exponential function of depth as shown in Fig. 1. The equation of motion has been taken from Biot's theory, and the displacement has been derived explicitly. The boundary condition has been imposed, and the closed form dispersion equation has been obtained. The numerical computations have been performed to show the influence of inhomogeneity parameter involved. It has been concluded that these parameters leave remarkable impacts on the phase velocity of Love wave propagation.

2 Statement of the Problem

The geometry of the problem consists of an inhomogeneous anisotropic crustal layer of finite thickness H overlying on a semi-infinite inhomogeneous isotropic medium. The wave propagates along x-direction with z-axis directing vertically downwards. The interface of the two medium is the plane $z = 0$. The upper boundary plane is assumed to be free (non-rigid). Following inhomogeneities have been assumed in the layer and the half space:

$$\text{Layer: } N = N_1\alpha^{(1+mz)},\ L = L_1\alpha^{(1+mz)} \text{ (Directional rigidities)},$$

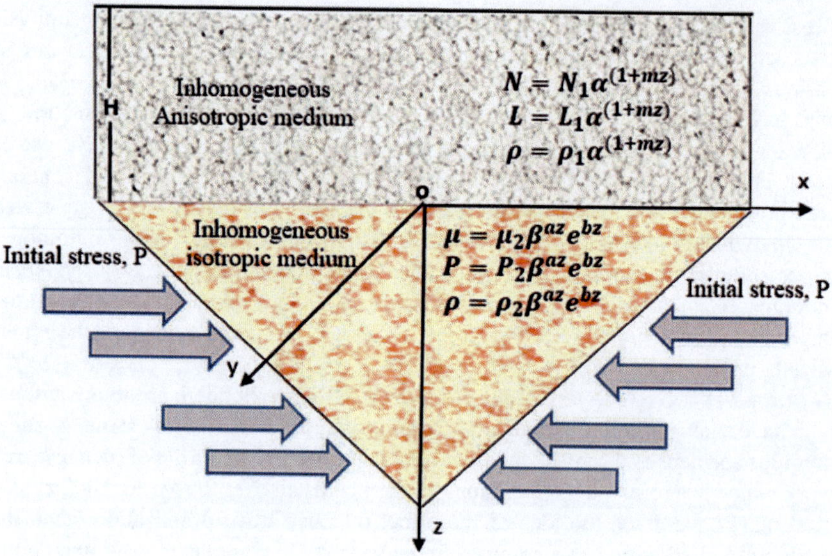

Fig. 1 Geometry of the problem

$$\rho = \rho_1\alpha^{(1+mz)}\text{(Density)}$$

where $N_1 = 3.99 \times 10^{10}$ N/m^2, $L_1 = 5.79 \times 10^{10}$ N/m^2, $\rho_1 = 2649$ kg/m^3.α, $m \in \mathbb{R}$ are constants with dimension equal to that of inverse of length.

$$\text{Half} - \text{Space: } \mu = \mu_2\beta^{az}e^{bz}, \rho = \rho_2\beta^{az}e^{bz}, P = P_2\beta^{az}e^{bz}$$

where $\mu_2 = 7.45 \times 10^{10}$ N/m^2, $\rho_2 = 3293$ $\frac{\text{kg}}{\text{m}^3}$. The values a, b and $\beta \in \mathbb{R}$, whose dimensions are equal to that of length inverse.

3 Solution for Non-homogeneous Anisotropic Layer of the Upper Medium

Let us consider the displacement components u, v and w in the direction of x, y and z, respectively. Using the general equation of motion as given by Biot [9] and considering the Love wave conditions, $u = 0$, $w = 0$ and $v = v_1(x, z, t)$

$$N\frac{\partial^2 v_1}{\partial x^2} + \frac{\partial}{\partial z}\left(L\frac{\partial v_1}{\partial z}\right) = \rho\frac{\partial^2 v_1}{\partial t^2} \tag{1}$$

For a wave propagation in x direction, we assume that

$$v_1 = V(z)\exp^{ik(x-ct)} \tag{2}$$

Using Eq. (2) in Eq. (1), we have

$$\frac{\mathrm{d}^2V}{\mathrm{d}z^2} + \frac{1}{L}\frac{\mathrm{d}L}{\mathrm{d}z}\frac{\mathrm{d}V}{\mathrm{d}z} + \frac{k^2}{L}\left(c^2\rho - N\right)V(z) = 0 \tag{3}$$

Now, we have

$$N = N_1\alpha^{(1+mz)}, L = L_1\alpha^{(1+mz)} \text{ and } \rho = \rho_1\alpha^{(1+mz)} \tag{4}$$

Substituting Eq. (4) in Eq. (3), we get

$$\frac{\mathrm{d}^2V}{\mathrm{d}z^2} + m\mathrm{Log}(\alpha)\frac{\mathrm{d}V}{\mathrm{d}z} + A_1V(z) = 0 \tag{5}$$

where $A_1 = k^2\left(\frac{c^2}{c_0^2} - \frac{N_1}{L_1}\right)$ and $c_0^2 = \frac{L_1}{\rho_1}$.

On solving Eq. (5) analytically, we have

$$v_1(z) = \Bigg(C_1\exp^{\frac{1}{2}z\left(-m\mathrm{Log}(\alpha)-\sqrt{-4A_1+m^2\mathrm{Log}(\alpha)^2}\right)} + C_2\exp^{\frac{1}{2}z\left(-m\mathrm{Log}(\alpha)+\sqrt{-4A_1+m^2\mathrm{Log}(\alpha)^2}\right)}\Bigg)\exp^{ik(x-ct)} \tag{6}$$

Equation (6) gives the displacement for non-homogeneous anisotropic layer.

4 Solution for the Lower Half Space

The lower medium has been considered as a non-homogeneous anisotropic half space under initial stress P in x-direction. The equation of motion as given by Biot [9] is

$$\frac{\partial s_{21}}{\partial x} + \frac{\partial s_{23}}{\partial z} - \frac{P}{2}\left(\frac{\partial^2 v_2}{\partial x^2}\right) = \frac{\partial^2(\rho v_2)}{\partial t^2} \tag{7}$$

where s_{ij} is the component of incremental stress in the half space; P is the initial compressive stress along x direction.

The stress–strain relations are

$$s_{21} = \mu\frac{\partial v_2}{\partial x}, s_{23} = \mu\frac{\partial v_2}{\partial z} \tag{8}$$

The variations for the rigidities and density for the lower half space are given by

$$\mu = \mu_2 \beta^{az} e^{bz},\ \rho = \rho_2 \beta^{az} e^{bz} \text{ and } P = P_2 \beta^{az} e^{bz} \tag{9}$$

Using the stress–strain relations and variations for rigidities and density in Eq. (7), we get

$$\left(1 - \frac{P_2}{2\mu_2}\right)\frac{\partial^2 v_2}{\partial x^2} + \frac{\partial^2 v_2}{\partial z^2} + (b + a\mathrm{Log}(\beta))\frac{\partial v_2}{\partial z} = \frac{\rho_2}{\mu_2}\frac{\partial^2 v_2}{\partial t^2} \tag{10}$$

Let us consider

$$v_2 = V(z)\exp^{ik(x-ct)} \tag{11}$$

Using Eq. (11) in Eq. (10), we have the ordinary differential equation

$$V''(z) + (b + a\mathrm{Log}(\beta))V'(z) + k^2\left(\frac{c^2\rho_2}{\mu_2} - 1 + \frac{P_2}{2\mu_2}\right)V(z) = 0 \tag{12}$$

where

$$A_2 = k^2\left(\frac{c^2}{c_1^2} - 1 + \zeta\right),\ c_1^2 = \frac{\mu_2}{\rho_2} \text{ and } \zeta = \frac{P_2}{2\mu_2} \tag{13}$$

On solving Eq. (12) analytically, we get

$$v_2(z) = \left(C_3\exp^{\frac{1}{2}z\left(-b-a\mathrm{Log}(\beta)-\sqrt{-4A_2+(b+a\mathrm{Log}(\beta))^2}\right)} + C_4\exp^{\frac{1}{2}z\left(-b-a\mathrm{Log}(\beta)+\sqrt{-4A_2+(b+a\mathrm{Log}(\beta))^2}\right)}\right)\exp^{ik(x-ct)} \tag{14}$$

Imposing the condition $v_2 \to 0$ as $z \to \infty$ in Eq. (14), it reduces to

$$v_2(z) = \left(C_3\exp^{\frac{1}{2}z\left(-b-a\mathrm{Log}(\beta)-\sqrt{-4A_2+(b+a\mathrm{Log}(\beta))^2}\right)}\right)\exp^{ik(x-ct)} \tag{15}$$

Equation (15) is the displacement in the inhomogeneous half space.

5 Boundary Conditions

(i) The stress vanishes at the free surface $z = -H$, so

$$(\tau_{yz})_1 = 0 \Rightarrow \frac{\partial v_1}{\partial z} = 0 \tag{16}$$

which results

$$\frac{1}{2}\left(-m\mathrm{Log}(\alpha) - \sqrt{-4A_1 + m^2\mathrm{Log}(\alpha)^2}\right)\exp^{-\frac{1}{2}H\left(-m\mathrm{Log}(\alpha)-\sqrt{-4A_1+m^2\mathrm{Log}(\alpha)^2}\right)}C_1$$
$$+\frac{1}{2}\left(-m\mathrm{Log}(\alpha) + \sqrt{-4A_1 + m^2\mathrm{Log}(\alpha)^2}\right)\exp^{-\frac{1}{2}H\left(-m\mathrm{Log}(\alpha)+\sqrt{-4A_1+m^2\mathrm{Log}(\alpha)^2}\right)}C_2 = 0 \quad (17)$$

(ii) The continuity of stress requires at the interface $z = 0$ requires

$$\left(\tau_{yz}\right)_1 = \left(\tau_{yz}\right)_2 \quad (18)$$

gives

$$\frac{1}{2}L_1\left(-m\mathrm{Log}(\alpha) - \sqrt{-4A_1 + m^2\mathrm{Log}(\alpha)^2}\right)C_1$$
$$+\frac{1}{2}L_1\left(-m\mathrm{Log}(\alpha) + \sqrt{-4A_1 + m^2\mathrm{Log}(\alpha)^2}\right)C_2$$
$$-\frac{1}{2}\mu_2\left(-b - a\mathrm{Log}(\beta) - \sqrt{-4A_2 + (b + a\mathrm{Log}(\beta))^2}\right)C_3 = 0 \quad (19)$$

(iii) At the interface $z = 0$,

$$v_1 = v_2 \quad (20)$$

We get,

$$C_1 + C_2 - C_3 = 0 \quad (21)$$

For the non-zero solution of Eqs. (17), (19) and (21), we must have

$$\begin{vmatrix} d_{11} & d_{12} & d_{13} \\ d_{21} & d_{22} & d_{23} \\ d_{31} & d_{32} & d_{33} \end{vmatrix} = 0, \quad (22)$$

where

$$d_{11} = \frac{1}{2}\left(-m\mathrm{Log}(\alpha) - \sqrt{-4A_1 + m^2\mathrm{Log}(\alpha)^2}\right)\exp^{-\frac{1}{2}H\left(-m\mathrm{Log}(\alpha)-\sqrt{-4A_1+m^2\mathrm{Log}(\alpha)^2}\right)}$$
$$d_{12} = \frac{1}{2}\left(-m\mathrm{Log}(\alpha) + \sqrt{-4A_1 + m^2\mathrm{Log}(\alpha)^2}\right)\exp^{-\frac{1}{2}H\left(-m\mathrm{Log}(\alpha)+\sqrt{-4A_1+m^2\mathrm{Log}(\alpha)^2}\right)}$$
$$d_{13} = 0$$

$$
\begin{aligned}
d_{21} &= \tfrac{1}{2}L_1\left(-m\text{Log}(\alpha) - \sqrt{-4A_1 + m^2\text{Log}(\alpha)^2}\right) \\
d_{22} &= \tfrac{1}{2}L_1\left(-m\text{Log}(\alpha) + \sqrt{-4A_1 + m^2\text{Log}(\alpha)^2}\right) \\
d_{23} &= -\tfrac{1}{2}\mu_2\left(-b - a\text{Log}(\beta) - \sqrt{-4A_2 + (b + a\text{Log}(\beta))^2}\right) \\
d_{31} &= 1 \\
d_{32} &= 1 \\
d_{33} &= -1
\end{aligned}
$$

6 Particular Cases

When there is no inhomogeneity involved in the layer and the half space, the dispersion Eq. (22) reduces to the classical equation as suggested by Love [3]. This can be shown theoretically as:

In the layer, at $z = -\frac{1}{m}$ $(m \neq 0)$, $N = N_1$, $L = L_1$, $\rho = \rho_1$. And, the half space, at $z = 0$, $\mu = \mu_2$, $\rho = \rho_2$. Equation (22) reduces to

$$
\tan\left(kh\sqrt{\frac{c^2}{c_0^2} - 1}\right) = \frac{\mu_1\sqrt{1 - \frac{c^2}{c_1^2}}}{L_1\sqrt{\frac{c^2}{c_0^2} - 1}}
$$

which is the classical equation of Love wave dispersion as described by Love [3].

7 Numerical Calculation and Graphs

In order to emerge with the effect of inhomogeneities involved in the layer and the half space, numerical computation of Eq. (22) has been performed with different values of parameters as shown in Table 1.

Table 1 Numerical computation

Figures	a/k	b/k	α/k	β/k	ζ	m/k	γ/k
1	1,2,3	0.1	2	0.5	1	5	1
2	1	0.1,0.2,0.3	2	0.5	1	5	1
3	1	0.1	1,2,3	0.5	1	5	1
4	1	0.1	2	0.5,1.0,1.5	1	5	1
5	1	0.1	2	0.5	1,2,3	5	1
6	1	0.1	2	0.5	1	5,10,15	1
7	1	0.1	2	0.5	1	5	1,2,3

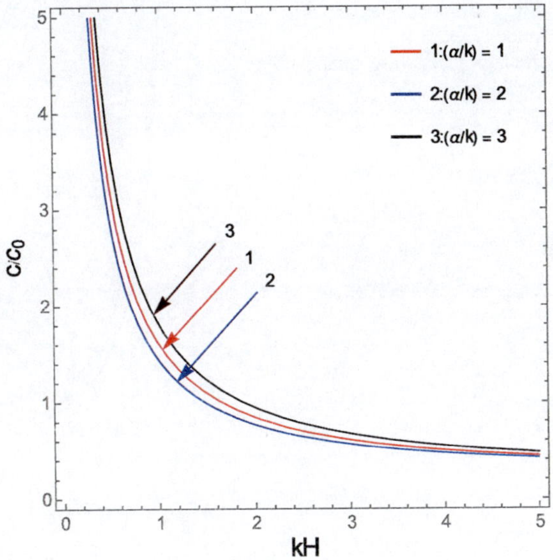

Fig. 2 Dimensionless phase velocity against dimensionless wave number for the parameter $\left(\frac{\alpha}{k}\right)$

Figure 2 studies the inhomogeneity parameter (α/k) associated with the rigidity, density and initial stress of the half space. The graph has been plotted for the dimensionless phase velocity against dimensionless wave number. It has been found that as the wave number increases, the phase velocity decreases consistently for all the values of (α/k). It can also be observed that magnitude of velocity fluctuates as the value of (α/k) goes from 1.0 to 3.0. This unusual behaviour deviates the trends of phase velocity and leaves a remarkable impact on the propagation behaviour of Love wave.

Figure 3 has been plotted to demonstrate the impact of (b/k) that appears in exponentiation with the modulus of rigidity and the density. The graph appears to be similar to Fig. 3 expect the periodic deviation that curve shows unlike to Fig. 3. It has been observed that as the magnitude of (b/k) increases from 0.1 to 0.3, the phase velocity decreases remarkably. It also appears that the curves are accumulating towards the lower and higher values of the wave number, giving rise to the critical points. At these points, although the magnitudes of (b/k) changes, but the phase velocity remains constant.

Figure 4 manifests the influence of (a/k) on the phase velocity of Love wave. The curves being quite apart from each other reveal the fact that increasing values of (a/k) deviate the phase velocity to a great extent. Unlike to Figs. 1 and 2, the curvature of the curves is on higher side, and the increasing value of (a/k) increases the phase velocity at a particular wave number. This shows that it has a remarkable effect on Love wave propagation.

Figure 5 represents the impact of increasing value of (β/k) on the phase velocity of Love wave. The curves are overlapped showing that although the values of (β/k) increase but the effects remained invisible.

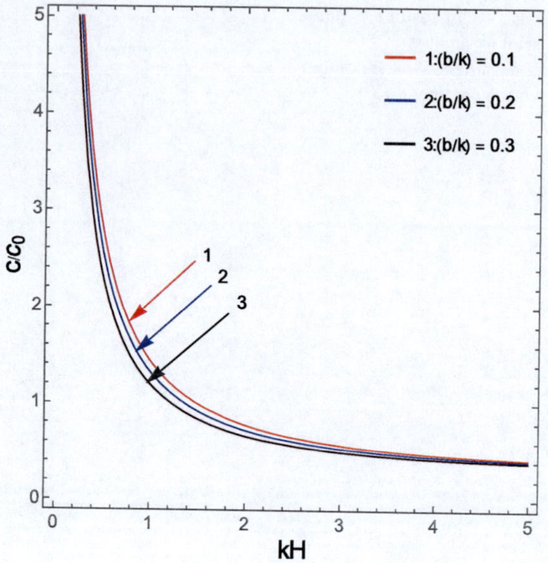

Fig. 3 Dimensionless phase velocity against dimensionless wave number for the parameter $\left(\frac{b}{k}\right)$

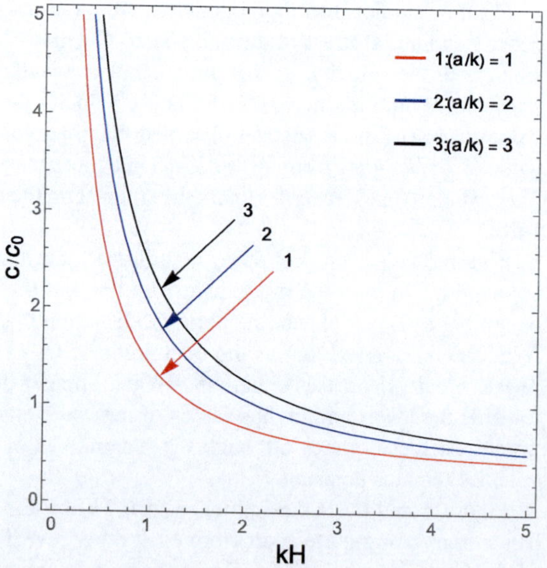

Fig. 4 Dimensionless phase velocity against dimensionless wave number for the parameter $\left(\frac{a}{k}\right)$

Figure 6 explains as to how does the inhomogeneity involved in the layer influences the phase velocity of Love wave. As the magnitude of (m/k) increases, the phase velocity also increases, whereas the phase velocity decreases as the wave

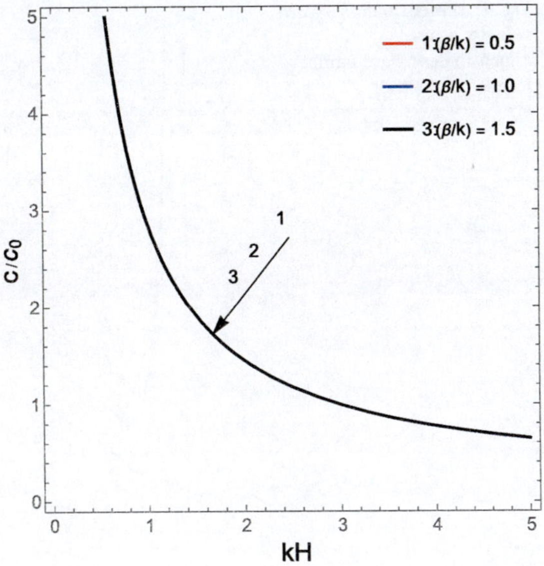

Fig. 5 Dimensionless phase velocity against dimensionless wave number for the parameter $\left(\frac{\beta}{k}\right)$

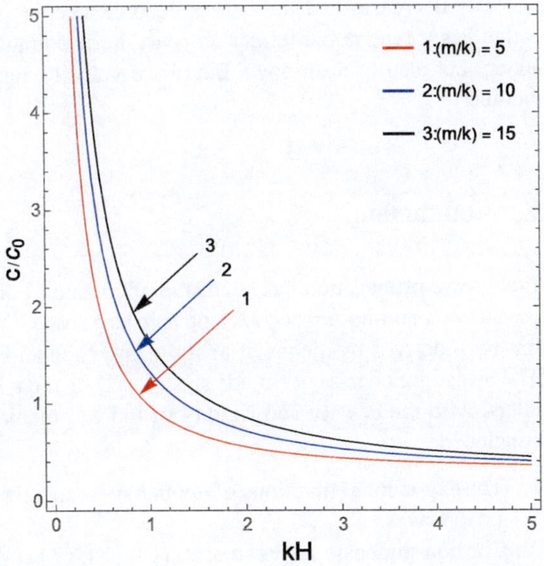

Fig. 6 Dimensionless phase velocity against dimensionless wave number for the parameter $\left(\frac{m}{k}\right)$

number increases. The graph being self-explanatory for the relevant impact that the inhomogeneity parameter (m/k) leaves on phase velocity.

Figure 7 has been plotted to show the effect of initial stress on the phase velocity of Love wave. As the magnitude of initial compressive stress increases, the phase

Fig. 7 Dimensionless phase velocity against dimensionless wave number for the parameter ζ

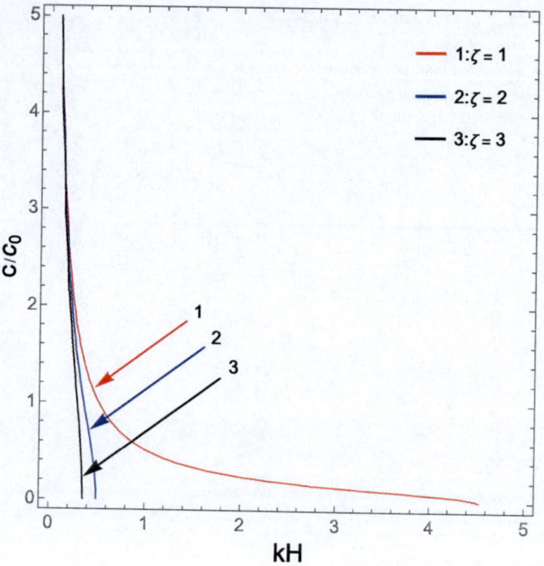

velocity decreases and the curves become steeper. This shows that the increasing value has a remarkable effect not only in deviating the phase velocity to a larger extent but also to drop down the phase velocity significantly at a particular wave number.

8 Conclusion

Love wave propagation has been studied in a non-homogeneous anisotropic crustal layer over a non-homogeneous isotropic half space. Closed form frequency equation has been derived by means of analytic approach and suitable boundary conditions. The investigation has been made specifically on the inhomogeneity parameters associated with the density and rigidity of the geo-media. Following points have been concluded:

1. The exponential functions of depth have a great bearing on the phase velocity of Love waves.
2. The non-dimensional parameters $\left(\frac{a}{k}\right)$, $\left(\frac{b}{k}\right)$, $\left(\frac{\alpha}{k}\right)$, $\left(\frac{\beta}{k}\right)$, $\left(\frac{m}{k}\right)$ and ζ have been found to have great bearing on the magnitude of Phase velocity. Some of the inhomogeneity parameters increases the magnitude of phase velocity while few of them decreases it. The unusual behaviour has also been found which ended up in recognizing possible critical points.
3. The theoretical studies of seismic waves are helpful in forecasting geophysical parameter during signal processing and seismic data analysis. The theoretical

study made here can be used to estimate the damage due to earthquake as well as in applications pertaining to geophysical prospecting.

Acknowledgements Authors extend their sincere thanks to SERB-DST, New Delhi, for providing financial support under Early Career Research Award with Ref. No. ECR/2017/001185. Authors are also thankful to DST New Delhi, for providing DST-FIST grant with Ref. no. 337 to Department of Mathematics, BITS-Pilani, Hyderabad Campus.

References

1. Shearer PM (2009) Introduction to seismology, 2nd edn. Cambridge University Press, Cambridge
2. Anderson DL (1961) Elastic wave propagation in layered anisotropic media. J Geophys Res 66:2593–2963
3. Love AEH (1911) Some problems of geodynamics. Cambridge University Press, London
4. Chattopadhyay A, Bandyopadhyay U (1986) Shear waves in infinite monoclinic crystal plates. Int J Eng Sci 24:1587–1596
5. Chattopadhyay A, Choudhury S (1990) Propagation, reflection and transmission of Magnetoelastic Shear waves in a self-reinforced medium. Int J Eng Sci 28:485–495
6. Mistri KC, Singh AK, Das A (2018) Attenuation and dispersion of SH-waves in a loosely bonded sandwiched fluid saturated porous layer. Soil Dyn Earthquake Eng 107:350–362
7. Vishwakarma SK, Panigrahi TR, Kaur R (2019) SH-wave propagation in linearly varying fibre-reinforced viscoelastic composite structure unintial stress. Arabian J Geosci 12:59
8. Love AEH (1944) A treatise on mathematical theory of elasticity. Dover Publication, New York
9. Biot MA (1940) The influence of initial stress on elastic waves. J App Phys 11:522–530
10. Chattaraj R, Sapan KS (2013) Love waves in the fiber-reinforced layer over a gravitating porous half-space. Acta Geophys 61(5):1170–1183
11. Nam NT, Merodio J, Ogden RW, Vinh PC (2016) The effect on the propagation of surface waves in a layered half-space. Int J Solids Struct 88(89):88–100
12. Shams M (2016) Effect of initial stress on Love wave propagation at the boundary between a layer and a half-space. Wave Motion 65:92–104
13. Bullen KE (1940) The problem of the Earth's density variation. Bull Seismol Soc Am 30(3):235–250
14. Ewing WM, Jardetzky WS, Press F (1957) Elastic waves in layered Media. McGraw-Hill Book Co., New York
15. Achenbach JD (1973) Wave propagation in elastic solids. North-Holland Publishing Company, Amsterdam, New York, Oxford
16. Aki K, Richard PG (2002) Quantitative seismology, 2nd edn. University Science Books, Sausalito, California
17. Pujol J (2003) Elastic wave propagation and generation in seismology. Cambridge University Press, Cambridge
18. Chapman C (2004) Fundamentals of seismic wave propagation. Cambridge University Press, Cambridge
19. Sari C, Salk M (2002) Analysis of gravity anomalies with hyperbolic density contrast: an application to the gravity data of Western Anatolia. J Balkan Geophys Soc 5(3):87–96
20. Wang CD, Chou HT, Peng DH (2017) Love-wave propagation in an inhomogeneous orthotropic medium obeying the exponential and generalized power law models. Int J Geomech 17(7):04017003

21. Kielczynski P, Szalewski M, Balcerzak A (2012) Effect of a viscous liquid loading on Love wave propagation. Int J Solids Struct 49:2314–2319
22. Gupta S, Vishwakarma SK, Majhi DK, Kundu S (2013) Possibility of Love wave propagation in a porous layer under the effect of linearly varying directional rigidities. Appl Math Model 37:6652–6660
23. Gupta S, Majhi DK, Kundu S, Vishwakarma SK (2013) Propagation of Love waves in non-homogenous substratum over initially stressed heterogeneous half-space. Appl Math Mech 34:249–258
24. Singh SS (2010) Love wave at a layer medium bounded by irregular boundary surfaces. J Vib Control 17:789–795
25. Kaur G, Singh D, Tomar SK (2019) Love waves in a nonlocal elastic media with voids. J Vib Control 0(0):1–14
26. Matthew MH, Tsai VC (2020) Perturbational and nonperturbational inversion of Love-wave velocities. Geophysics 85(1):19–26
27. Furumura T, Kennett B (2020) Propagation of distinct Love-wave pulses from regional to teleseismic distances in continental and oceanic environments. Geophys J Int. https://doi.org/10.1093/gji/ggaa028
28. Piotr K, Marek S, Andrzej B, Krzysztof W (2020) Dispersion curves of love waves in elastic waveguides loaded with a Newtonian liquid layer of finite thickness. Arch Acoust 45(1):19–27

Earthquake Induced Ground Motion Mitigation Using Geotextile Encased Infill Material System

S. P. Vijay Kumar, S. Ganesh Kumar, and Elias Anju Mary

Abstract Occurrence of earthquakes creates combination of horizontal and vertical ground motions. When these ground motions interact with saturated or partially saturated soils, soil liquefaction occurs. To mitigate the effects of ground motions, there are conventional isolation systems available for isolating and improving the performance of foundation structure. However, assessment of these isolation barrier systems in liquefiable soils was limited. In the present research, an attempt was made to develop an effective and economical barrier model for mitigating ground motions induced by an earthquake. The barrier system was installed in a liquefiable deposit to evaluate both isolation and liquefaction resistance characteristics. For developing barrier system, commercially available polyurethane (PU) foam which is known for its absorption characteristics was used. The experiments were performed on saturated ground model having 40 and 60% relative density and they were tested under repeated shaking conditions with and without barrier reinforcement. Experimental results showed that the developed PU foam isolation barrier system placed inside liquefiable soil performs well in mitigating earthquake borne ground motions, improves seismic resistance of soil and minimizing settlement compared to unreinforced ground bed.

Keywords Isolation barrier · Infill material · Ground motions · Geotextile

1 Introduction

Occurrence of earthquake induced ground motions always poses problems to the safety of infra-structures. Particularly, when structures are located in liquefiable deposit, the failures were mainly due to (i) interaction between foundation structure

S. P. Vijay Kumar (✉)
IIT(ISM) Dhanbad, Dhanbad 826004, Jharkhand, India
e-mail: spvijay.18mt0390@agp.iitism.ac.in

E. Anju Mary
Karunya Institute of Technology and Sciences, Coimbatore 641114, Tamil Nadu, India

S. P. Vijay Kumar · S. Ganesh Kumar · E. Anju Mary
CSIR-CBRI Roorkee, Roorkee 247667, Uttarakhand, India

T. G. Sitharam et al. (eds.), *Seismic Hazards and Risk*, Lecture Notes in Civil Engineering 116, https://doi.org/10.1007/978-981-15-9976-7_23

and the generated ground motions under earthquake induced shaking and (ii) generation of excess pore water pressures which reduces strength of saturated soil deposit. Presently, the adopted conventional isolation systems are intended for mitigating horizontal acceleration response only. However, within the vicinity of earthquake epicenter, there are reports available regarding failures of structures due to combined vertical and horizontal ground motions generated during earthquake. When buildings are located near to epicenter, the structure experiences a combined horizontal and vertical ground acceleration responses which create vertical deformation and failure of structure [1]. The response of vertical component of ground motion is found to be higher in near fault regions, higher magnitude seismic events and in shorter period events relative to the horizontal component [2]. Also, vertical ground motions are associated with higher frequency oscillations and hence there are limited hysteretic energy dissipation mechanisms available for estimating vertical inelastic response [3]. In addition to the above, studies on the use of energy absorbing isolation barrier system in liquefiable soils was also limited.

Considering the above challenges, the present research aims at developing an effective economical solution for mitigating both horizontal and vertical acceleration response induced by an earthquake. In addition, the impact of developed isolation barrier model in improving seismic resistance of liquefiable deposit was also evaluated. For experimental study, the barrier system was installed in saturated sand deposit having 40 and 60% density subjected to dynamic loading conditions. The isolation barrier was made with commercially available PU foam which was widely accepted especially in shaking table tests as an absorbing boundary. Since, the study involves installation of barrier within soil bed, possibility of clogging can happen due to its porous characteristics. Geotextile was encased around the PU foam barrier to minimize clogging and to prevent soil-foam intrusion. Then, soil bed with isolation barrier was placed on uniaxial shaking table and experiments were performed. The effect of model ground with and without isolation barrier was studied and presented.

2 Motivation and Objective

The primary objective of this research is to develop an efficient isolation barrier system to ensure the safety of existing structure against earthquake induced ground motion conditions. Usually PU foams have been used in shaking table experimental studies as an absorbing boundary member. Considering its absorbing and damping characteristics, PU foam was selected and designed as an isolation barrier system. The beneficial absorbing characteristics motivated to use foam barrier model inside the liquefiable deposit to test its dynamic performance under repeated shaking events. The developed foam barrier was installed as a confining member to confine foundation model and their performance was evaluated for combined liquefaction and ground motion mitigation characteristics. For experimental study, poorly graded sand was used to model liquefiable ground having 40 and 60% density. Since the finer particles in the sand can cause clogging, geotextile was wrapped around the barrier and installed.

All the experimental investigations were performed with Uni-axial shaking table and for study both sinusoidal motions having 0.1, 0.2 and 0.3 g loading with 5 Hz frequency and scaled down Uttarkashi earthquake motion was applied to the model ground. It was observed that the shaking of different loading intensities provided in uniaxial direction generated both horizontal and vertical components of acceleration in which the existence of vertical acceleration was confirmed by the variation of pore pressure readings along the depth of prepared model ground. Since the application of acceleration loading can generate combined horizontal and vertical acceleration amplitudes, the isolation barrier model was developed similar to inverted T-shaped barrier system. This shape offers effective anchorage and also to absorb vertical ground motions generated due to shaking.

All the experimental tests were conducted in sub-sequential acceleration loading conditions to evaluate the isolation performance of the system under repeated earthquake events. Tests were performed with and without confined isolation barrier model to evaluate the effect of generated ground motion characteristics and liquefaction response. Through shaking table studies, efficiency of the system under sequential acceleration loading was studied to evaluate the performance of developed isolation barrier in ground motion mitigation and in liquefaction resistance.

3 Methodology

For the experimental studies, poorly graded sand was selected. Soil sample was collected from solani river bed in Roorkee, Uttarakhand. Laboratory experiments such as specific gravity, sieve analysis, relative density and permeability tests were carried out to determine the soil properties. A model tank of dimension 1.4 m × 1 m × 1 m was used for model ground preparation. Saturated ground model having 600 mm height with 40 and 60% density was prepared and tested with and without isolation barrier under repeated acceleration loading conditions.

3.1 Preparation of Un-reinforced Ground Condition

The saturated ground model was prepared by sand pluviation technique. Desired relative density for soil was achieved by first adding the water required for complete saturation of soil and then pouring the soil into the tank through a conical hopper having 60° inverted cone at bottom from a specific height as calculated. From the height of fall experiments, the height of pouring has been decided to be 110 and 160 mm for achieving 40 and 60% relative densities respectively.

The soil was filled in three layers of 200 mm height for achieving uniformity in sand bed preparation and for monitoring pore water pressure generated during shaking; piezometers were fixed at a height of 200 and 400 mm from the bottom of the tank. For estimating acceleration response, two accelerometers were placed at 150 and 300 mm depth respectively from bottom. For measuring foundation displacements with and without barrier system, a reduced scaled foundation footing model was embedded in the soil bed. Then, the prepared ground bed was subjected to sequential acceleration of 0.1, 0.2 and 0.3 g loading. A constant excitation frequency of 5 Hz was selected for performing the tests with aforementioned loading intensities in order to simulate medium to high seismic conditions pertaining in the ground model. Higher base motion frequency tends to increase the acceleration amplitudes which thereby results higher incremental pressures in the ground model [5]. Test with an increased acceleration loading was performed only after complete dissipation of generated pore water pressure from previous acceleration loading.

3.2 *Isolation Barrier System*

Poly urethane foam was well known for its high resilience, abrasion and vibration absorption characteristics. Considering their vibration absorption characteristics, in this study PU Foam was primarily selected as an isolation barrier.

Preparation of PU Foam isolation barrier

The PU foam was procured in liquid form. It consists of two components, resin and hardener. Both had a density of 1.2 g/cc. The procured PU foam had a capacity of expanding up to 30 times its original volume. Considering the mixing and application losses involved, a nominal volume expansion factor of 25 times was fixed. The desired quantity of foam liquid required for the fabrication of barrier was prepared by mixing resin and hardener in a ratio of 1:1 and poured inside the inverted T-shaped mould. To prevent clogging, geotextile material was reinforced on the surface of the barrier system.

Two inverted T-shaped anchored model having dimensions of Heel: 200 × 500 × 50 mm, Toe: 100 × 500 × 50 mm and Stem: 50 × 500 × 300 mm was prepared for the study. The prepared isolation barrier was placed at 500 mm c/c spacing and installed inside the tank at 300 mm height from bottom and then soil filling was carried out. Figure 1 showing the complete details of the isolation barrier system installed inside the ground deposit. For measuring foundation response under shaking, scaled down isolated square footing having 150 mm size was placed. Positioning of isolation barrier was placed in such a way that, the heel portion and stem portion absorbs vertical and horizontal ground motions and also confines the foundation model placed inside.

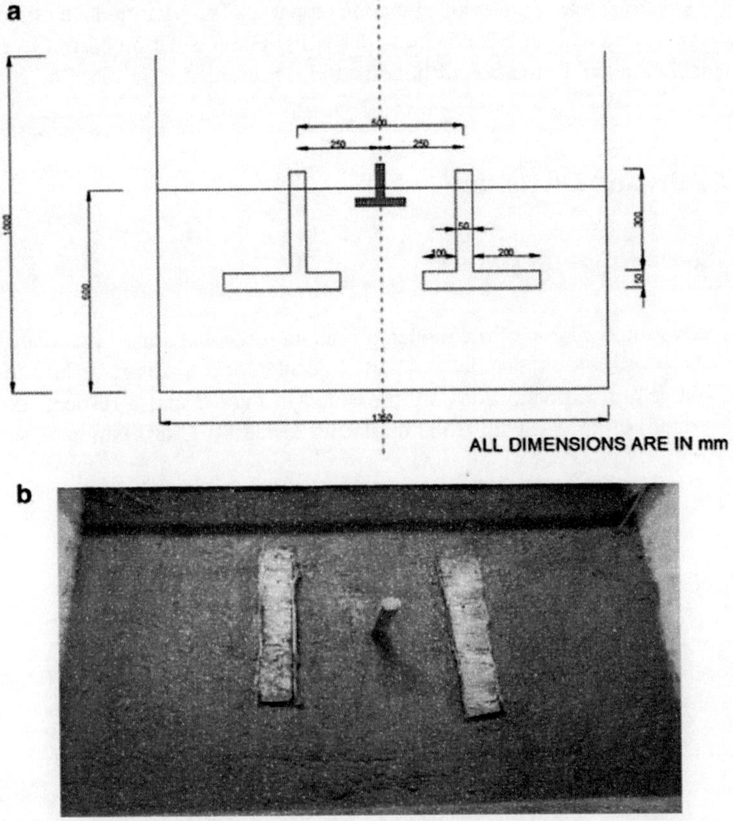

Fig. 1 PUF isolation barrier system. **a** Side view. **b** Top view

4 Experimental Testing

For experimental testing, scaled down ground motions of Uttarkashi earthquake and sinusoidal wave loading having 0.1, 0.2 and 0.3 g acceleration intensities with 5 Hz frequency was applied to both unreinforced and reinforced ground. The soil with and without isolation barrier system was tested for the scaled down earthquake loading in order to observe the response of reinforced and unreinforced ground model to the real time earthquake condition where the loading condition is non-linear. The Uttarkashi earthquake (1991) was selected due to its violent intensity (IX) and closer locality to the site where the soil sample is taken. For measuring acceleration response with isolation barrier system, three accelerometers are placed inside the model tank i.e. two within the confined area at 100 mm height and at 400 mm depth and one outside the barrier area at 100 mm depth from top ground surface respectively. Using piezometers, the generated excess pore water pressure was measured. The effects of foundation settlement, acceleration response, pore pressure response with and without isolation system are evaluated and compared. The performance of isolation barrier is assessed and the effectiveness of the PU Foam isolation barrier in ground motion isolation and mitigation of liquefaction is presented.

5 Results and Discussion

5.1 Acceleration Response

The acceleration response of the model ground under scaled down Uttarkashi earthquake conditions and sinusoidal sequential acceleration loading for both unreinforced and reinforced conditions are presented in Figs. 2 and 3 respectively. The model ground reinforced with isolation barrier tested for Uttarkashi earthquake of

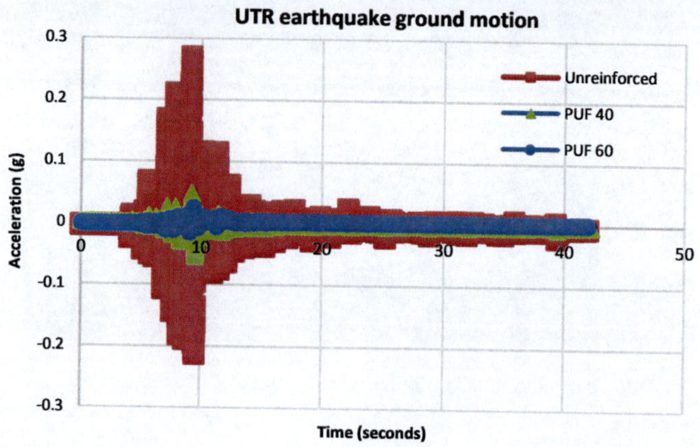

Fig. 2 Mitigation of Uttarkashi earthquake scaled ground motion by isolation barrier

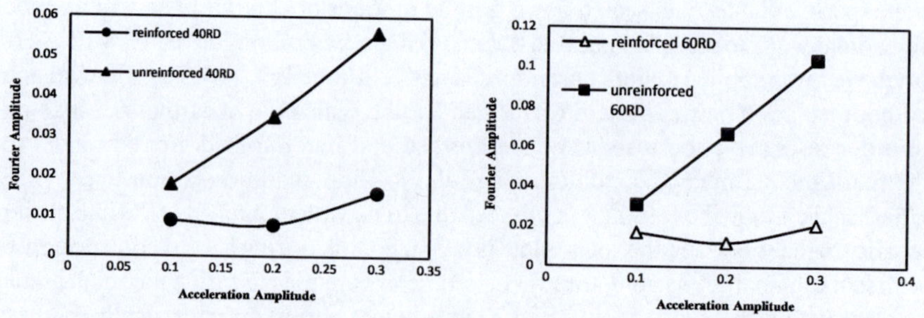

Fig. 3 Peak acceleration response with and without isolation barrier under repeated sinusoidal acceleration loading

PGA: 0.31 and was found to provide vibration damping of about 83% (PGA: 0.052) under 40% relative density and about 90% damping (PGA: 0.029) under 60% relative density conditions (Fig. 2).

Similarly, reinforced ground model subjected to sequential acceleration loading of 0.1, 0.2 and 0.3 g at 5 Hz frequency was also compared with unreinforced conditions. It was evaluated that peak fourier amplitude near the footing model reinforced with PU foam condition is reduced by 75% under 40% relative density and 70% under 60% relative density when compared to that of unreinforced (Fig. 3).

5.2 *Pore Water Pressure*

The effect of pore water pressure under sub-sequential incremental loading for both un-reinforced and reinforced conditions was discussed. The prepared ground model was subjected to the sequential amplitude sinusoidal loading incrementally i.e. 0.1, 0.2 and 0.3 at 5 Hz frequency. The generated pore water pressure was estimated by means of two glass piezometers located at 200 and 400 mm from ground level. The mean effective stress at bottom is higher than that at top due to the higher overburden pressure experienced at the bottom layers. Thus, the top portion of ground model is more susceptible to re-liquefaction occurrence when subjected to repeated shaking events. The variation of excess pore water pressure in top piezometer with time for 40 and 60% relative density of soil bed are presented (Fig. 4).

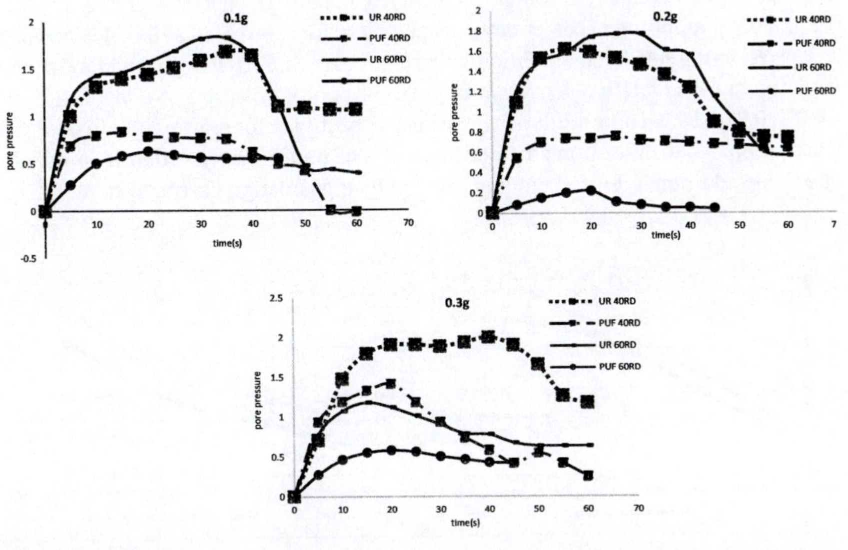

Fig. 4 Variation of pore pressure with and without isolation barrier for 40 and 60% Relative densities under sequential incremental loading

It can be clearly seen that the peak pore water pressure generated in reinforced condition is much lower when compared to that of unreinforced condition. However, the time for attainment of peak occurs fast in case of PUF barrier reinforcement. This may be due to the confinement provided by foams on two sides of the foundation model and non-presence of effective drainage system. However, no foundation failure was observed compared to unreinforced conditions. The studies with sequential loading validate that the isolation barrier is effective in dissipating the generated excess pore water pressure under repetitive acceleration conditions.

5.3 *Variation in Relative Density*

After each shaking event, the relative density of the soil tends to increase due to soil grain rearrangement. In case of repeated un-drained shaking events, generation of water from bottom to top disturbs the soil deposition making more compact rearrangement thus increasing density. The sequential variation in relative density was evaluated using digital static cone penetrometer. Digital Cone Penetration test (DCPT) was conducted after each acceleration intensity and the increase in the cone penetration resistance was evaluated at different depths (10, 20, 30, 40 and 50 cm) of soil ground model. To compare the effect of soil compression with and without barrier isolation reinforcement under repeated acceleration loading, the variation of relative density is evaluated and presented (Fig. 5).

For the prepared 40% RD ground model, the variation in mean relative density of the sand deposit attained is found to be approximately 4, 3 and 9% greater for 0.1, 0.2 and 0.3 g amplitude accelerations respectively for reinforced soil deposits when compared to unreinforced condition whereas in case of 60% RD ground model, the variation in mean relative density is found to be approximately 5, 4 and 7% greater for 0.1, 0.2 and 0.3 g amplitude accelerations respectively for reinforced soil deposits when compared to unreinforced ground model. The improvement in density is mainly due to the additional reinforcement provided by the isolation barrier system.

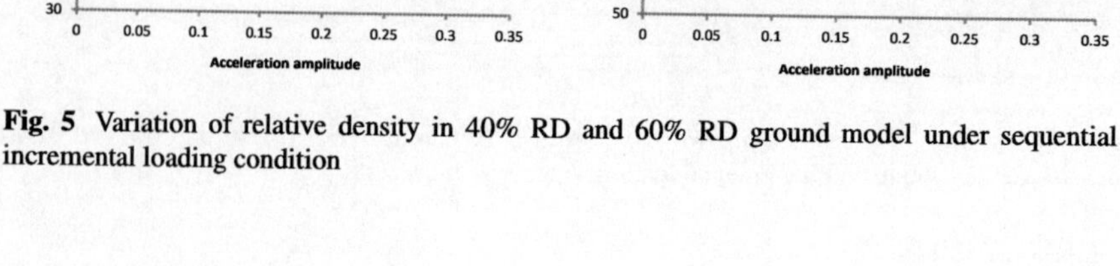

Fig. 5 Variation of relative density in 40% RD and 60% RD ground model under sequential incremental loading condition

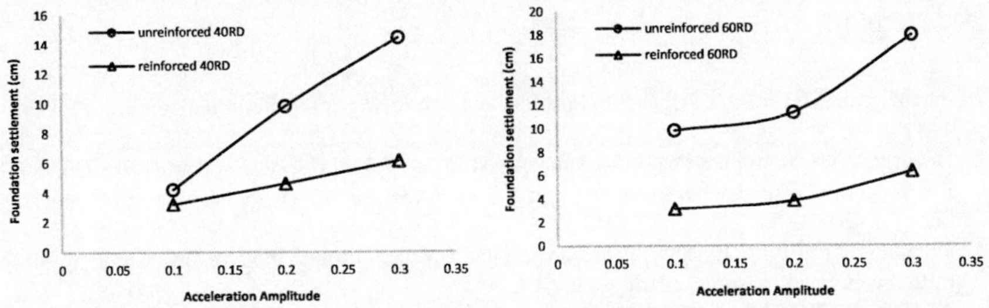

Fig. 6 Foundation settlement in 40% RD and 60% RD ground model under sequential incremental loading condition

5.4 *Foundation Settlement*

A scaled foundation model was embedded at center of the prepared ground model. The foundation settlements with respect to incremental sequential amplitude accelerations under both reinforced and unreinforced soil deposits are evaluated and presented below (Fig. 6). Due to the effective confinement, enhanced absorption and dissipation characteristics provided by the PUF barriers on both sides of foundation, foundation settlement was found to be minimum compared to unreinforced conditions.

It can be seen that for ground prepared with 40% density, the foundation settlement in the PUF isolation barrier reinforced model is about 65% less when compared to that of unreinforced. In case of 60% density ground condition bed, the foundation settlement in reinforced model is reduced by about 77%. The reduction in foundation settlement suggests that, isolation barrier system not only effective in mitigating ground motion response but also confines foundation structure and improve stability under repetitive dynamic loading.

6 Conclusion

Based on the obtained experimental test results, the following conclusions were made.

- The selected isolation barrier system is effective in mitigating vertical deformation and failure in the structure during dynamic loading.
- The isolation barrier improves seismic resistance of liquefiable soil deposits.
- The developed isolation barrier confines foundation member effectively, minimize ground motion interaction thereby minimizing settlement and improving safety of foundation structure.

References

1. Papazoglou AJ, Elnashai AS (1996) Earthquake engineering and structural dynamics. 25:1109–1137
2. Bozorgnia Y, Campbell KW (2004) The vertical-to-horizontal response spectral ratio and tentative procedures for developing simplified V/H and vertical design spectra. J Earthquake Eng 8(2):175–207
3. Kunnath SK, Erduran E (2008) Effect of near-fault vertical ground motions on seismic response of highway overcrossings. J Bridge Eng 13(3)
4. Al-Hussaini TM, Ahmad S (1991) Design of wave barriers for reduction of horizontal ground vibration. J Geotech Eng 117(4):616–636
5. Krishna AM, Latha GM (2007) Seismic response of wrap-faced reinforced soil retaining wall models using shaking table tests. Geosynthetics Int 14(6):355–364

Modified Tuned Liquid Damper for Vibration Control of Short Period Structures

Dhirendra Kumar Pandey and Sudib Kumar Mishra

Abstract Tuning of structures of short periods to the long period sloshing liquid of a tune liquid damper (TLD) possess difficulty. However, proper tuning of the tuned damper devices is must for the desired control efficiency. To overcome this difficulty, a new implementation is proposed contrary to the conventional rigidly connected TLD with relatively shallow water depth. Presently, flexible pads with lower time period of vibration are inserted between the structure and the TLD, for which, the depth of the liquid in water tank may allow to be relatively high. The frequency of the container (inclusive of impulsive mass of the liquid) vibration is tuned to the fundamental frequency of the structure, which is observed to allow sufficient sloshing. The elastomeric pads are designed for adequate tuning. The performance of the proposed implantation studied numerically. Two short period structures are chosen (0.3 and 0.5 s) for the present study. Six ground motions representing the various seismic hazard levels from SAC ground motion data base are employed as input excitation. The numerical study reveals that the damper system is effectively reducing the response of structure yet it allows a relatively larger depth of water in the tank that can be harnessed for storage purposes as well.

Keywords Tuned liquid damper · Short period structures · Deep water tank

1 Introduction

Tuned liquid damper (TLD studied for controlling the response of the tall structure subjecting to random loading in past [1–4]. Modification in conventional TLD is also suggested to enhance the performance of the TLD. Some modifications are described briefly herein. These modifications are mostly done by altering the tank geometry. Enhanced preference of the damper is demonstrated by inserting the semicircular block [5] and by installing the floating roof on the liquid [6]. A study on sloped bottom TLD is done by the Gardarsson et al. [7]. The nonlinear shallow water model of the

D. K. Pandey (✉) · S. K. Mishra
Indian Institute of Technology Kanpur, Kanpur 208016, India
e-mail: dkpandey@iitk.ac.in

T. G. Sitharam et al. (eds.), *Seismic Hazards and Risk*, Lecture Notes in Civil Engineering 116, https://doi.org/10.1007/978-981-15-9976-7_24

TLD was suggested by the Sun et al. [8]. Yu et al. [9] suggested the nonlinear tune mass damper (TMD) model of water sloshing by matching the energy dissipation. All these studied were done for suppressing the response of the long period structures. However, short period structures are also susceptible for the seismic loading. TLDs have their own inherent limitation in finding the practically feasible length for the tuning of the damper to short period structures. To overcome these limitations, Ghosh et al. [10] proposed a modified tune liquid column damper in which the damper is attached by the spring to the structure, and the whole damper assembly is tuned with the structure. A similar study is done by the Benerji et al. [11], by putting TLD on a TMD and demonstrated that TMD system is increasing the acceleration at the base of TLD. It is also demonstrated that the increased base acceleration of TLD is enhancing the damper performance by enhanced liquid sloshing and wave breaking. However, the performance of the damper is not demonstrated for the short period structure. Benerji et al. [12] suggested a higher depth ratio by violating the shallow water depth ratio requirement of the TLD, for the short period structures. Nayak and Thakre [13] studied the seismic performance of the existing water tank by performing the nondestructive testing. However, the performance of the damper is compromised by insufficient tuning. Pandey et al. [14] suggested a new type of implementation in which shallow water TLD is placed on the flexible elastomeric pads and whole damper system is tuned with structure and the performance of the damper is demonstrated for the suppressing the acceleration response of the short period structure. Viewing the study done [14], a similar work but relaxing the shallow water depth requirement of the TLD is suggested herein. In this study, deep water TLD (depth ratio 0.6) is used, and the frequency of container mass (inclusive of impulsive mass) is tuned with the frequency of the structure by adopting a suitable geometric and the elastic properties of the elastomeric pads. Then the damper parameters are optimized and the damper performance is demonstrated by employing the damper parameters obtained from the optimization.

2 Modeling of the Damper

The equation of motions is derived by adopting the linear model of water tank suggested by Housner et al. [15]. The performance of the damper is demonstrated by obtaining the solution of the system of equations numerically. The optimum damper parameters are obtained by the parametric variation performed on the damper parameters. Finally, the performance of the damper is illustrated by subjecting, a suite of selected ground motions to the employed structure.

The damper is modeled by following the Housner's water tank model [15]. The liquid mass is bifurcated into the impulsive mass (mass which vibrates along with the liquid) m_i and liquid sloshing mass m_s. In the present implementation, the container and the impulsive mass is tuned with the structure by adopting the suitable dimensions of the elastomeric pad. The adopted model of the structure damper system with idealized linear sloshing of the liquid is shown in Fig. 1. The tank wall is assumed

Fig. 1 Idealized model of the structure damper system

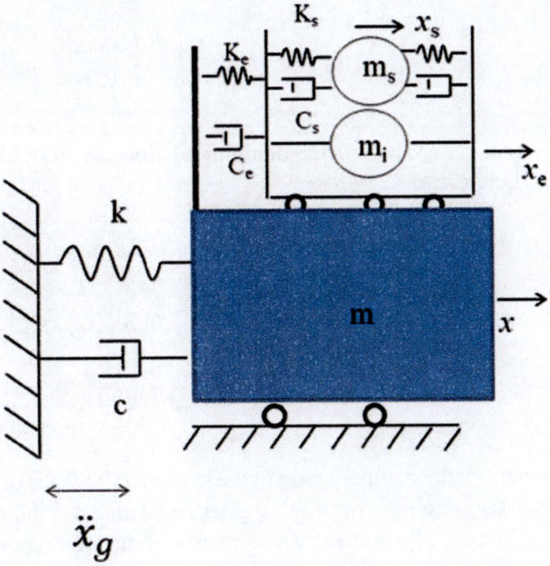

rigid, which is a reasonable assumption for the concrete and prestressed water tanks. The system of the equations of the vibration of the proposed structure damper system may be expressed as;

$$[M]\{\ddot{X}\} + [C]\left\{\dot{X}\right\} + [K]\{X\} = -[M]\{r\}\ddot{u}_g \tag{1}$$

Here $\{X\}$ is the displacement vector associated with the sloshing mass, impulsive mass and the structure, respectively. These degrees of freedoms are defined with respect to ground and may be written as,

$$\{X\} = \left\{x_s\ x_e\ x\right\}^T \tag{2}$$

Matrix, $[M]$ depicted the mass matrix and given as,

$$[M] = \begin{bmatrix} m_s\ m_s\ m_s \\ m_s\ m_e\ m_e \\ m_s\ m_e\ M \end{bmatrix} \tag{3}$$

where m_s is the mass of the liquid contributing in sloshing, m_e is the total mass liquid and the container at the base of the elastomeric pad and M is the mass of short period structure inclusive of the damper mass.

Matrix, $[C]$ is the viscous damping matrix and given as,

$$[C] = \begin{bmatrix} c_s & 0 & 0 \\ 0 & c_e & 0 \\ 0 & 0 & c \end{bmatrix} \tag{4}$$

where c_s, c_e and c is the damping of sloshing liquid mass, elastomeric pad and the structure and written as,

$$c_s = 2\varsigma_s m_s \omega_s \tag{5}$$

$$c_e = 2\varsigma_e m_e \sqrt{k_e / m_e} \tag{6}$$

$$c = 2\varsigma M \omega \tag{7}$$

where ς_s the damping ratio of the (assumed 0.5%) is of the sloshing liquid mass, ς_e is the damping ratio of the elastomeric pad and ς is the damping ratio of the structure.

Matrix, $[K]$ is the stiffness matrix of the structure damper system and given as,

$$[K] = \begin{bmatrix} k_s & 0 & 0 \\ 0 & k_e & 0 \\ 0 & 0 & k \end{bmatrix} \tag{8}$$

where k_s, k_e and k is the stiffness of sloshing mass, elastomeric pad and the structure, respectively. $\{r\}$ is influence vector and may be written as,

$$\{r\} = \{0\ 0\ 1\}^T \tag{9}$$

$\ddot{u}_g$ is base acceleration input of the earthquake. The stiffness of the elastomeric pads is chosen so that the frequency of the container (inclusive of impulsive mass) and the frequency of the structure can be tuned.

3 Selection of the Ground Motion

The ground motions which are selected in this study are chosen from the SAC ground data base, in which the ground motions are categorized in three different hazard bins named as design basis earthquake (DBE), maximum considerable earthquakes (MCE) and the serviceability design earthquake (SDE). Total six (two from each hazard bins) earthquake motions are selected for the study. The properties of the selected ground motions are illustrated in Table1. The response spectra of the acceleration of the selected ground motions are illustrated in Fig. 2. The average response acceleration of selected earthquake motions is also shown in Fig. 2 with the thicker

Table 1 Properties of the selected ground motions

S. No.	EQ. details	Magnitude	Dominant period (s)	PGA (g)
1	la03 Imperical Valley, 1979	6.5	0.16	0.39
2	la06, Imperical Valley, 1979	6.5	0.30	0.23
3	la24, Loma Prieta, 1989	7.0	0.72	0.47
4	la26, Northridge, 1994	6.7	0.30	0.94
5	la42, Coyte Lake, 1979	5.7	0.20	0.33
6	la46, Kern, 1952	7.7	0.54	0.16

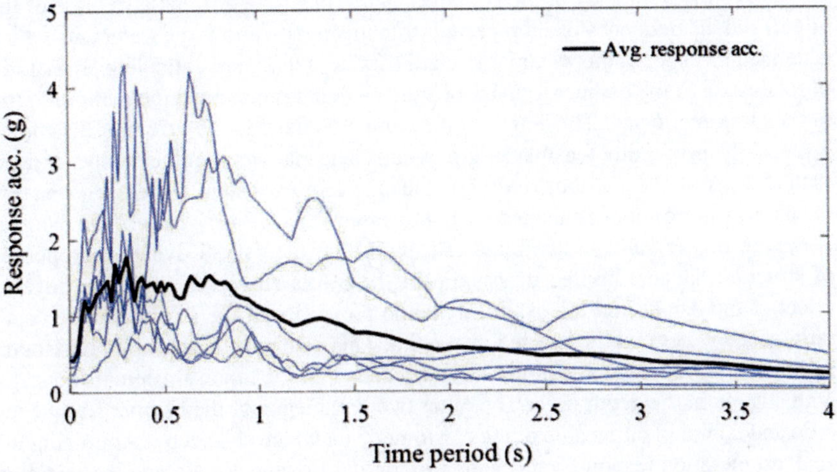

Fig. 2 Response acceleration spectra of ground motions

line. It is observed from the figure the dominate period of the selected ground motion is varying from the 0.2 to 0.9 s. This is clearly indicating that these ground motions are vulnerable to short period structures. It is also observed that some ground motions are showing significant spectral ordinate at a higher time period (around 1.5 s). This type of selected motion will be helpful in judging the performance of the damper when the structure is subjected to the excitations of the long period as well. It is observed from Table 1 that the dominate period of the selected ground motions is varying from the 0.16 s. to the 0.72 s. This variation in the dominate period of the chosen ground motions is selected to judge the robustness of the damper system in suppressing the response of the structure when subjected to ground motions of wide range of frequency content. The peak value of ground acceleration (PGA) of the selected ground motions is varying from 0.16 to 0.94 g. This variation is chosen to judge the performance of the proposed damper with the varying level of the excitation level.

4 Selection of the Structure and Damper Parameters

Two short period structures of the period of the 0.3 and 0.5 s are chosen for the present study. These selected periods fall in the category of the short periods structures, which typically ranges from 0.2 to 0.9 s. The damping ratio of both the structures is chosen as 1%. This deliberate choice of the low damping ratio is to judge the actual response reduction capacity of the damper, which significantly suppressed in the structure of the higher damping ratio. The dynamic properties of the selected structures are listed in Table 2. The important damper parameters that need to be selected to judge the performance of the are the length of the water tank, frequency ratio which defines as the ratio of the frequency of the container mass (inclusive of impulsive mass of the liquid) and the frequency of short period structure, depth ratio of the water tank which is defined as depth of the water and length of water tank, mass ratio defined as, ratio of total mass of the damper to mass of short period structure and the damping ratio of the elastomeric pad. The length of the damper is fixed to the 0.56 m. This choice is due to the practically feasible length of tank. The mass ratio of the damper is fixed as the 5%. From the previous studies on the tune damper system, it is established that the damper performance enhances with increased value of mass ratio of the damper. However, higher value of the mass ratio will alter the overall dynamic properties of structure. Hence, limited mass ratio of the 5% is chosen to avoid the adverse effect of the damper on the overall dynamic properties of the structure. The depth ratio of the water tank is selected as the 0.6. This value is chosen as the maximum depth ratio beyond which the linearized frequency of the liquid sloshing is constant with change in the depth ratio. The other two parameters of the damper (frequency ratio and damping ratio value of the elastomeric pads) are chosen by minimizing the peak acceleration response ratio with varying the frequency ratio and the damping of the elastomeric pads. The variations of the peak acceleration ratio by the varying value frequency ratio, the damping value of elastomeric pad, the damping of the structure and the mass ratio of the damper are illustrated in Fig. 3a–d, respectively. It is apparent from Fig. 3a that the peak acceleration ratio is not uniform at a particular frequency ratio and for different ground motion. This variation may be due to the fact that the different ground motions poses variability in their frequency and the PGA level. The average value (average of responses obtained from the selected ground motions) of the peak acceleration ratio is also plotted in Fig. 3a–d. From the average value of the peak acceleration ratio, it is observed that the optimum frequency ratio is 0.95 for both the structures. However, this value is not same for all the ground motions. It is also observed from Fig. 3a that the performance efficiency of the

Table 2 Details of the selected buildings and the damper parameters

S. No.	Structure parameters		Damper parameters			
	Time per. (s)	Damp. rat (%)	Mass rat (%)	FR	EDR (%)	Tank length (m)
1	0.3	1	5	0.95	5	0.56
2	0.5	1	5	0.95	5	0.56

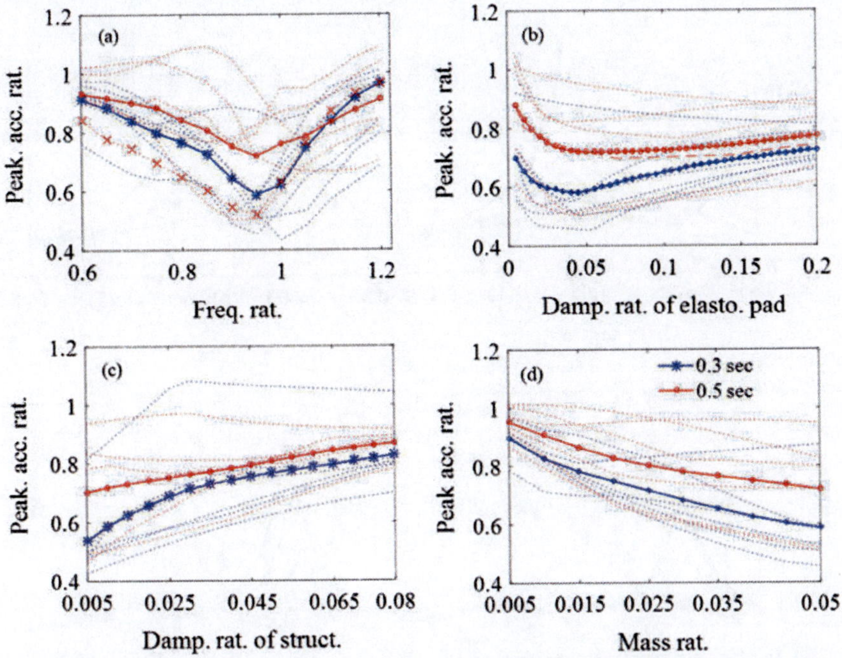

Fig. 3 Peak acceleration ratio of the structure with varying value of **a** frequency ratio, **b** damping ratio of the elastomeric pad, **c** damping of structure and **d** mass ratio of damper

damper in suppressing acceleration response is better for the structure having time period of the 0.3 s in comparison to the structure having time period 0.5 s. It is apparent from Fig. 3b that the damping ratio of the elastomeric pad corresponding to the minimum average peak acceleration ratio is around 5% for both the structures. However, this optimum value of the damping ratio of the elastomeric pad is not true for all the ground motions. It is apparent from Fig. 3c that that the performance of the damper is compromised for the higher damping ratio of the structures. This type of observation is also reported in the work of the Benerjee et al. [11]. The damper performance with varying level of the mass ratio is shown in Fig. 3d. It is seen from Fig. 3d that the efficiency of the damper is enhanced with the increased value of the mass ratio. However, 5% mass ratio is restricted in this present study. This restriction is in fact that the higher mass ratio will alter the overall dynamic properties of the structure, and also the higher mass ratio of the damper may lead to the higher damper mass, which possesses difficulty in handling and maintenance of the damper.

The optimum parameters obtained from the previous discussion are further verified by the obtaining the response function of the structures by subjecting the structure to a sine sweep function.

The response function for the structures having period of 0.3 and 0.5 s for varying value of excitation ratio which may be defined as ratio of the frequency of excitation

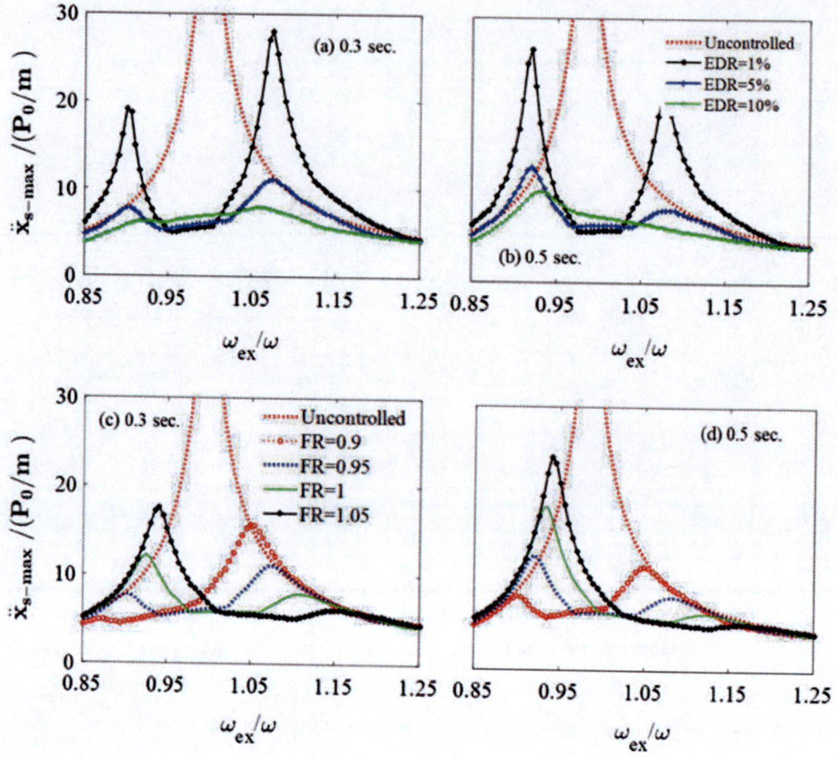

Fig. 4 Response function of the structure with varying value of damping ratio of the elastomeric pad for the structure of time period **a** 0.3 s, **b** 0.5 s with varying value of frequency ratio for the structures of time period, **c** 0.3 s, and **d** 0.5 s

of the sine sweep and the frequency of structure, and the varying damping value of the elastomeric pads is shown in Fig. 4a, b, respectively. The response function is plotted for the elastomeric pad damping ratios (EDR) values of 1, 5 and 10%. It is seen from Fig. 4a, b that the optimum value of the damping ratio of the elastomeric pad for both the structures is 10%. This optimum value deviate from the optimum values of the EDR obtained from the structure subjecting to recorded suite of ground motions. A similar plot for the frequency ratio (FR) values of the 0.9, 0.95, 1 and 1.05 is also plotted in Fig. 4c, d. It is seen from Fig. 4c, d that the optimum value of the FR is 0.95 for both the structure. This obtained value is similar to the optimum FR obtained previously. Based on the previous discussion, the adopted damper and structure parameters are listed in Table 2.

5 Performance of the Damper

The control efficiency of the damper is obtained by subjecting selected short period structures to suite of the selected ground motion. The damper parameters are chosen as discuss in Sect. 4. The response quantities that need to be obtained are the acceleration and the displacement. The obtained acceleration and displacement response of the structure excited by the selected set of the ground motions are listed in Tables 3 and 4, respectively. It is apparent from Table 3 that control efficiency in reducing peak acceleration varies from 12 to 54% for 0.3 s structure and from 5 to 49% for the 0.5 s structure whereas efficiency in reducing the rms acceleration ranges from 44 to 63% for 0.3 s and 25 to 57% for the 0.5 s structure. It is also observed that the effect of excitation level in the performance of the damper might not be established from the present investigation because for some ground motions, the controlled efficiency is higher even though PGA level of the ground motion is lower. However, generally the control efficiency of the damper should improve with the enhance excitation level since it will enhance liquid sloshing and wave breaking. But for the recorded ground motions, the effect of frequency content that of the ground motions on the response of the structure cannot be overruled.

In this study, much of the selected ground motion possess the dominate period from the 0.3 to 0.5 s. It is also observed that the tune liquid damper slightly increases the structure period. It may possible that the slight increase in the time period may bring the structure period in the range of the dominate period of the ground motion. Hence, this increase in the time period may hamper the damper performance significantly. The acceleration time histories of the 0.3 s structure subjecting to the three ground motions la03, la06 and the la42 are illustrated in Fig. 5a–c, respectively. A similar plot

Table 3 Acceleration response of the structure

S. No.	EQ. No.	Per. (s)	Peak acc. (m/s^2)			r.m.s acc. (m/s^2)		
			Uncon	Con	Effic. (%)	Uncon	Con	Effici. (%)
1	la03	0.3	8.47	7.40	12.69	1.75	0.92	47.26
		0.5	9.33	6.71	28.16	2.42	1.17	51.36
2	la06	0.3	14.37	6.53	54.53	2.66	1.07	59.94
		0.5	6.13	5.06	17.46	1.37	0.88	35.55
3	la24	0.3	10.30	5.99	41.78	2.68	1.47	44.92
		0.5	20.30	16.16	20.39	6.83	5.06	25.96
4	la26	0.3	42.25	22.29	47.23	14.41	5.33	63.03
		0.5	28.09	14.57	48.15	11.55	4.90	57.55
5	la42	0.3	18.69	9.50	49.16	4.54	1.86	58.97
		0.5	6.72	6.38	5.16	1.68	1.17	30.22
6	la46	0.3	5.44	3.15	42.03	1.19	0.55	54.26
		0.5	9.57	4.84	49.47	1.59	0.89	43.71

Table 4 Dispacement response of the structure

S. No.	EQ. No.	Per. (s)	Peak. disp. (cm)			r.m.s disp. (cm)		
			Uncon	Con	Effici (%)	Uncon	Cont	Effici. (%)
1	la03	0.3	2.74	1.71	37.38	0.42	0.26	38.93
		0.5	6.00	4.90	18.24	1.57	0.81	48.52
2	la06	0.3	3.36	1.60	52.34	0.61	0.24	60.37
		0.5	3.63	3.07	15.51	0.88	0.54	38.13
3	la24	0.3	3.06	1.95	36.12	0.73	0.53	27.88
		0.5	4.64	3.90	15.80	4.64	3.90	15.80
4	la26	0.3	9.64	5.55	42.40	3.33	1.31	60.53
		0.5	18.10	12.95	28.44	7.67	3.44	55.17
5	la42	0.3	3.74	2.12	43.41	1.01	0.39	61.43
		0.5	3.38	2.95	12.72	1.04	0.71	32.03
6	la46	0.3	1.31	1.02	22.34	0.28	0.15	48.18
		0.5	5.58	3.60	35.47	1.03	0.62	39.93

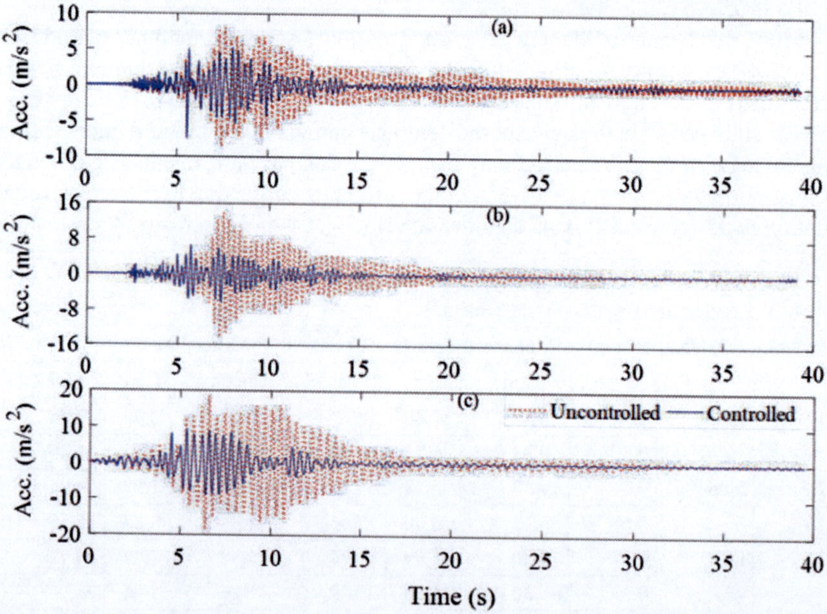

Fig. 5 Acceleration response time history of the structure when subjected to **a** la03, **b** la06 and **c** la 42 ground motions

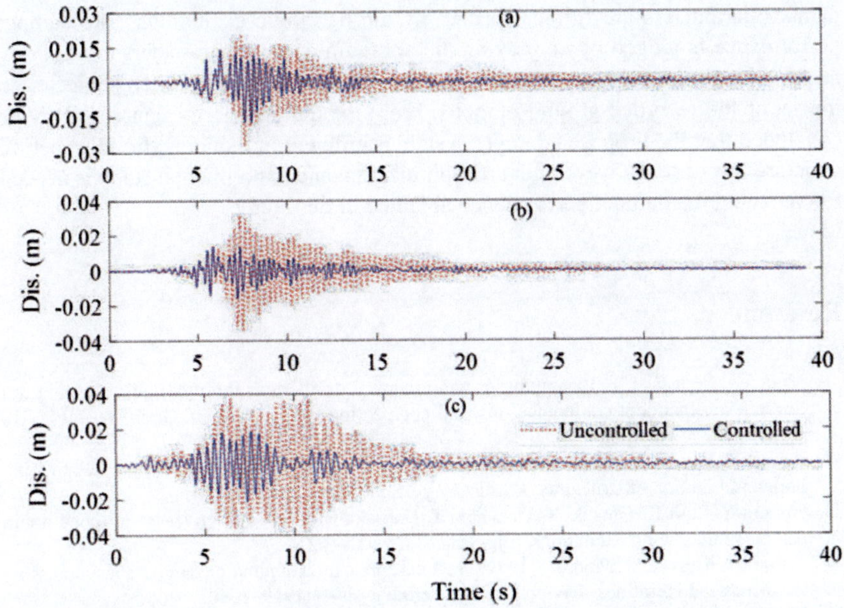

Fig. 6 Displacement response time history of the structure when subjected to **a** la03, **b** la06 and **c** la 42 ground motions

of the displacement time histories is plotted in Fig. 6a–c. It is observed from Table 4 that the control efficiency of the damper in reducing the displacement response varies significantly varies significantly among the ground motions. It may be concluded from Tables 3 and 4 that proposed damper system is efficient in reducing response of the structure. It is observed from Figs. 5 and 6 that proposed damper system is efficient in reducing acceleration and displacement response of the short period structure significantly.

The damper performance may further improved by selecting the damper parameters more accurately by employing the optimization techniques. The damper performance needs to be verified experimentally as well. However, this study is restricted for the numerical analysis only.

6 Conclusions

A new implementation in the conventional tune liquid damper is proposed by inserting the flexible elastomeric pads between the TLD and the structure. The system of the equations of the motion of the proposed structure damper system is derived. The optimum damper parameters are obtained by the parametric study of the structure response subjected to the suite of recorded ground motions. The obtained optimum damper parameters are further verified by the harmonic excitations. The damper performance is judged by employing the optimum damper parameters and solving the coupled structure damper equations of motions numerically when subjected to the set of the recorded ground motions. From the damper performance, it may be concluded that the proposed damper system is efficient in reducing the short period structure response. However, general optimization and experimental study is needed for verifying the damper performance obtained in this study.

References

1. Modi VJ, Welt F (1987) Damping of wind induced oscillations through liquid sloshing. In: Seventh international conference on wind engineering, vol 5, Aachen, Germany, July 6–10 (1987)
2. Fujino Y, Sun LM, Pacheco BM, Chaiseri P (1992) Tuned liquid dampers for suppressing horizontal motion of structures. J Struct Eng ASCE 118(10):2017–2030
3. Tamura Y, Fujii K, Ohtsuki T, Wakahara T, Kohsaka R (1995) Effectiveness of tuned liquid dampers under wind excitations. Eng Struct 17(5):609–621
4. Tamura Y, Kohsaka R, Modi VJ (1992) Practical application of nutation damper for suppressing wind-induced vibrations. J Wind Eng Ind Aerodyn 41–44:1919–1930
5. Modi VJ, Akinturk A (1998) An efficient liquid sloshing damper for vibration control. J Fluid Struct 12(8):1055–1071
6. Ruiz RO, Taflanidis AA (2001) Characterization and design of tuned liquid dampers with floating roof considering arbitrary tank cross-sections. J Sound Vib 368:36–54
7. Gardarsson S, Yeh H, Reed D (2001) Behavior of sloped bottom tuned liquid damper. J Eng Mechan ASCE 127(3):266–271
8. Sun LM, Fujino Y, Pacheco BM, Chaiseri P (1991) Modeling of tuned liquid damper (TLD). In: Proceedings of 8th international conference on wind engineering, pp 1883–1894, IAWE, London, Canada
9. Yu J, Wakahara T, Reed D (1999) A nonlinear numerical model of the tuned liquid damper. Earthquake Eng Struct Dynam 28:671–686
10. Ghosh A, Basu B (2004) Seismic vibration control of short period structures using the liquid column damper. Eng Struct 26(13):1905–1913
11. Banerji P, Samanta P (2011) Vibration control of structures using hybrid mass liquid damper. Eng Struct 33:1291–1301
12. Banerji P, Murudi M, Shah AH, Popplewall N (2000) Tuned liquid dampers for controlling earthquake response of structures. Earthquake Eng Struct Dynam 29(5):587–602
13. Nayak CB, Thakare SB (2019) Seismic performance of existing water tank after condition ranking using non-destructive testing. Int J Adv Struct Eng 11:395–410
14. Pandey DK, Sharma MK, Mishra SK (2019) A compliant tuned liquid damper for controlling seismic vibration of short period structures 132:405–428
15. Housner GW (1963) The dynamic behavior of water tanks. Bull Seismol Soc Am 53(2):381–387

Numerical Study on Vibration Screening Using Trench Filled with Sand–Crumb Rubber Mixture

Rahul Barman, Abir Sarkar, and Debjit Bhowmik

Abstract The present study investigates numerically the effectiveness of open trench (OT) barriers as well as in-filled trench (IFT) barriers using sand–crumb rubber (SCR) mixture as vibration isolators to minimize the transmission of ground vibration. The numerical analysis is carried out considering an active screening using a square trench barrier surrounding the vertically oscillated square foundation which is acting as a dynamic source, and the trench is filled with SCR mixture. In most cases, the vibratory energy which affects the nearby structure is carried by the Rayleigh wave. The vertical dynamic displacements are estimated at different distances along the ground surface to determine the amplitude reduction factor (ARF). In the present study, the numerical analysis is carried out using a three-dimensional (3D) finite element analysis (FEA). The effect of various influencing factors such as the depth of the trench, width of the trench, distance of the trench from the source, exciting frequency of the applied load and mix-proportion of sand–crumb rubber mixture on the ARF is investigated in the present study. The efficiencies of the open trench (OT) and in-filled trench (IFT) are estimated from the value of ARF.

Keywords Vibration screening · Sand–crumb rubber mixture · Amplitude reduction factor · Acoustic impedance ratio · Vibration barrier

1 Introduction

One of the major problems in the field of soil dynamics is to stop or at least minimize the transmission of ground vibration generated by high-speed machines, blasting, moving traffic load or high-speed railway. These sources of vibration may cause

---

R. Barman (✉) · A. Sarkar · D. Bhowmik
National Institute of Technology, Silchar, Assam, Silchar, Assam, India
e-mail: rb.dinhata@gmail.com

A. Sarkar
e-mail: abirsarkar2007@gmail.com

D. Bhowmik
e-mail: debjitbhowmik@gmail.com

T. G. Sitharam et al. (eds.), *Seismic Hazards and Risk*, Lecture Notes in Civil Engineering 116, https://doi.org/10.1007/978-981-15-9976-7_25

disturbance to adjacent structures and disrupting the operation of nearby sensitive equipment. Sometimes, though the ground vibration is not causing any damage to the nearby structure, the intensity of vibration may be intolerable to the nearby sensitive equipment. Therefore, it is necessary to control the transmission of the energy of vibration through the propagation of surface waves by means of vibration screening. Effective vibration screening can be achieved by proper diffraction, scattering and interception of surface waves [1, 2]. There are two types of screening—(1) active screening and (2) passive screening. In the case of active screening, the trench is provided surrounding the foundation that is the source of the disturbance. In the case of passive screening, the trench is provided surrounding the sensitive equipment far from the source of vibration. From the literature [1–11], it is cleared that the efficiency of the vibration screening depends on the depth of the trench. The depth of the trench depends on the Rayleigh wavelength (λ). The wave velocity in the medium is required to be controlled for the effective screening of the surface wave. The acoustic impedance ratio (A_r) defines as the product of the density of the material and the velocity of the Rayleigh wave, which indicates how fast waves can travel through the material. In general, the open trench (OT) is the most effective method as the air has very low A_r [1, 12–14], but at the same time, it is very unsafe as the issues of the wall instability is a big problem for the construction. Hence, an in-filled trench (IFT) is preferred as a better alternative. Crumb rubber is a waste material derived from used tires of vehicles and has very low A_r. Nowadays, as the recycling of the waste tires is a big concern, the use of sand–crumb rubber mixture as filling material in the IFT may be beneficial both from design and environmental point of view. In the present study, it is aimed to investigate the suitability and efficiency of the IFT barrier.

2 Problem Statement and Materials

2.1 Methodology

A numerical simulation is carried out to study the behavior of OT and IFT barrier. Uniform harmonic loading of constant amplitude as shown in Fig. 1 is applied on a square foundation (1 m × 1 m) on the ground surface. The applied load can be defined by the function, $P(t) = P_0 \sin(\omega t)$, where $P(t)$ is the load intensity in kN/m^2 at time t sec, P_0 = initial load intensity in kN/m^2, ω is the circular frequency ($\omega = 2\pi f$), and f is the operating frequency. The square foundation is considered to be situated on linear elastic, homogeneous sub-soil deposits. In this numerical study, the initial pressure amplitude is considered as $P_0 = 1$ kN/m^2 with an operating frequency, $f = 10$ Hz for the dynamic loading time of 10 s. The self-weight of the foundation is considered to be 10 kN/m^2 on the base of the foundation. The square trench is simulated at a distance, $l = 0.4\lambda$ from the center of the foundation, which is the source of the disturbance. The trench is filled with SCR mixture in case of IFT. The

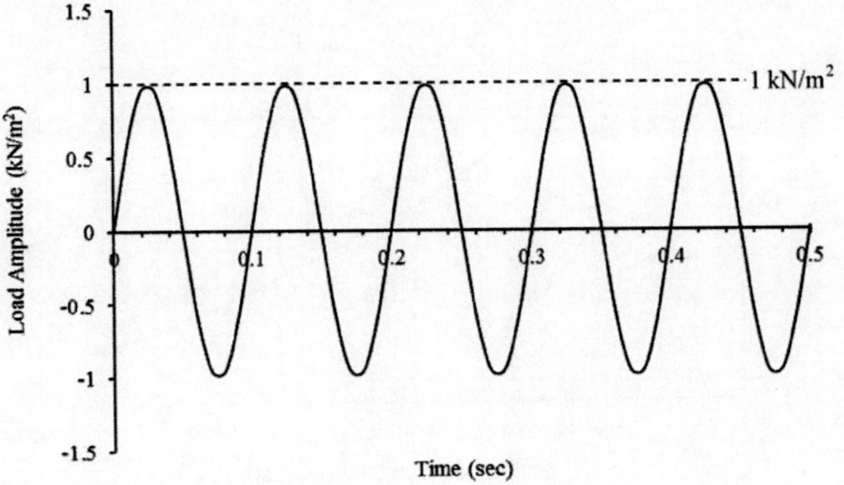

Fig. 1 Dynamic input sinusoidal loading

main focus is to determine the efficiency of vibration screening (active) for both OT and IFT in terms of amplitude reduction factor (ARF). The vertical displacement amplitudes of different pick-up points are estimated at different distances from the source of the disturbance.

Amplitude Reduction Factor (ARF) The amplitude reduction factor (ARF) is defined as the ratio of vertical amplitudes measured at a particular pick-up point considering the trench in the model to that without considering the trench [1]. The efficiency of the OT and IFT is depending on the ARF of the barrier. The efficiency of the vibration barrier increases with the decrease of ARF.

$$\text{ARF} = \frac{\text{Vertical displacement amplitude with trench}}{\text{Vertical displacement amplitude without trench}} \tag{1}$$

$$\text{The efficiency of Trench} = (1 - \text{ARF}) \times 100\% \tag{2}$$

The low A_r material has lower ARF but high efficiency for the in-filled trench of vibration screening.

2.2 Material Properties

Soil properties are determined by conducting different laboratory tests on undisturbed soil samples collected from the site situated near the New Library Building at National Institute of Technology Silchar. Soil is considered as a linear elastic, homogenous and isotropic deposit for the numerical analysis using PLAXIS 3D [15]. The groundwater

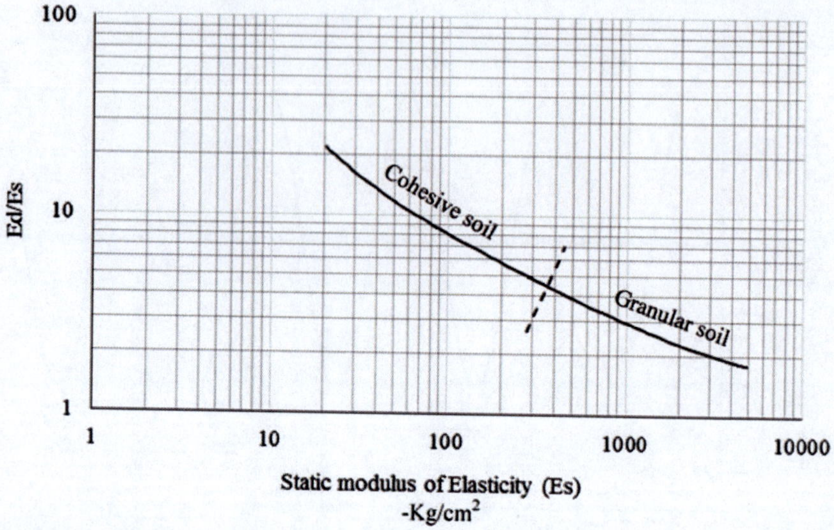

Fig. 2 Dynamic and elastic moduli of elasticity [16]

table has been considered at great depth. The dynamic properties of soil are estimated from the static Young's modulus measured from the initial slope of the stress–strain curve of the unconfined compression strength test using different well-established relationships. The dynamic Young's modulus (E_d) with respect to static Young's modulus (E_s) is determined from empirical curves proposed by Alpan [16] from Fig. 2. The primary wave velocity, shear wave velocity and Rayleigh wave velocity have been calculated using Eqs. 3, 4 and 5, respectively [17]. Various properties of the soil considered in the numerical model are given in Table 1. For the simulation of IFT barrier, the SCR mixture is also considered as a linear elastic material in PLAXIS 3D [15]. Dutta [18] carried out an experimental investigation to find out different dynamic properties of the Barak river sand and crumb rubber (SCR) mixture of different proportions. The maximum and minimum dry densities of Barak river sand are determined as 17.7 and 14.2 kN/m^3, respectively. The specific gravity is found to be 2.64 for the Barak river sand. The crumb rubbers are mixed with sand at different percentages like 10, 20, 30, 40, 50 and 70% by weight of sand for 35 and 65% relative densities (D_r).

$$V_P = \sqrt{\frac{(1-\mu)E}{(1+\mu)(1-2\mu)\rho}} \tag{3}$$

$$V_S = \sqrt{\frac{E}{2(1+\mu)\rho}} \tag{4}$$

Table 1 Material properties considered for the soil deposit

Property	Unit	Value
Dry unit weight (γ_b)	kN/m^3	21.7
Water content (w_w)	%	16.25
Specific gravity (G_s)	–	2.59
Cohesion (c_u)	kN/m^2	60.01
Friction angle (ϕ)	(°)	27.3
Primary wave velocity (V_P)	m/s	337.3
Shear wave velocity (V_S)	m/s	180.3
Rayleigh wave velocity (V_R)	m/s	167.27
Shear modulus (G)	kN/m^2	7.191 x 10^4
Dynamic Young's modulus (E_d)	kN/m^2	1.84 x 10^5
Wavelength (λ)*	m	16.7
Acoustic impedance ratio ($A_r = V_R \times \rho$)	kg m^{-2} s^{-1}	3.63 x 10^5
Rayleigh wave co-efficient (α)	–	3.11
Rayleigh wave co-efficient (β)	–	0.007958
Poisson's ratio (μ)	–	0.3

*The Rayleigh wavelength (λ) is calculated corresponding to the operating frequency, $f = 10$ Hz

$$V_R = \left(\frac{0.87 + 1.12\mu}{1 + \mu}\right) V_s \tag{5}$$

$$\lambda = \left(\frac{VR}{f}\right) \tag{6}$$

where ρ is the density of the soil.

Dutta [18] determined different dynamic properties of SCR mixtures using resonant column apparatus (RCA) at the low strain level. The A_r for different SCR mixtures at 35% D_r is lower compared to those of the SCR of 65% D_r as shown in Fig. 3, calculated from the results obtained by Dutta [18]. Hence, in this study, properties of the SCR mixtures at 35% D_r reported by Dutta [18] are used as shown in Table 2.

3 Finite Element Analysis (FEA)

Finite element analysis (FEA) is carried out for a model of one-fourth of the total geometry (to reduce the complexity of calculation) of the soil mass considered for the analysis. The discretization of the model is done using ten nodded tetrahedral elements using PLAXIS 3D [15] as shown in Fig. 5. The average element size for

Fig. 3 Variation of A_r with the percentage of crumb rubber [18]

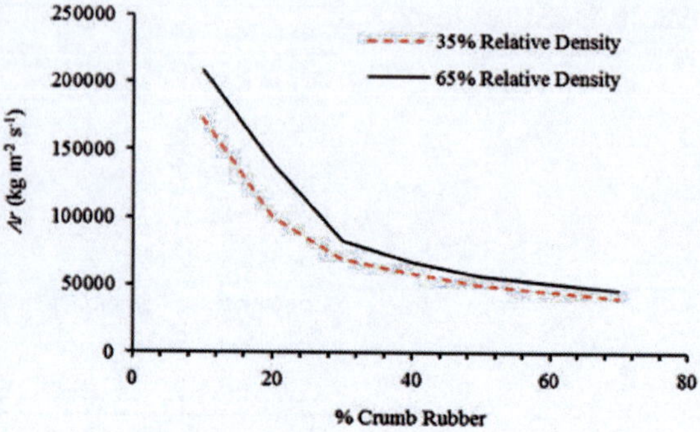

Table 2 Basic and dynamic properties of SCR mixture at 35% D_r [18]

Sample	% Crumb rubber	Unit Wt. (kN/m^3)	Young's modulus (kN/m^2)	Shear modulus (kN/m^2)	A_r (kg m^{-2} s^{-1})
SCR10	10	13.34	59,350	22826.42	172913.11
SCR20	20	11.51	22,970	8833.09	99865.10
SCR30	30	10.14	12,270	4719.23	68539.99
SCR40	40	9.05	10,020	3855.62	58530.21
SCR50	50	8.13	8248	3172.33	50325.35
SCR70	70	6.72	6448	2479.69	40440.21

this analysis is considered as per the criteria of wave propagation as proposed by Kramer [19]. The model boundaries should be situated at an adequate distance from the source of disturbance to reduce the boundary effect due to possible reflections. The size of the total soil mass considered in the present analysis as 142 m × 142 m × 118 m (height). The absorbent boundaries are adopted to avoid spurious reflections [20]. The square trench is provided at 0.4λ from the source of disturbance, and width (w) of the trench is considered as 0.05λ as shown in Fig. 4. The pick-up points are kept at a distance (x) of 0.5λ, 1.0λ, 1.5λ, 2.0λ and 3.5λ from the source. Similarly, the depth (d) of the trench is varied as 0.4λ, 0.6λ, 0.8λ and 1.0λ. The exciting frequency is considered for this analysis as 10 Hz for dynamic excitation time of 10 s. The time step has been considered in the present FEM analysis according to Eq. 7 as suggested by Valliappan and Murti et al. [21].

$$\Delta t \leq \frac{\text{Average Element Size}}{\text{Velocity of slowest propagating wave}} \tag{7}$$

The physical damping caused by the viscous effect has been taken as Rayleigh damping. The damping can occur for viscous medium, friction and plasticity of the soil deposit. In most cases, the damping ratio of the soil lies within 4–6%. In this study, the damping ratio is assumed 5% for soil deposit as it is very common for

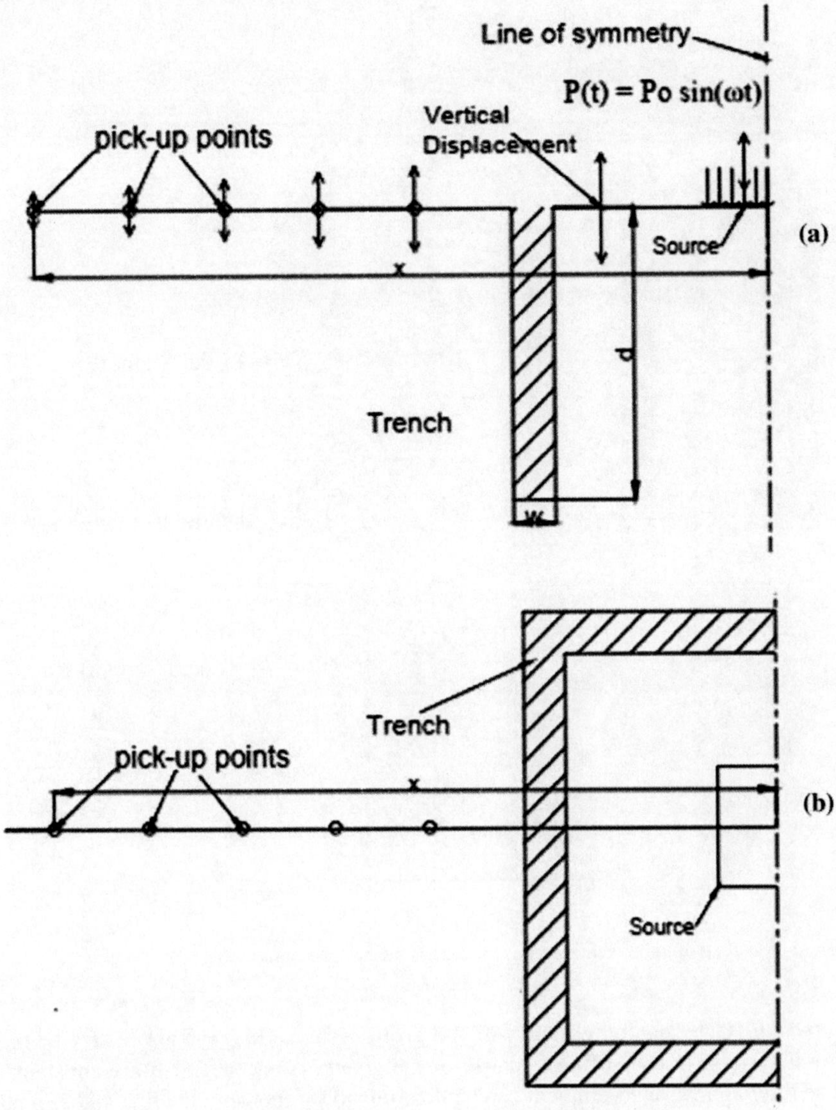

Fig. 4 Typical setup for the numerical model. **a** Sectional view and **b** Top view

vibration screening problems [3, 22]. The Rayleigh damping is proportional to the mass and stiffness of the system [23] which can be written as shown in Eq. (8).

$$[C] = \alpha[M] + \beta[K] \tag{8}$$

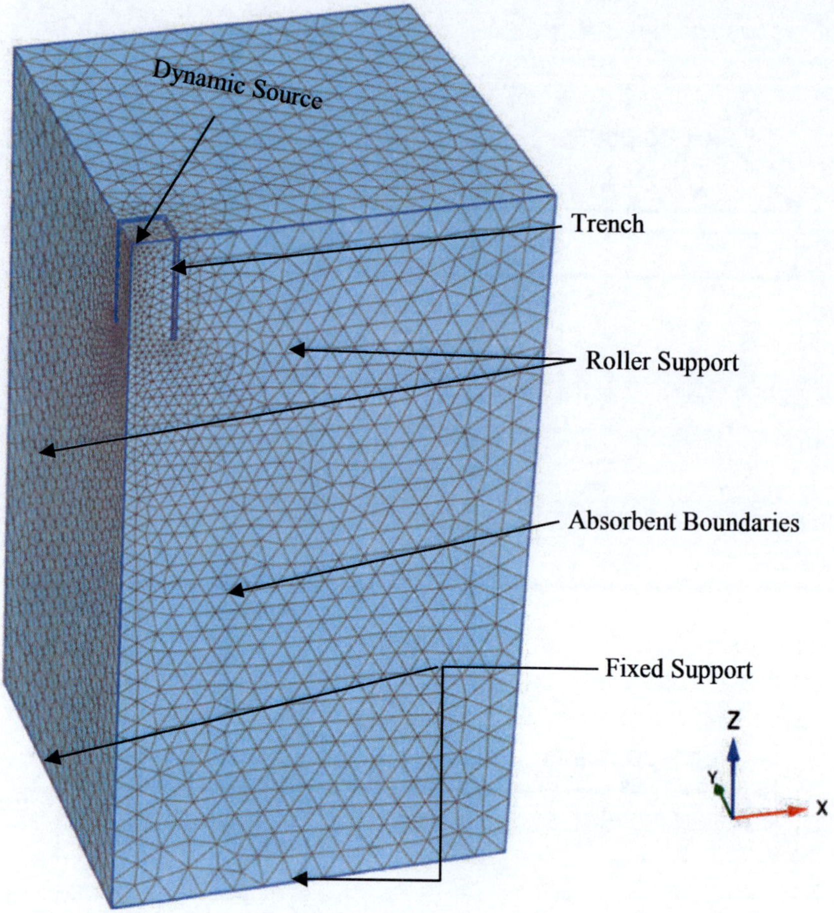

Fig. 5 Meshing of finite element analysis and boundary conditions

in which [C] = damping matrix of the given system; [M] = mass matrix of the given system; [K] = stiffness matrix of the given system; α and β are constants. The Rayleigh wave co-efficient can be determined by choosing the first and second natural frequency of the soil deposit, which is calculated in the PLAXIS property automatically. The Rayleigh wave coefficients are given in Table 1.

4 Result and Discussion

The ARF values for both OT and IFT barriers have been estimated using 3D FEA. The ground displacement has been estimated at different pick-up points. In Figs. 6

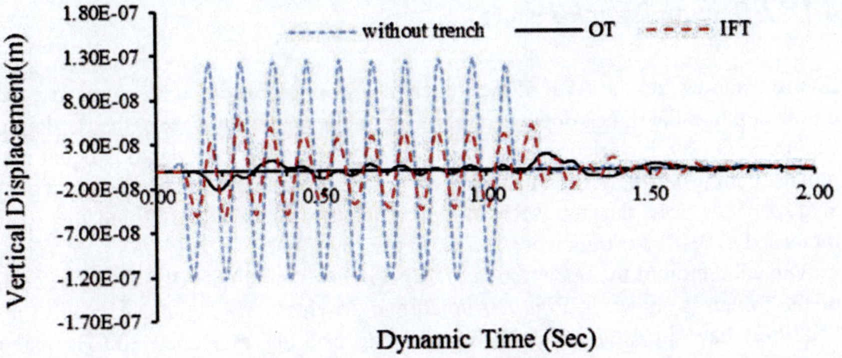

Fig. 6 Vertical displacement variation of OT and IFT at nearest pick-up point ($x = 0.5\lambda$, $f = 10$ Hz, $d = 1.0\lambda$)

and 7, the vertical displacements with respect to time have been shown comparing no trench condition, OT and IFT for the first pick-up point at a distance of 0.5λ and the far most pick-up point at 3.5λ, respectively. The vertical displacement amplitude for no trench is much greater than the OT and IFT. Figure 6 shows that the OT barrier is more effectively screening the vibration compared to the IFT barrier near the source. However, it is seen that at a greater distance the effectiveness of the OT barrier is only marginally higher than that of the IFT barrier from Fig. 7.

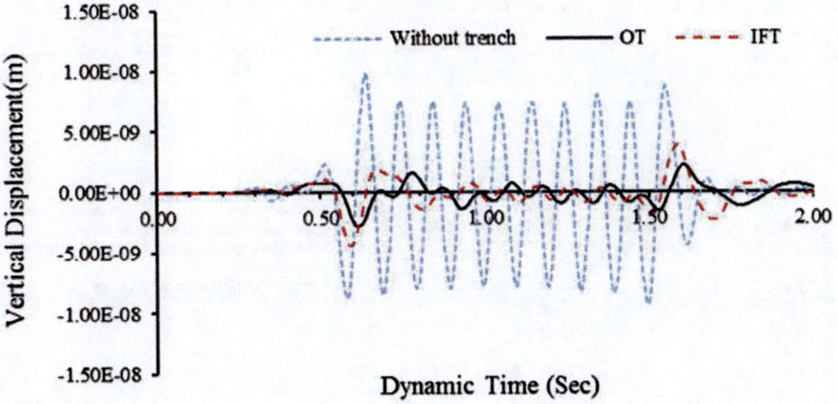

Fig. 7 Vertical displacement variation of OT and IFT at far most pick-up point ($x = 3.5\lambda$, $f = 10$ Hz, $d = 1.0\lambda$)

4.1 Effect of Trench Depth

Figure 8 shows that the ARF of the trench barrier is decreasing with increasing the trench depth as the trench depth is depending on the wavelength (λ) of the Rayleigh. The trench depth is varied from 0.4λ to 1.0λ. When the trench depth is equal and greater than 1.0λ, the ARF of the trench is found minimum. From the numerical analysis, it is seen that the ARF of trench depth of 1.0λ is 0.169 and the ARF increased to 0.481 for trench depth 0.4λ in the case of OT.

The validation of the open trench of this numerical study of ARF with the depth of the trench is shown in Fig. 9. Ahmad and Al-Hussaini [24] and Ahmad et al. [3](1996) have investigated ARF by applying exciting frequency 300 Hz using boundary element method. Majumder and Rajesh [25], Bose et al. [8], Alzawi and El-Naggar [5] have also investigated ARF of the open trench using finite element method.

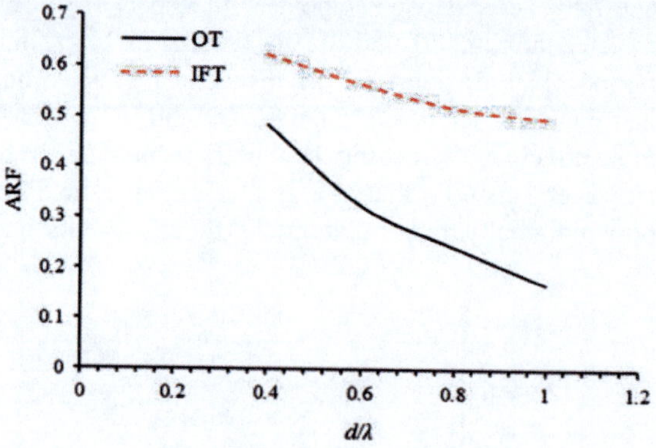

Fig. 8 ARF variation with depth of trench ($x = 0.5\lambda$, $f = 10$ Hz)

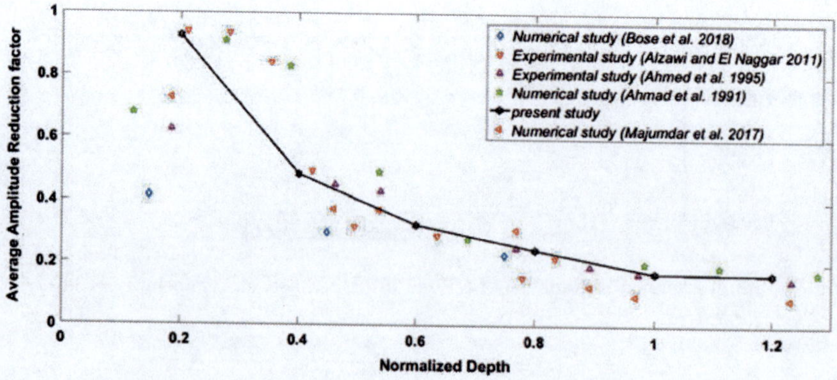

Fig. 9 Comparison of ARF variation with depth with previous studies

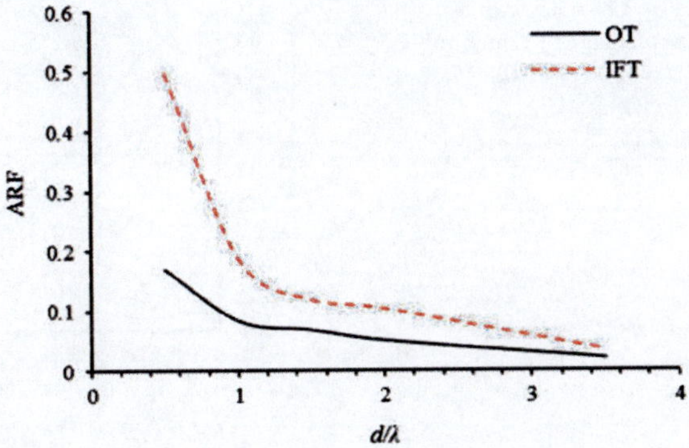

Fig. 10 ARF variation with distance of the pick-up point ($d = 1.0\lambda, f = 10$ Hz)

4.2 *Effect of the Distance of Pick-up Point*

The pick-up points are set up at a distance of 0.5λ, 1.0λ, 1.5λ, 2.0λ and 3.5λ from the source of the disturbance. The amplitude reduction factor has determined by increasing the distance of the pick-up point from the source shown in Fig. 10.

4.3 *Effect of the Exciting Frequency*

The frequency is inversely proportional to the wavelength, and so for the constant depth of the trench, the ARF decreases with the increase of exciting frequency (f) as shown in Fig. 11. It is clear that higher frequency gives higher efficiency of the trench barrier.

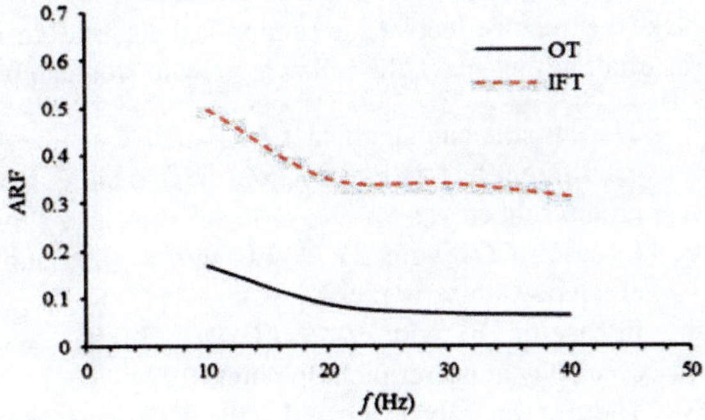

Fig. 11 ARF variation with exciting frequency ($x = 0.5\lambda$, $d = 1.0\lambda$)

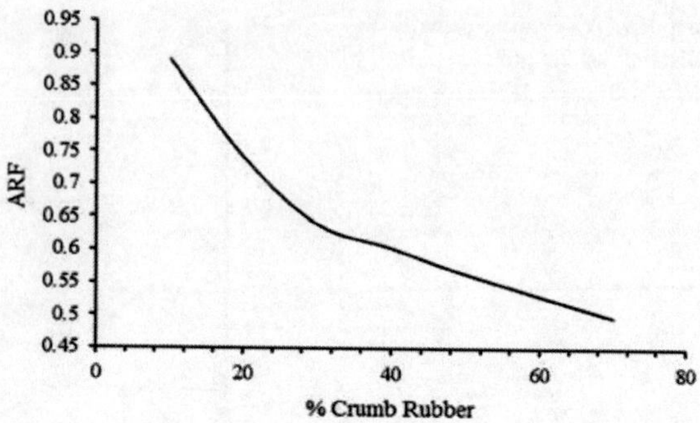

Fig. 12 ARF with percentage of crumb rubber ($x = 0.5\lambda$, $d = 1.0\lambda$, $f =$ 10 Hz)

4.4 *Effect of the Percentage of Crumb Rubber in SCR Mixture*

The ARF decreases when the percentage of crumb rubber is being increased from 10 to 70% in the SCR mixture (Fig. 12). With increasing the quantity of crumb rubber in SCR mixture, the A_r of SCR decreases and the efficiency of IFT increases.

5 Conclusion

The present FEM investigation shows that the open trench is very effective for vibration screening. But in the case of OT, it causes a major problem in that region, as well as issues, wall instability problems during monsoon. That is why the IFT trench should be provided instead of OT as a vibration isolation system. Though IFT has pretty high ARF and low efficiency with comparing that of the OT, it does not cause any major problem. The SCR mixture can be used as a screening material for IFT. To make the trench effective and economical, the SCR mixture can be used as screening material. In this study, the following conclusion can be made:

i. The suitable and significant ARF of IFT can be achieved with proper analysis of different factors like d/λ, x/λ, w/λ, exciting frequency (f) and percentage of crumb rubber.
ii. In case of OT with $d/\lambda = 1.0$, $w/\lambda = 0.05$ and $l/\lambda = 0.4$, it gives 80–90% efficiency at nearer pick-up point (0.5λ).
iii. In case of IFT with $d/\lambda = 1.0$, $w/\lambda = 0.05\lambda$ and $l/\lambda = 0.4$, it gives 50–60% efficiency at nearer pick-up point (0.5λ).
iv. The present analysis shows that the ARF decreases with the increase of frequency from 10 to 40 Hz. IFT with exciting frequency 40 Hz provides 65–70% efficiency at the nearer pick-up point (0.5λ) for $d/\lambda = 1.0$, $w/\lambda = 0.05\lambda$ and $l/\lambda = 0.4$.
v. The percentage of crumb rubber for the SCR mixture is one of the important factors of this analysis. With the increase of the crumb rubber, the efficiency of

the trench increases but the load-carrying capacity of the soil decreases. So the percentage of crumb rubber should be provided such that it must balance the efficiency of the trench and the load-carrying capacity of the sub-soil.

References

1. Woods R (1968) Screening of surface waves in soils. J Soil Mechan Found 94:951–979
2. Woods RD, Bernett NE, Sagesser R (1974) Holography, a new tool for soil dynamics. J Geotech Eng ASCE 100(11):1231–1247
3. Al-Hussaini TM, Ahmad S (1996) Active isolation of machine foundations by in-filled trench barriers. J Geotech Eng 122:288–294. https://doi.org/10.1061/(ASCE)07339410
4. Alzawi A, Hesham El Naggar M (2011) Full scale experimental study on vibration scattering using open and in-filled (GeoFoam) wave barriers. Soil Dyn Earthq Eng 31(3):306–317. https://doi.org/10.1016/j.soildyn.2010.08.010
5. Alzawi A, El-Naggar MH (2011) Full-scale experimental study on vibration scattering using open and in-filled (GeoFoam) wave barriers. Soil Dyn Earthq Eng 31:306–317. https://doi.org/10.1016/j.soildyn.2010.08.010
6. Dasgupta B, Beskos DE, Vardoulakis IG (1990) Vibration isolation using open or filled trenches, Part 2:3-D homogeneous soil. Comput Mech 6(2):129–142
7. Majumder M, Ghosh P (2016) Numerical study on a novel vibration screening technique using intermittent Geofoam. GeoChina 2016, 25–27th July 2016, Shandong, China
8. Bose T et al (2018) Efficiency of open and infill trenches in mitigating ground-borne vibrations. J Geotech Geoenviron Eng 144(8):1–11. https://doi.org/10.1061/(ASCE)GT.1943-5606.0001915
9. Beskos DE, Dasgupta B, Vardoulakis IG (1986) Vibration isolation using open or filled trenches. Comput Mech 1(1):43–63. https://doi.org/10.1007/BF00298637
10. Saikia A, Das UK (2014) Analysis and design of open tench barriers in screening steady-state surface vibrations. Earthq Eng Eng Vib 13(3):545–554
11. Banerjee, P. K., Ahmad, S. & Chen, K. Advanced application of bem to wave barriers in multi-layered three-dimensional soil media. Earthq. Eng. Struct. Dyn. 16, 1041–1060 (1988).
12. Babu GLS, Srivastava A, Rao KSN, Venkatesha S (2011) Analysis and design of vibration isolation system using open trenches. Int J Geomech 11(5):364–369. https://doi.org/10.1061/(ASCE)GM.1943-5622.0000103
13. Celebi E, Firat S, Beyhan G, Cankaya I, Vural I, Kirtel O (2009) Field experiments on wave propagation and vibration isolation using wave barriers. Soil Dyn Earthq Eng 29(5):824–833. https://doi.org/10.1016/j.soildyn.2008.08.007
14. Segol G, Lee PCY, Abel JR (1978) Amplitude reduction of surface waves by trenches. J Eng Mech Div ASCE 104(3):621–641
15. PLAXIS 3D (2018) Reference manual
16. Alpan I (1970) The geotechnical properties of soils. Earth Sci Rev 6(1):5–49. https://doi.org/10.1016/0012-8252(70)90001-2
17. Brinkgreve RBJ (2006) Plaxis: finite element code for soil and rock analyses: 2D-Version 8.5: (User's Guide), Balkema, Delft, The Netherlands
18. Dutta D (2016) Study of dynamic behaviour of sand and sand-crumb rubber mixture using resonant column apparatus. M.Tech. Thesis, Department of Civil Engineering, National Institute of Technology, Silchar
19. Kramer SL (1996) Geotechnical earthquake engineering. Prentice-Hall International Series in Civil Engineering and Engineering Mechanics, Prentice-Hall, New Jersey
20. Lysmer J, Kuhlmeyer RL (1969) Finite dynamic model for infinite media. J Eng Mech Div Proc ASCE 95(4):859–877

21. Valliappan HS, Murti V (1984) Finite element constraints in the analysis of wave propagation problems. UNICIV Rep. No. R-218; School of Civil Engineering, University of New South Wales, Australia
22. Alzawi A, El-Naggar MH (2009) Vibration scattering using geofoam material as vibration wave barriers. In: GeoHalifax, 52nd Canadian geotechnical conference
23. Chowdhury I, Dasgupta SP, Computation of rayleigh damping coefficients for large systems
24. Ahmad S, Al-Hussaini TM (1991) Simplified design for vibration screening by open and in-filled trenches. J Geotech Eng ASCE 117(1):67–68. https://doi.org/10.1061/(ASCE)0733-9410
25. Majumder M, Ghosh P, Rajesh S (2017) An innovative vibration barrier by intermittent geofoam-a numerical study. Geomech Eng

A Study on Trenching Techniques for Vibration Isolation: An Overview

Pankaj Bariker and **Sreevalsa Kolathayar**

Abstract This paper presents a detailed review of past research studies on trenching techniques for vibration isolation. With rapid urbanization and fast-growing infrastructures in developing countries like India, it is very likely for many structures subject to vibrations from earthquakes, traffic movements, or machine vibrations. Though the trench technique for vibration screening is an established technique, there has been a lot of advancements in vibration screening in recent years. This paper presents an overview of the developments in trenching techniques for vibration screening from its inception to the latest advancements. The studies have been categorized into laboratory element/model tests, field tests, and numerical analyses. The review of studies highlights that the open trenches perform better isolation for shallow depths whereas the infilled trenches perform better than open trenches when the normalized depth of trench is deeper.

Keywords Urbanization · Machine vibrations · Screening · Isolation

1 Introduction

In the present day, India is developing faster with a focus on infrastructure development leading to the emergence of Smart cities, Amruth cities and Express-ways, Bullet trains and Industries with modern equipment. These projects will generate vibrations either by vehicular movement or by machine operations that later affect the adjacent structures. This modernization will lead to an increase in the land value of the place that will not allow using more places for tackling the vibration effect. Hence, by utilizing the trench techniques one can isolate adjacent structures both in active and passive ways.

Adoption of open trench techniques will lead to the cutting down of the medium through which the vibrations are traveled, whereas the infilled trench techniques lead to the vibration reduction by their ability to reduce shear wave velocity. It will also

P. Bariker (✉) · S. Kolathayar
National Institute of Technology Karnataka, Surathkal, India
e-mail: pankajbariker@gmail.com

T. G. Sitharam et al. (eds.), *Seismic Hazards and Risk*, Lecture Notes in Civil Engineering 116, https://doi.org/10.1007/978-981-15-9976-7_26

encourage deeper trench by providing suitable passive resistance thereby avoiding the caving in of the side walls. Trenching techniques for vibration screening can be adopted in terms of open and infilled trenches as investigated by Beskos et al. [4] modeling the open and infilled trenches under plane strain conditions by BEM, for the motion of the rigid foundations and surface blasting.

Woods [25] presented the concepts of the trenching techniques attenuating the surface waves through which most of the wave energy is transferred to soil medium by making a preliminary attempt in this field. Bo et al. [5] performed a comprehensive finite element analysis showing the significance of thin trenches as wave energy barriers, i.e., attenuating the transmission of vibration to below subsoil. Coulier et al. [7] studied the effectiveness of the stiff wave barrier by using jet grouting to inject cement slurry into the soil below, increasing the effectiveness of the trench in isolation. The screening performance of multiple trenches was observed to be better for isolation than a single trench for vibrations induced by trains as concluded by Younesian and Sadri [26] and Estroff and Adam [10]. The deficiency of adopting the most generic method to select different trench configurations based on the vibration input and output requirements was first dealt by Hojjat [13] developing the technique that used dual infilled trenches along with artificial neural network.

2 Experimental Studies on Open Trenches for Vibration Isolation

Saibaba Reddy et al. [21] dealt with controlling vibration at source by conducting a series of block vibration tests with active isolation trench by adopting geophones to record the data of ground motions both in lateral and longitudinal directions as shown in Fig. 1. The parameters like embankment depth and dimensions of the trench were studied at different operating frequencies. It was concluded that the vertical vibrations of the block significantly reduces by 50% when embedded within the ground and also the transfer of the lateral vibrations to the ground was reduced by the provision of the open trench barriers.

The infilling materials being rice husk, quarry dust, and natural soil which are readily available as byproducts. They concluded that infilled materials in the trench will reduce the vertical motion of the block because of side friction between block and infilling material. Out of these, natural soil showed a significant reduction in vertical motion of the block and transmission of the horizontal vibrations is reduced by about 58% and more when trenches are filled with infilling materials out of which the rice husk gave the maximum reduction because it has almost nil shear wave velocity.

Ashref Alzawi and Naggar [3] dealt with the ground vibrations generated by machines which had caused unfavorable effects on the adjacent structures from annoyance to structural damage. A full-scale experimental study was performed to

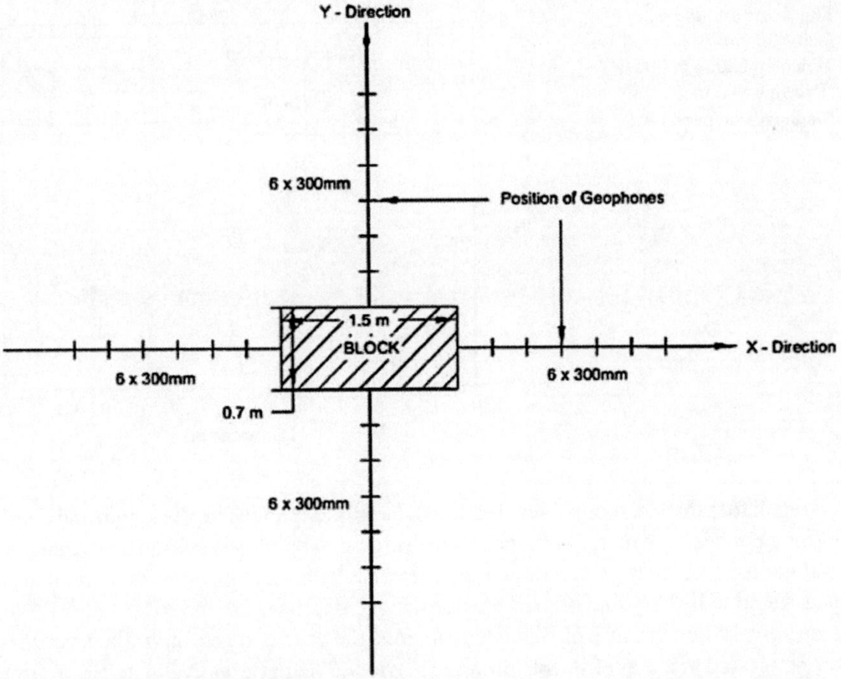

Fig. 1 Plan showing location of ground vibrations observations (Drawn by Author after Saibaba Reddy et al. [21])

check the performance of open trenches and also to study the influence of the geometry of the trench wall and location of the vibrating source from the point of the study. The effectiveness of the wave barriers is studied based on the reduction in the soil particle velocities by changing the operating frequency and location of the vibrating source. From experimental data, it is concluded that the open trenches reduce the transmitted vibration to about half the original and more. The barriers are generally found to be more effective when the depth of trench (D) $D > 0.6$ They worked on the vibration screening by geofoam-infilled trenches in comparison with open trenches along with the parametric study over the influence of wall geometry and distance of the source of vibration. It is concluded that the geofoam barrier was found out to be an alternative for vibration screening which reduces by about 68% and more. It is found that the effectiveness of the deep trenches decreases when provided with open trenches and remains unchanged with the geofoam barrier. Even, same observations were found in the finite element package ABAQUS and validated.

Ulgen and Toygar [24] highlighted the attenuation of ground-borne vibration and its effects on frequency, distance of the source from the trench, layering of the strata, and irregular geometry of soil profile by conducting a series of full-scale experiments. The variation of shear wave velocity at the site, A 0.8 m wide and 4.5 m depth trench was made, studied for open case thereafter water, and geofoam as the infilling material

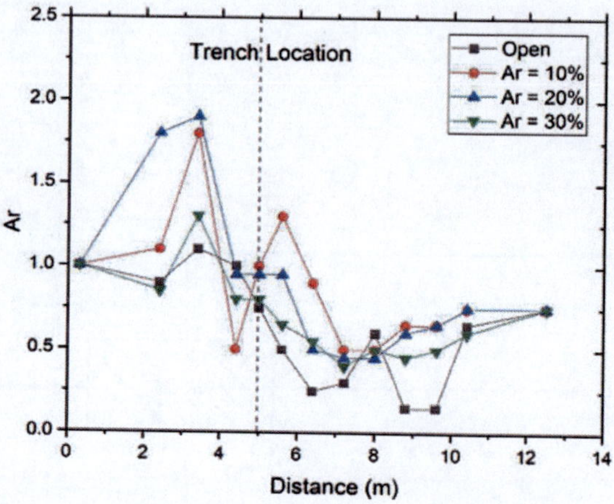

Fig. 2 Amplitude reduction factor for source at 5 m from trench operating at 30 Hz (Drawn by Author after Madhavisefat et al. [16])

in trench for which response were recorded. Results depicted that the length and depth of the trenches have major influence on the screening effectiveness and open trenches and geofoam filled trenches shown the better isolation and the amplitude reduction was found to be 67% and higher, the normalized depth was found to be 1 and 1.5 m, respectively, and because of stability problems, the open and geofoam-filled trenches are suggested to use as efficient soft wave barriers. The same was validated even from numerical and previous literature.

Madhavisefat et al. [16] presented the effectiveness of open and infilled trenches for vibration isolation by conducting a series of full-scale filed experimental tests. SRM is the sand–rubber mixture used as the infilling material which is lightweight, highly energy absorbing, and ecofriendly material. The vibrations created at the source are picked up by a series of geophones situated after trenches at regular intervals of 2.1 and 1.0 m before trench and at every 0.8 m after trench, where trench is situated at 5.0 m from the source. The source was operating at different frequencies ranging from 30 to 90 Hz. The results confirmed that the effectiveness of the SRM was found to be close to open trenches leading to the reduction of attenuation by about 60–70% by increasing the depth of trenches and is as shown in Fig. 2.

3 FEM and FDM Studies on Trenching Isolation Technique

Ahmad et al. [2] dealt with the vibrations caused by machine foundations that are detrimental to adjacent structures and nearby sensitive instruments. A three-dimensional algorithm study was done on the parametric study and the effectiveness of open trench as screening mediums with two geometrical foundation shapes. And it was highlighted that the amplitude reduction ratio not only depends on the depth of the trench but also on the size of the footing and the distance of the trench. The width of the trench has lesser influence near the source but vanishes as the trench movies away from the source. The normalized optimum depth (*D*) depends on the source dimension (B_f) and normalized length of barrier from source (*L*).

Adam [1] presented the theoretical investigation over the reduction of vibrations induced by railway movement on the building, in which he had modified the previously used boundary element and finite element algorithm and applied for the direct time domain. It was concluded that the provision of open trenches will reduce the structural vibrations due to railway passage, the open trenches are said to perform better for shallow depth trenches. Increasing the distance between the open trench and the building decreases the reduction as a result of scattered horizontal motions and accelerations. This behavior is less pronounced in the case of using the infilled trench. The mass density and poisons ratio of the infilling material has its effect on the vibration isolation.

Sivakumar Babu et al. [23] conducted a field vibration test at the site and validated with the finite difference software FLAC to suggest an effective screening system for vibration. The various parametric studies were performed by providing gravel-bed with rubber pad and a cutoff trench for vibration screening. It is noted that just by proving the rubber pad and gravel pit, the vibrations are controlled adequately and the cutoff trench is only needed to bring down the vibration level below the tolerance limits and the same was validated with the experimental test.

Ju and Li [15] performed a 3D finite element analysis to study the effect of open trenches in vibration screening filled with different water levels that have zero shear wave velocity. It was highlighted that the isolation contribution of the open trenches filled with water is more than open trenches as shown in Fig. 3 because the incident wave from the water and diffracted wave from the bottom of the trench have different wave phases. Hence, a trench can reduce wave transmission in two important directions leading to effective isolation and also it is concluded that the mat foundations significantly reduce the horizontal vibration.

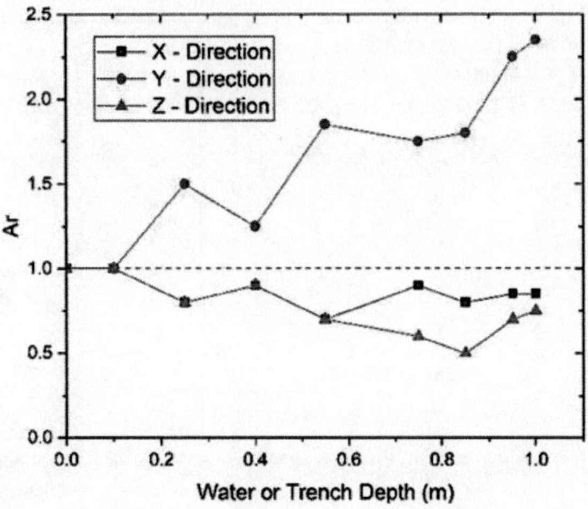

Fig. 3 Illustration of the model with different domain axis and comparison ratio changing in different directions (Drawn by Author after Ju and Li [15])

Das and Kumar [8] made an attempt to study the effect of rectangular open trenches in isolating the vibrations numerically by finite element package PLAXIS, in which the key geometrical features like width, depth, and distance from the vibrating source in terms of the normalized depth are studied. The study is conducted for varying depths of the trench and machine at operating frequencies and different trench dimensions. It is concluded that the vibration screening efficiency is primarily governed by the normalized depth of barrier and the width of the trench is influenced very less and this study did not show the effect of the width of the trench in isolating vibration. The simplified design formulae and rage of its applicability are shown in Table 1.

Duzgun [9] made an attempt to do parametric studies to assess the dynamic properties of soil and to check the effectiveness of trench in isolating vibrations on soil for 2D problems by coupling finite and infinite elements. It is amended that the provision of open or infilled trenches will reduce structural vibrations. Essentially open trenches perform better than infilled ones for vibration isolation as shown in Fig. 4. Isolation is directly dependent on the depth of the trench and location of the trench from the source vibration creating. And one can use the coupling of finite and infinite elements to solve isolation problems.

Table 1 Simplified design formulae and range of applicability (Tabulated by Author after Das and Kumar [8])

CASE	Vibration component	Trench location (m)	Range of W (m)	Formulae recommended
Active	Vertical	1	Lesser than 0.6	$0.28D^{-0.44}$
Passive	Vertical	5	Lesser than 0.6	$0.18D^{-0.95}$
Active	Horizontal	1	Lesser than 0.4	$0.43D^{-0.59}$
Passive	Horizontal	5	Lesser than 0.4	$0.37D^{-0.71}$

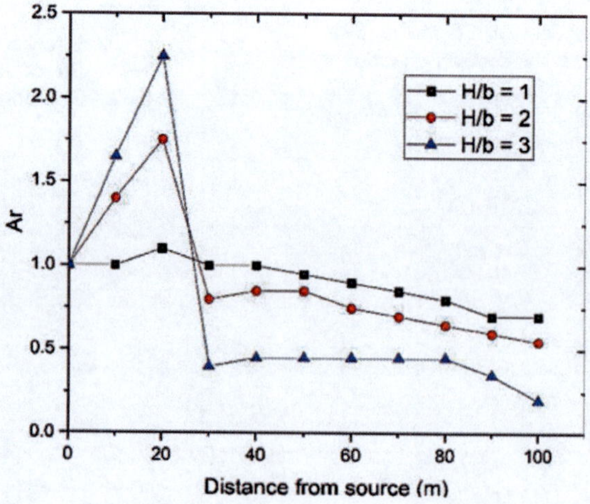

Fig. 4 Effect of H/b ratio on isolation (Drawn by Author after Duzgun [9])

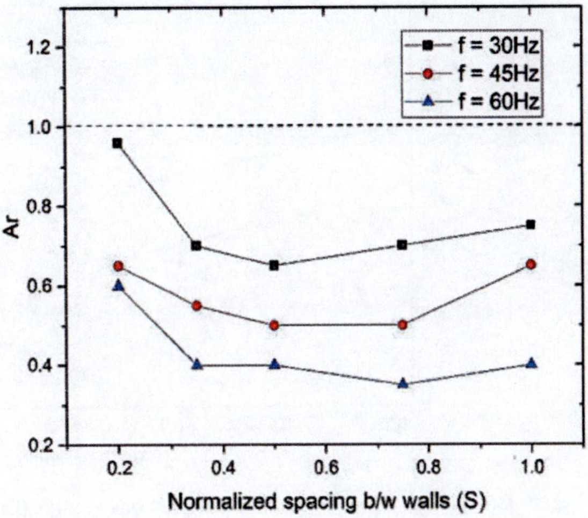

Fig. 5 Average amplitude reduction ratio with normalized spacing between walls at the different operating frequencies (Drawn by Author after Bose et al. [6])

Bose et al. [6] proposed and presented mitigation of ground-borne vibrations by 2D and 3D finite element models using PLAXIS for investigating the isolation effectiveness for trenches (i.e., open and infilled). Numerical studies shown that for open trenches normalized depth will influence the effectiveness of isolation and is very shallow, whereas for infilled trenches, it was observed that the low-density materials perform well and their performance is highly sensitive in nature, finally the study was extended to study the effectiveness of geofoam-filled trenches that are found to be better performing for vibrations generated by train movements as shown in Fig. 5.

Feng et al. [11] tried to device the findings of buried trenches in vibration screening by the 2D frequency-domain finite element method as an axisymmetric model in COMSOL, in which he stated the influence of geotechnical parameters normalized by Rayleigh wavelength (L_R) and estimated screening capacity of the buried trench by simplified design formula. Program outcomes depicted that the buried depth and trench depth influence the screening effect whereas that of the width of the trench is lesser. The optimum location of the trench from vibrating source and buried depth was found to be $5L_R$ and $0.1L_R$, respectively. The top space between buried trench and ground will contribute to some diffracted vibrations to some extent.

Jayawardana et al. [14] made an attempt in achieving higher vibration isolation generally for the cases like passage of trains by adopting dual trenching techniques which thereby reducing the unrealistic depths of trench to be adopted when provided with single trench and the results are also cross-checked with that of finite element model in ABAQUS for present work and past literature, which depicted the close aggregation lead to validation of the work. Since consideration of many geotechnical parameters that are controlling the effectiveness of the isolation in analyzing, an attempt of developing artificial neural network (ANN) is done for this case as shown

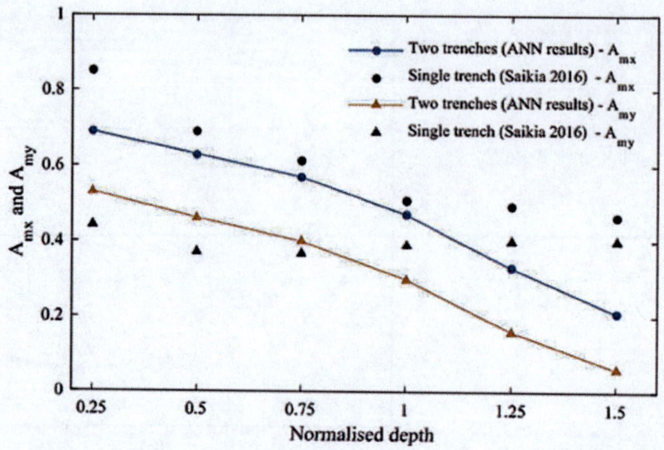

Fig. 6 Comparison of ANN results with previous studies (Drawn by Author after Jayawardana et al. [14])

in Fig. 6, which can be further extended to a wider range with further simulations to develop a generic prediction model.

4 Recent Developments

Disturbances get transmitted to the immediate surrounding body, i.e., soil medium as vibration waves called surface waves and body waves. Since surface waves carry almost all the wave energy, attenuating the same is the primary requirement. Adoption of the trench in the path of the wave transmission will cause isolation or screening of the vibration [19].

Vibration screening for buildings fitted with sensitive instruments is always difficult. Hence, a case study is conducted to study the effectiveness of vibration isolation by trenching techniques, and finally, after noting and analyzing the vibrations of internal and external sources, it was concluded that the trenching techniques reduce the amplitudes of vibrations to the permissible extent [22].

Trench barriers can be used for decay of the surface waves efficiently in perspective of complex band structures [18], on conducting a numerical model using finite elements concepts on a physical model and formulations, that validated through good correlations with previous studies and also conducted the parametric study to evaluate the effect of different geometrical parameters and damping due to infilling material, which is confirmed from both domains, i.e., frequency and time.

Structural vibrations caused by the passage of traffic will disturb the comfort of people in nearby buildings. Zeng et al. [27] Proposed the isolation of the backfill zone of the foundation pit (IBF) method to reduce the building vibration using a new isolation material that has higher axial stiffness and lower shear stiffness (HALS). Two four-story test buildings were constructed and tests are conducted onsite, which reduced the vibration by about 31–53%, indicating better screening performance.

A theoretical study on the efficiency of ground vibrations screening by a row of piles as passive barriers in a 3D analysis emerged as an equivalent method to that with an infilled trench, by increasing the pile size or by decreasing the pile spacing. Hence, the principle of modeling of the barrier for vibration transmission by the row of piles by an infilled trench is discussed, that highlighted the effectiveness of the trenching techniques in screening by Gao et al. [12].

The most generic techniques need to be employed for the problems like trenching methods for attenuating vibrations which have a lot of influencing factors that cannot be used while analyzing manually, hence development of ANN models will lead to the next level of base in this regard as done by Jayawardana et al. [14] and Hojjat [13]. Attempts for using fuzzy logic to study the effectiveness of the trenches in vibration isolation with a set of methods to characterize and quantify different variable parameters that the effectiveness of trench isolation will depend upon if, it is taken care of then, will help in improving the efficiency of the trenching techniques for attenuation problems with the concepts of Ross [20].

5 Conclusions

This paper reviewed sufficient literature on trenching techniques (i.e., open and infilled trenches) in vibration screening or isolation. From this, it is highlighted that the open trenches perform better isolation for shallow depths whereas the infilled trenches perform better than open trenches when normalized depth of trench is deeper, that too the infilling materials with lesser shear wave velocity and lesser friction materials contribute more in vibration isolation like geofoam, the width of trench does not influence the vibration screening much. But, when the vibration sources are machine foundations, minimum embedment is to be provided for machine foundation below ground and placing the foundation block over a gravel bed, rubber pad, and sand–rubber mixture (SRM) leads to lesser vibration transmission to the surrounding soil.

Many developed finite element software packages listed in the above review had resulted in closer behavior to the full-scale experimental results, hence before adopting any infilling materials for isolation, its behavior and effectiveness in screening the vibration can be checked by modeling in the available software packages, so that the better performance is achieved. Dual or multi-infilled trench techniques can be adopted when there is a chance of larger vibration transmission. Artificial neural network (ANN) can be developed for better performance and knowledge aggregation when planned to adopt multiple trenching techniques.

References

1. Adam MA (2002) Reduction of train induced building vibrations using open and infilled trenches. In: International conference on civil and architectural engineering. ICCAE, Cairo, Egypt
2. Ahmad S, Al-Hussaini TM, Fishman KL (1996) Investigation on active isolation of machine foundations by open trenches. J Geotech Eng ASCE 454–461
3. Ashref Alzawi M, Naggar EI (2011) Full scale Experimental study on vibration scattering using open and in-filled. Soil Dyn Earthq Eng 306–317
4. Beskos DE, Dasgupta B, Vardoulakis IG (1986) Vibration isolation using open or filled trenches. Comput Mech 1:11–24
5. Bo Q, Ali L, Irini DM (2014) Numerical study of wave barrier and its optimization design. Finite Element Anal Des 84:1–13
6. Bose T, Choudary D, Sprengel J, Ziegler M (2018) Efficiency of open and infill trenches in mitigating ground-borne vibrations. J Geotech Geoenviron Eng ASCE 04018048-1 to 04018048-11
7. Coulier P, Cuellar V, Degrande G, Lombaert G (2015) Experimental and numerical evaluation of the effectiveness of a stiff wave barrier in the Soil. Soil Dyn Earthq Eng 77:238–253
8. Das AS, Kumar U (2014) Analysis and design of open trench barriers in screening. Earthq Eng Eng Vib 13(3):545–554
9. Duzgun OA (2015) Efficiency of trenches on vibration isolation under time dependent loads. Period Polytech Civil Eng 133–142
10. Estroff V, Adam M (2005) Reduction of train-induced building vibrations by using open and infilled trenches. Comput Struct 83:11–24
11. Feng S, Li J, Zhang X, Chen Z, Zheng Q, Zhang D (2019) Numerical analysis of buried trench in screening surface vibration. Soil Dyn Earthq Eng 126:105822–1 to 105822-6
12. Gao G-Y, Jian S, Jun Y (2015) Theoretical analysis of a row of piles as passive barriers and an equivalent in-filled trench model. Springer J 22:1919–1928
13. Hojjat A (2001) Neural networks in civil engineering. Comput Aid Civil Infrastruct Eng 16:126–142
14. Jayawardana P, Thambiratnam DP, Perera N, Chan T (2019) Dual in-filled trenches for vibration mitigation and their predictions using. Soil Dyn Earthq Eng 122:107–115
15. Ju SH, Li HC (2011) 3D analyses of open trench barriers filled with water. J Geotech Geoenviron Eng ASCE 1114–1120
16. Madhavisefat E, Salehzadeh H, Heshmati AA (2017) Full-scale experimental study on screening effectiveness of SRM-filled trench barriers. Geotechnique 201–214
17. Orehov VV, Nagendra Moghaniou R, Negahdar Hassan (2012) Investigation effects of trench barrier on the reducing energy of surface waves in soils. 15 WCEE LISBOA 2012:106–116
18. Pu X, Shi Z (2020) Broadband surface wave attenuation in periodic trench barriers. J Sound Vibr 468. Article no. 115130
19. Pu X, Shi Z, Xiang H (2018) Feasibility of ambient vibration screening by periodic geofoam-filled trenches. Soil Dyn Earthq Eng 104:228–235
20. Ross TJ (2005) Fuzzy logic with engineering applications. Wiley, Hoboken
21. Saibaba Reddy E, Rama Sastri K, Abishekar Paul M (2001) An experimental investigation on trench isolation. In: International conferences on recent advances in geotechnical earthquake engineering and soil dynamics. Scholars' Mine 2.41.1–2.41.6
22. Sitharam TG, Sebastian Resmi, Fazil Febin (2018) Vibration isolation of buildings housed with sensitive equipment using open trenches—case study and numerical simulations. Soil Dyn Earthq Eng 115:344–351
23. Sivakumar Babu GL, Srivastava A, Nanjunda Rao KS, Venkatesha S (2011) Analysis and design of vibration isolation system using open trenches. Int J Geomech ASCE 15:364–369
24. Ulgen D, Toygar O (2015) Screening Effectiveness of open and infilled wave barriers: a full-scale Experimental study. Construct Build Mater 12–20

25. Woods RD (1968) Screening of surface waves in soils. University of Michigan, Ann Arbor, IP-804
26. Younesian D, Sadri M (2014) Performance analysis of multiple trenches in train-induced wave mitigation. J Low Freq Noise Vibr Active Control 33:47–63
27. Zeng Y, Pan P, Zhang D, Yang J (2020) Experimental study of isolation in the backfill zone of the foundation pit (IBF) method to reduce ground-borne vibration in buildings. Eng Struct 202. Article no. 109740

Blast-Induced Ground Motion in Geo-media

Harshada Sharma, Sharang Totekar, A. K. Darpe, S. P. Singh, and Vasant Matsagar

Abstract The increase in the activities of accidental as well as terrorist-initiated blast explosion cause cataclysmic effects on occupants and structures. The underground, as well as above-ground explosion, results in several hazards; one such hazard is a blast-induced ground motion (BIGM). It is important to predict the intensity of vibration due to the BIGM to safeguard the personnel and structures. In an event of a blast, the ground vibrations are measured based on parameters such as peak particle velocity (PPV), peak ground acceleration (PGA), and principle frequency content (PF). These parameters quantify the ground motion characteristic depending on intensity, arrival time, and wave duration; whereas, the ground vibration is usually expressed in terms of the PPV. In the case of underground blast, the prediction of the BIGM due to large-scale explosion depends upon the charge weight, radial distance, depth of burst (DoB), site constants, and type of propagating medium. The prediction of the PPV includes various empirical relations, code-based equations, predictor equation using regression analysis based on experimental data, artificial neural network (ANN) models, and finite element (FE) based numerical simulation methods. The differences in the estimated PPV values may occur due to discontinuities and the presence of water table in the geo-medium. Herein, the PPV due to an underground explosion is estimated using the abovementioned approaches. Also, the FE-based numerical simulations are carried out to investigate the influence of different geo-media like granite, limestone, sandstone, and sandy loam for prediction of the BIGM parameters such as PPV, PGA, and PF. The simulation-based estimate of the PPV for different geo-media is compared with the abovementioned approaches for various scenarios. The numerical simulations confirmed the ground shock attenuation with an increase

---

H. Sharma · S. Totekar
School of Interdisciplinary Research (SIRe), Indian Institute of Technology (IIT) Delhi, New Delhi 110016, India

A. K. Darpe · S. P. Singh
Department of Mechanical Engineering, Indian Institute of Technology (IIT) Delhi, New Delhi 110016, India

V. Matsagar (✉)
Department of Civil Engineering, Indian Institute of Technology (IIT) Delhi, New Delhi 110016, India
e-mail: matsagar@civil.iitd.ac.in

T. G. Sitharam et al. (eds.), *Seismic Hazards and Risk*, Lecture Notes in Civil Engineering 116, https://doi.org/10.1007/978-981-15-9976-7_27

in the radial distance as the PPV shows reduction in the amplitude with an exponential decay. The charge shape and DoB also show significant influence on the PPV estimation. The PGA and PF estimates based on the empirical and code-based equations show good resemblance with those predicted using the numerical simulations. The comparison of various prediction models has been presented, helping researchers and designers to select the adequate PPV predictor equation or approach based on scenarios and site conditions.

Keywords Blast-induced ground motion · Underground explosion · Peak particle velocity · Peak ground acceleration

1 Introduction

The accidental or manmade blast activities above or below the ground surface release enormous energy, which generates compressive and tensile types of waves. The high pressure and impulse possessed by these waves induce ground shock which causes particles to vibrate in the medium through which it travels. The particle motion leads to the ground vibration which is generally called the blast-induced ground motion (BIGM). The shock wave travels radially outward in all direction from the point of explosion. In near-field region, the amplitude of vibration is significantly high which decays down as it travels away from the detonation point. The blast-induced waves reduce rapidly to a point, where the induced compressive strength of waves is less than the failure compressive strength of the geo-media [1]. The intensity of vibration can be predicted based on charge weight, radial distance, i.e., distance from the point of explosion to the point of interest and scaling laws.

In recent years, the blast-induced ground motion (BIGM) is gaining more attention of the researchers as the behavior of geo-media under blast loading has disastrous effects on the nearby structures. In the near-field region of blast, the geo-media properties do not affect the wave propagation or intensity of vibration; however, in far-field region of blast, the geo-media properties affect the wave propagation [2], and also, the wave propagates with constant velocity [3]. Thus, the induced vibrations are significant as they may cause damage to the nearby structures. The blast-induced vibrations must be assessed from safety point of view to prevent damage to facilities. This can be monitored by predicting the ground motion characteristics such as peak particle velocity (PPV), peak ground acceleration (PGA), peak ground displacement (PGD), and principle frequency content (PF). Several empirical formulae have been proposed by different researchers for the prediction of the peak particle motion [4] which is generally expressed in terms of the PPV and are influenced by the characteristics of rock mass, geological structures, blast design parameters, and explosive characteristics [5, 6]. Most of the empirical formulas available for predicting the PPV are a function of explosive quantity, distance between detonation point and point of interest, and the site constants which define site and geological conditions. The different explosive materials are usually expressed in terms of Trinitrotoluene

(TNT) equivalency as standard explosive material. The most of the empirical or code-based relations generally use TNT equivalency to represent explosive weight and are based on spherical shape charge.

The wave propagation in a medium depends upon the intensity of explosion, shape and location of charge, density of medium, seismic velocity, and geotechnical parameters such as elastic modulus, shear modulus, bulk modulus, and compressive strength of soil or rocks. The decay in wave amplitude happens due to geometric and hysteretic attenuation which depends on the energy dissipation mechanism. The geometric attenuation is caused due to propagation of the wave in radially outward direction causing increase in the radial distance. This leads to spread of the energy over a larger surface area.

The hysteresis effect is experienced due to dissipation of energy through fracturing or fragmenting the medium in which explosion-induced high-energy wave propagates. The dissipation of energy depends on the strength and geotechnical properties of the medium to cater the transmitted stress with or without undergoing plastic deformations. Herein, the blast-induced ground motion (BIGM) is estimated and compared for different geo-media using empirical relations, code-based equations, and numerical simulations. The PPV estimated using numerical simulations is also compared with the experimental data presented in the literature. The study also investigates wave propagation parameters and directional component of motion for various geo-media.

2 Underground Blast—Theory and State-of-the-Art

2.1 *Stress Wave Propagation*

Underground explosions have more devastating abilities than the equivalent intensity air bursts, as it produces significantly higher pressure and impulse which dissipates slower in the geo-media in comparison to air. Also, point of detonation with respect to medium also plays significant role in generating higher stress due to coupling between the explosion and the medium. Detonation of explosive releases tremendous energy in the form of compressive wave resulting in ground shock which propagates in the medium causing particle motion. This motion of the particles induces ground vibration which is referred as blast-induced ground vibration. The ground motion is characterized based on stress wave propagation in the medium, and its intensity is usually expressed in terms of the peak particle velocity (PPV) along with other components such as PGA, PGD, and PF associated with it. The explosion near or under the ground surface causes surface as well as body waves. The body waves consist of primary (*P*) waves which are induced due to compression and dilation of soil or rock with particle motion parallel to the direction of wave propagation, while secondary (*S*) waves are induced due to distortion or shearing of the soil or rock with particle motion perpendicular to the direction of propagation of wave. The *P* waves are set up due to isotropic component of transient stress pulse while S waves are induced due to deviatoric component of the stress pulse. The body waves govern the ground shocks in case of the buried explosion in the near-field region, while surface wave dominates at far-field region for the buried explosions. The particle near the ground surfaces experiences the circular motion which generates Rayleigh (R) waves. These surface waves play a vital role in case of the surface or shallow explosion, while the body waves are significant in case of the buried explosion.

The seismic velocity (C) is generally expressed in terms of density (ρ) of the medium through which it propagates and elastic modulus (E) of the soil or rock as expressed in Eq. 1,

$$C = \sqrt{\frac{E}{\rho}} \tag{1}$$

The arrival of these waves is termed as direct induced ground shock in underground blast scenario, which primarily depends on the radial distance and seismic velocity which in turn depends upon the propagation medium as expressed in Eq. 2. The arrival time varies drastically depending on the ground media as the seismic velocity shows significant variation for soil, clay, weathered, or jointed rock and hard rock.

$$t_{ag} = \frac{12{,}000R}{C} \tag{2}$$

where t_{ag} is the arrival time, R is the radial distance, and C is the seismic velocity. The wave arrival time is high when direct induced shock wave propagation is through soils, while in the rocks it is the minimum.

2.2 *Peak Particle Velocity (PPV)*

When explosive charge is detonated in a rock medium, the released energy results in generation of ground shock with high compressive wave. This wave propagates with particle motion, which induces ground vibrations. The motion of the particles is possible in three dimensions, i.e., vertical and radially horizontal directions. These particles possess velocity which causes vibration, and the maximum velocity with which the particles vibrate is termed as peak particle velocity (PPV). The ground vibration intensity depends primarily on the charge weight (W), distance between point of interest and point of detonation, i.e., standoff distance (R) radially, scaling laws, site and geological constants.

Code-based relations

The PPV can be estimated using several code-based or empirical relations based on the abovementioned parameters, as presented in the literature. The prediction of vertical and horizontal component of the PPV has been presented in Eqs. 3 and 4 based on scaling laws using UFC-3-340-02 [7] guidelines, which consider that the particles vibrate with same velocity regardless of the direction and medium in which the wave travels. The Indian Standard, I.S. 6922 [5] predicts the PPV due to underground explosion for two types of geo-media based on site constants, charge weight, and radial distance, as indicted in Eq. 5.

$$V_v = \frac{150}{Z^{1.5}} \tag{3}$$

$$V_h = V_v \tag{4}$$

$$\text{PVV} = K_1 \left(\frac{W^{\frac{2}{3}}}{R} \right)^{1.25} \tag{5}$$

where V_v is the velocity in vertical direction while V_h is the velocity in horizontal direction, Z ($= R/W^{1/3}$) is the scaled distance, and K_1 is the constant for soil and rock.

Empirical relations

The estimation of the blast-induced ground motion (BIGM) due to the direct induced shock is a complex phenomenon as it depends on several factors. Due to a lack of complete understanding of the rock or soil beneath, it is challenging to evaluate precisely rock or soil properties; however, the uncertainties involved in the geo-medium for the prediction of the BIGM also make it a difficult task. The experimental evaluations for prediction of the ground vibrations also pose several restrictions due to safety, cost, instrumentation, and administrative hurdles. Considering these difficulties, the empirical models or relations provide a reasonable estimate of the BIGM. Several empirical relations proposed by different researchers help to predict peak particle velocity (PPV) based on site constants, geotechnical parameters, and the scaling laws. Most of the empirical relations are based on scaling laws which correlate the radial distance and charge weight. Equations 6–9 represent various empirical relations or predictor equation based on artificial neural network (ANN) or regression method [8], Weibull method, reciprocal method, and power method [9] respectively, presented in the literature to predict the PPV due to underground explosions.

$$\text{PPV} = 7.8008814 + 0.833482 Q_{\text{max}} - 0.07484 D \tag{6}$$

$$\text{PPV} = 25.680 - 24.335\, \mathrm{e}^{-359.4402 D^{-2.69}} \tag{7}$$

$$\text{PPV} = \frac{1}{0.0071 D - 0.0218} \tag{8}$$

$$\text{PPV} = 374.947D^{-1.2057} \tag{9}$$

where PPV is the peak particle velocity, Q_{max} is the maximum charge per delay, and D is the distance of detonation point from point of interest.

2.3 Numerical Model

Underground explosion involves complex physics due to uncertainties and intricate behavior of the explosives and propagation media. The behavior of the underground medium may not remain linearly elastic due to release of high energy during explosion which results in large deformation, temperature, and high strain rate. This phenomena can be modeled using coupled analysis involving principles of computational fluid dynamics (CFD) and solid mechanics using general purpose finite element (FE) analysis tools such as ANSYS-AUTODYN [10]. The numerical methods typically employ multi-material arbitrary Lagrangian Eulerian (MMALE) approach to simulate explosive events such as blast. In the present study, the ground vibration caused due to the underground blast event is predicted using three-dimensional (3D) finite element (FE) tool, ANSYS-AUTODYN based on the MMALE formulations. The detailed FE modeling procedure has been described in the following sections systematically.

Material modeling of air and explosive

The constitutive property (law) of air has been modeled using ideal gas equation of state (EoS) as expressed in Eq. 10,

$$P = (\gamma - 1)\frac{\rho}{\rho_0}E \tag{10}$$

where P is the current absolute pressure in air, E is the initial internal volumetric energy density, ρ is the current mass density, ρ_0 is the initial mass density of the ideal gas (0.001225 g/cm^3), and γ (= 1.4) is an adiabatic exponent for air.

The detonation of explosive material has been modeled using Jones-Wilkins-Lee (JWL) equation of state (EoS) which is able to simulate the hydrodynamic computation of a wide range of explosive expansions. The JWL equation is able to simulate the most types of high explosives (HE) using Eq. 11, while constants used in the JWL equation have been presented in Table 1.

Table 1 Material properties for modeling of explosive

Density (g/cm^3)	A (kPa)	B (kPa)	R_1	R_2	w
1.63	3.73×10^8	3.747×10^6	4.15	0.9	0.35

$$P = A\left(1 - \frac{w\eta}{R_1}\right)e^{\frac{-R_1}{\eta}} + B\left(1 - \frac{w\eta}{R_2}\right)e^{\frac{-R_1}{\eta}} + w\rho e \quad (11)$$

where $\eta = \rho/\rho_0$, ρ_0 is reference density and ρ density of explosives. The values of the constants A, B, R_1, R_2, and w for many common explosives have been determined from dynamic tests and reported in the literature.

Material strength model

The rock mass is a model in the Lagrangian mesh domain which can simulate the dynamic behavior under a high rate of loading due to the explosion. The dynamic properties of the rock mass significantly alter the stress wave propagation in the rock mass. A piecewise linear Drucker-Prager strength model is used to simulate the dynamic and inelastic behaviors of the rock mass under the extreme events such a blast. The stress deviator in terms of invariants can be written as depicted in Eq. 12,

$$F = \sqrt{J_2} - \alpha I_1 - k = 0 \quad (12)$$

where J_2 is the second invariant of the stress deviator; I_1 is the first invariant of the stress tensor, $I_1 = (\sigma_1 + \sigma_2 + \sigma_3)$, $J_2 = 0.5S_{ij}S_{ij}$; α and k are the material constants. The studies have confirmed that the strength of the rock increases linearly with the logarithm of the loading rate. The linear equation of state (EoS) is considered to represent initial elastic behavior of the intact rock. Here, in the present study, properties used for granite rock mass are: the mass density of 2620 kg/m^3, the elastic modulus of 75 GPa, the uniaxial compressive strength of 250 MPa, and the average seismic velocity of the rock considered is around 5660 m/s [11]. The properties used for limestone considered are: elastic modulus of 77 MPa, mass density of 2600 kg/m^3, and Poisson's ratio of 0.25 [12]. The properties used for sandstone are: density of 2638 kg/m^3, unified compressive strength as 98 MPa, Poisson's ratio of 0.23 [13].

3 Results and Discussion

The numerical simulations have been carried out using multi-material arbitrary Lagrangian Eulerian (MMALE) approach to predict the ground vibration induced due to underground explosion. The rock medium is modeled in Lagrangian domain while explosive and air are modeled in Eulerian domain with ideal gas and JWL as equation of state (EoS). The model is validated using experimental data [11] available in the literature which shows reasonable agreement with the estimated PPV with variation of about 15–18%. The variation in the PPV estimate is attributed to uncertainties in the rock medium considering large-scale test and wide range of Rock Quality Designation (RQD), i.e., 75–100% in contrast to the intact rock model in simulations. The PPV in various geo-media have been estimated numerically and compared with those estimated using the code-based relations, empirical relations, and experimental results presented in the literature. The following sections describe the PPV estimation in different geo-media along with arrival time, wave attenuation, and directional component of the PPV.

3.1 Peak Particle Velocity in Varying Geo-media

The blast-induced ground vibration is usually expressed in terms of peak particle velocity (PPV) which represents maximum velocities with which particles travel/oscillate in medium as body or surface waves depending on buried or surface explosion. Herein, the PPV estimated using the numerical simulations are compared with the code-based relations, empirical relations or predictor equations, and experimental results available in the literature for various geo-media. The estimated PPV using the abovementioned approaches signify the effects of wave propagation, attenuation in various geo-media, and variation with scaled distances used to represent different blast scenarios. Figure 1 shows comparison of the PPV estimates for various geo-media such as granite (Gr), limestone (LS), and sandstone (SS) estimated using the code-based guidelines, empirical relations, and numerical simulations. The PPV estimates show drastic variation in the amplitudes for the lower scaled distance below 2 m/kg$^{1/3}$, i.e., in near-field region, while it decays down significantly to minimal variation in far-field region beyond 5 m/kg$^{1/3}$. The UFC-3-340-02 [7] shows invariably higher estimates regardless of the propagation medium as it predicts equal amplitude

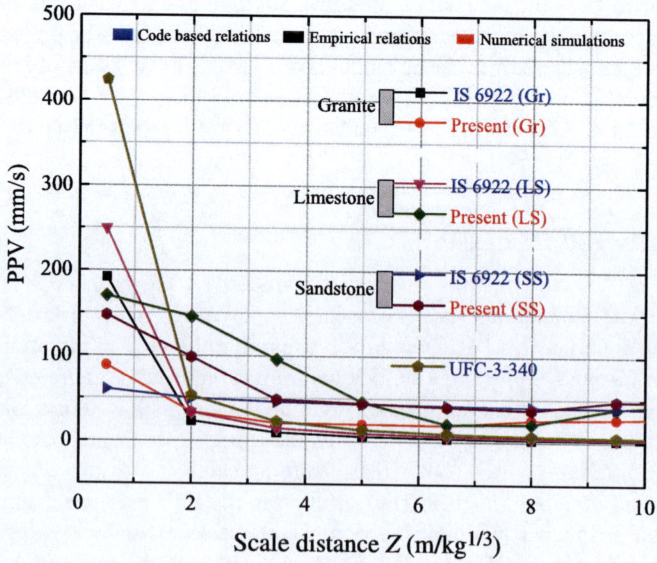

Fig. 1 PPV comparison estimated with various methods in different geo-media for varying scaled distance (*Z*)

for all soils or rocks. The prediction using I.S. 6922 [5] distinguish between soft rock or soils and hard rock using site constants as depicted in Eq. 5, which shows some resemblance with the prediction using numerical simulations and empirical relations presented in the literature. The limestone (LS) shows higher stress transmission with significantly higher PPV in comparison to sandstone (SS) and granite (Gr) in the near-field region, while a reverse trend is exhibited in far-field region. This signifies importance of various geo-media in wave propagation or transmission, as well as in attenuation of the shock wave based on radial distance and energy dissipation based on strength and geotechnical properties of the medium.

3.2 *Wave Transmission and Attenuation*

The seismic velocity has been estimated using Eq. 1 based on density and stiffness of the soil or rock to understand influence of geo-media in transmitting stress wave. The particle motion appears to be high in case of the rock with maximum for the limestone followed by granite and sandstone, while soils exhibit relatively lower particle motion. The loose soils show poor stress transmission in comparison to clay, saturated soils, and dense soils. The variation in the seismic velocity for different geo-media has been compared and presented in Table 2. The upper bound estimates in case of the rock signifies intact rock with RQD above 75%, while lower bound relates to the highly weathered or disintegrated rock with RQD less than 25%. The soft rock or medium quality rock with RQD in the range of 25–75% exhibits the stress transmission or particle motion between upper and lower bounds. The similar inference can be made for various soil types regarding the upper and lower bounds of seismic velocity. The loose or poorly graded soils exhibit least stress transmission, however clay or saturated soils show higher transmission. Nonetheless, such stress transmission is significantly lower in comparison with the rocks. The saturated soils show higher stress transmission due to the presence of water. The impedance of media plays a vital role in the stress transmission or energy dissipation. This signifies

Table 2 Upper and lower bound values of seismic velocities for various geo-media

Soil type	Density (kg/m^3)	Seismic velocity (m/s)	
		Lower bound	Upper bound
Granite (Gr)	2620	244	4572
Sandstone (SS)	2638	914	4267
Limestone (LS)	2600	2133	6400
Loose soil	1712	–	183
Dense soil	1800	184	900
Saturated soil	1600–1700	901	1920
Clay	1900	1921	2590

poor attenuation or dissipation of the transmitted wave in case of hard or intact rocks with higher transmission, while loose or poorly graded soils exhibit higher wave attenuation with lower wave transmission. The wave attenuation is significantly influenced by the strength and geotechnical properties of the geo-media as energy dissipation through hysteresis depends on the energy loss in fracturing the rock or plastic deformation of the soil. The plastic deformation and fracture of rock or soil in turn depend upon characteristic and ultimate strength of the geo-media. Thus, prediction of the ground vibration induced due to underground explosion is highly complex considering the physics involved in the explosion event, the uncertainties and non-linear behavior of soil or rock media. The variation in the soil properties significantly alters the wave motion or propagation; moreover, the soil beneath may be composed of layers or variation in RQD in all three principle directions, i.e., vertical, transverse/radial directions.

3.3 *Wave Arrival Time*

The arrival time of shock wave at the point of interest or structure depends on radial distance and seismic velocity which is based on density and stiffness of the medium in which the wave propagates. The arrival time of shock wave is depicted in Eq. 2. The large variation in seismic velocity has been observed within and across various geo-media as presented in Table 2, which influences the arrival time significantly and proves to be challenging to distinguish between super seismic and out running region in case of a surface burst. The arrival time is higher for loose soils or highly weathered rock having low seismic velocity, while intact rock or saturated soil exhibits faster transmission of the shock wave with lower arrival time as depicted in Fig. 2. The hard rocks, saturated soils, and clay possess higher seismic velocities which show faster wave propagation as plotted against the scaled distance, while loose or granular soils and weathered, or soft rocks show higher attenuation with significant energy dissipation resulting in higher arrival time. The arrival time has been compared for the rocks such as granite, limestone, sandstone, and soils or clays with loose, dense, and saturated conditions. It is noted that loose or poorly graded soils or weathered rocks can be used effectively to attenuate and delay shock wave arrival in reducing the catastrophes in the extreme events such as blast.

3.4 *Directional Component*

In an underground blast event, the ground shock or stress wave travel radially outwards in all directions, with generation of either body waves and/or surface waves depending upon the depth of explosion, i.e., buried burst or surface burst. The shock wave propagation can alter with the same geo-media, based on site geometry, explosive shape, and the presence of uncertainties such as voids, water table, etc., while in

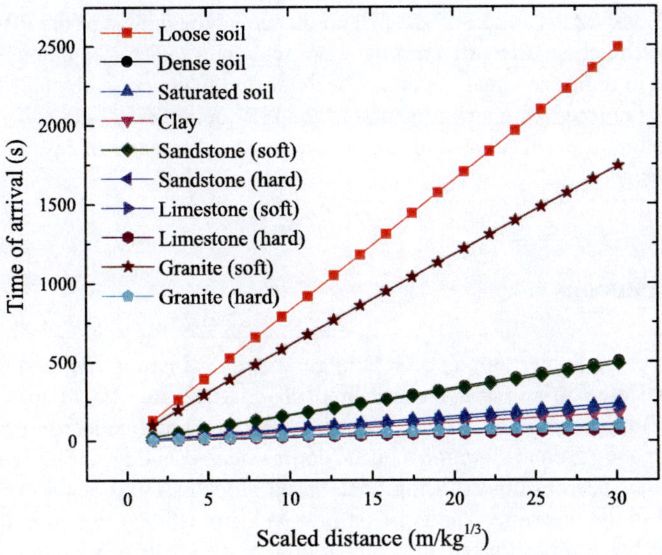

Fig. 2 Wave arrival time for underground blast scenario in different geo-media

layered medium the propagation depends upon the reflection and transmission at the interface of the two mediums. Table 3 depicts the comparison of vertical and horizontal components of peak particle velocity (PPV) estimated using numerical simulations and with experimental data [11] presented in the literature for the granitic rock medium. The comparison of vertical component of simulated and experimental data shows close resemblance, considering the intact rock medium in numerical simulation in contrast to experimental investigation which specifies RQD in the range of 75–100% and other uncertainties in the rock beneath for such large field test. The vertical and transverse components estimated using the numerical simulations and experimental results presented in the literature show variation of about 30–55% for the considered blast scenarios with average value of about 37 and 40% for the experimental [11] and simulation-based estimates. However, code-based guidelines, such

Table 3 Comparison of vertical and horizontal velocity components

Charge weight, *W* (kg)	Scaled distance, *Z* (m/kg$^{1/3}$)	Velocity (m/s)			
		Wu et al. [11]		Present study	
		Vertical	Horizontal	Vertical	Horizontal
10,000	3.713	0.140	0.060	0.105	0.036
10,000	3.787	0.080	0.018	0.065	0.021
10,000	4.755	0.075	0.020	0.055	0.019
10,000	4.898	0.090	0.033	0.07	0.023
10,000	7.180	0.045	0.025	0.034	0.019

as UFC-3-340-02 [7] and I.S. 6922 [5], predict equal magnitude of the PPV regardless of the direction. Also, most of the empirical and predictor equations based on the ANNs or regression analysis do not account for the direction in the PPV prediction. This can lead to higher estimation of the PPV in horizontal direction resulting in overdesign of the facilitates or may prove to be detrimental in case of sensitive facility design.

4 Conclusions

Blast-induced ground motion (BIGM) in geo-media is investigated, and numerical simulations are carried out using the multi-material arbitrary Lagrangian Eulerian (MMALE) approach. Detailed modeling of the geo-media and explosives is presented for investigating ground vibration induced due to underground explosion. The ground vibration has been estimated using code-based guidelines and empirical relations presented in the literature. The Peak Particle Velocity (PPV) estimation in various geo-media has been estimated and compared with the abovementioned approaches to investigate variation of the PPV for various scaled distances, wave propagation or transmission, arrival of the shock wave, and its attenuation in different geo-media. The following conclusions are drawn based on the abovementioned investigation.

1. The numerical simulations using the MMALE approach for estimation of the PPV show good agreement with the experimental results presented in the literature.
2. The UFC-3-340-02 [7] leads invariably to higher estimation of the PPV in comparison to I.S. 6922 [5] estimation. However, effect of geo-media is captured by the empirical relations and numerical simulations which differ from the UFC-3-340-02 [7] guidelines.
3. The limestone shows higher stress transmission with significantly higher PPV in comparison to the sandstone and granite in the near-field region, while the reverse trend is exhibited in the far-field region.
4. The loose or poorly graded soils or weathered rocks can be used effectively to attenuate and delay the shock wave arrival in reducing the catastrophes during accidental or manmade explosions.
5. The vertical and transverse components of the estimated PPV using numerical simulations and experimental results presented in the literature shows average variation of about 40% and 37%, respectively; whereas, code-based guidelines predict equal magnitude of the PPV regardless of the direction.

References

1. Esen S, Onederra I, Bilgin HA (2003) Modelling the size of the crushed zone around a blasthole. Int J Rock Mech Min Sci 40(4):485–495

2. Dowding CH (1985) Blast vibration monitoring and control, vol 297. Prentice-Hall, Englewood Cliffs
3. Silva-Castro JJ (2012) Blast vibration modeling using improved signature hole technique for bench blast. Ph.D. Dissertation, University of Kentucky, USA
4. Langefors U, Kihlstrom B (1963) The modern technique of rock blasting. Wiley, New York, USA
5. Indian Standard IS (1973) Criteria for safety and design of structures subjected to underground blast. Bulletin No. IS-6922, Bureau of Indian Standards (BIS), New Delhi
6. Kumar R, Choudhary D, Bhargava K (2016) Determination of blast-induced ground vibration equations for rocks using mechanical and geological properties. Rock Mech Rock Eng 8(3):341–349
7. Unified Facilities Criteria (UFC) 3-340-02 (2008) Structures to resist the effects of accidental explosions. U.S. Department of Defense, USA
8. Parida A, Mishra MK (2015) Blast vibration analysis by different predictor approaches—a comparison. Proced Earth Planet Sci 11:337–345
9. Ray S, Dauji S (2019) Ground vibration attenuation relationship for underground blast—a case study. J Instit Eng Ser A, 1–13
10. ANSYS-AUTODYN (2005) User's manual, version 6.1, Century Dynamics Horsham, UK
11. Wu C, Lu Y, Hao H, Lim WK, Zhou Y, Seath CC (2003) Characterization of underground blast induced ground motions from large scale field tests. Shock Waves 13(3):237–252
12. Ainalis D, Ducarne L, Kaufmann O, Tshibangu J, Verlinden O, Kouroussis G (2018) Improved analysis of ground vibrations produced by manmade sources. Sci Total Environ 517–530
13. Yilmaz O, Unlu T (2013) Three dimensional numerical rock damage analyses under blasting load. Tunn Undergr Space Technol 38:266–278

Phase Velocity of Love-Type Waves in a Periodic Irregular Anisotropic-Layered Substratum

Tapas Ranjan Panigrahi, Sumit Kumar Vishwakarma, and Rupinderjit Kaur

Abstract The present study has been carried out to investigate the transference of Love-type surface waves in an anisotropic heterogeneous earth-crust lying over an isotropic layer and an isotropic inhomogeneous bedded half-space. The rigidities and densities have been assumed varying as a function of depth in both the media. These variations are exponential in the layer while in the half-space, they are the product of linear algebraic function and an exponential function of depth. Following the theory of elastic waves, the mathematical model has been established. Separation of variable has been employed to obtain the displacement in the layer and the half-space. Intrinsic boundary conditions have been imposed to obtain the dispersion equation. It has been found that the dispersion equation is a function of inhomogeneity parameters as well as the initial stress and the term that governs the irregularity of the structure. Graphical representations have been performed to exhibit the findings, which clearly shows the significant impact of these parameter on the phase velocity of Love-type surface waves.

Keywords Love-type wave · Initial stress · Heterogeneous half-space · Dispersion equation · Phase velocity · Corrugated surface

1 Introduction

Over the last few years, the study of propagation behaviour of Love waves in an elastic medium with layered structure has been of great interest. Many seismologists have been following the theory given by Love [1], presented a thorough and comprehensive outputs on seismic wave propagation in an isotropic and anisotropic medium. It was

T. R. Panigrahi (✉) · S. K. Vishwakarma · R. Kaur
Department of Mathematics, BITS-Pilani, Hyderabad Campus, Hyderabad 500078, India
e-mail: p20170412@hyderabad.bits-pilani.ac.in

T. G. Sitharam et al. (eds.), *Seismic Hazards and Risk*, Lecture Notes in Civil Engineering 116, https://doi.org/10.1007/978-981-15-9976-7_28

investigated by Bullen [2], the density in different layers within the earth varies at different rates. He has approximated that on going down from 413 to 984 km, it varies quadratically, whereas the density varies as a linear function of depth parameter from 984 km to the central core. In various technological applications, the problem on non-homogeneity has been assumed in unbounded space with variations in rigidities and densities varying linearly with a certain discontinuous jump. Sari and Salk [3] have explained the propagation of the wave in a medium hyperbolic variation in the density of layer with depth as a hyperbolic function. Likewise, different authors have taken different forms of variations like harmonic, linear, quadratic and exponential, simulating the variations in the density and other geological parameters inside the earth. For this reason, the problem of surface waves in heterogeneous-layered media over a half-space is of great interest in various applications.

It was British mathematician A .E. H. Love in 1911 discovered an important surface wave named as Love wave in his honour. This wave is a horizontally polarized wave, resultant of SH-waves that are tracked near the surfaces. Wolf [4] has investigated Love waves in layers with an irregular boundaries, whereas scattering of seismic waves generated by an irregular seabed has been studied by Santos and Tadeu [5]. References may be given to Zhang and Chen [6], Singh [7], Poonia and Mada [8], Mistri et al. [9] and Liu et al. [10] for their constructive works towards surface wave propagation. It was Aki [11] who has used Love wave to study the earthquake mechanism, while Noyer [12] studied the effect of variation in layer thickness on Love waves. Wang and Tang [13] investigated finite-difference modelling of elastic wave propagation, whereas Du et al. [14] studied Love wave propagation in layer magneto-electro-elastic structure with initial stress. Following the theory of Du et al. [15] analysed as to how Love waves behave in a pre-stressed piezoelectric layered structure. In addition to that, recently, Gupta et al. [16] and Vishwakarma et al. [17–19] investigated the influence of non-homogeneity and linearly varying initial stresses on the phase velocity of Love waves and SH-waves.

Up to now, very few attempts have been made to study the propagation of Love-type waves in a periodic irregular earth's crustal layer over a bedded half-space. In the present study, the dispersion relation for the Love-type wave propagating in a periodic irregular anisotropic inhomogeneous substratum over an irregular anisotropic half-space has been obtained in the closed-form. Rigidities, densities and initial stress have been assumed to vary as a function of depth. Closed-form dispersion equation has been deduced by the help of intrinsic boundary conditions. It has been observed that phase velocity of Love-type waves is greatly affected by the fluctuation of wavenumber, periodic irregularity, inhomogeneity parameters and initial stress. In order to analyse these effects more profoundly, numerical computation and graphical demonstration have been carried out. Moreover, a comparative study have been made to discuss some of the important points of the problem, and classical results are deduced as special cases of the problem.

2 Geometry and Statement of the Problem

In the present problem, layered substratum has been taken over a half-space with x- and y-axis horizontally perpendicular and positive z-axis in a vertical direction towards the center of the earth as shown in Fig. 1. The inhomogeneities associated with the rigidity, density, and initial stress have been given as (Gubbins [20])

Medium 1: $N = N_1\alpha^z$, $L = L_1\alpha^z$, $\rho = \rho_1\alpha^z$, $z, \alpha \in R$, where $L_1 = 5.98 \times 10^{10}$ N/m^2, $N_1 = 7.34 \times 10^{10}$ N/m^2, $\rho_1 = 3195\,\text{kg/m}^3$.

Medium 2: $\mu = \mu_2$, $\rho = \rho_2$, where $\mu_2 = 7.45 \times 10^{10}$ N/m^2, $\rho_2 = 3293\,\text{kg/m}^3$.

Medium 3: $\mu = \mu_3(1 + az)e^{bz}$, $\rho = \rho_3(1 + az)e^{bz}$ and $P = P_3(1 + az)e^{bz}$, $z, a, b \in R$ where $\mu_3 = 6.02 \times 10^{10}$ N/m^2, $\rho_3 = 3364\,\text{kg/m}^3$.

where N and L are directional rigidities towards the vertical and the horizontal directions, ρ is the density of the medium, μ and P are the rigidities and initial stress, respectively.

3 Solution for Non-homogeneous Anisotropic Layer (Medium 1)

Let us consider u, v, and w are the displacement components along the x-, y- and z-direction, respectively. Discussing the general equation of motion and using the Love wave condition, i.e. $u = 0$, $v = v_1(x, z, t)$, and $w = 0$, the equation of motion can be written as (Biot [21])

$$N\frac{\partial^2 v_1}{\partial x^2} + \frac{\partial}{\partial z}\left(L\frac{\partial v_1}{\partial z}\right) = \rho\frac{\partial^2 v_1}{\partial t^2} \tag{1}$$

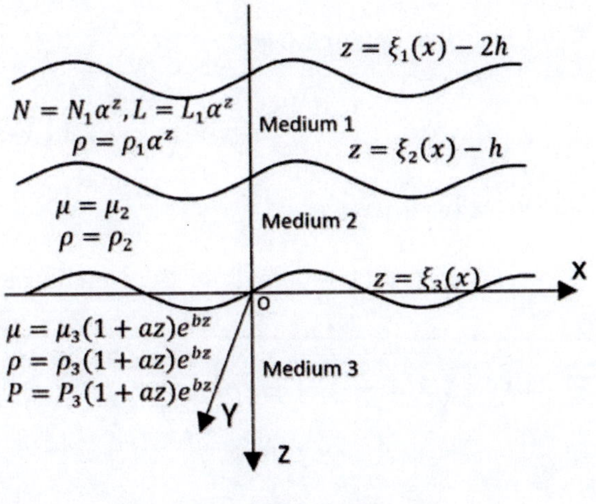

Fig. 1 Geometry of the problem

For a wave propagating along x-direction, we assume

$$v_1 = V(z)\mathrm{e}^{(ik(x-ct))},\ k > 0,\ c > 0 \tag{2}$$

Using Eq. (2) in Eq. (1), we get

$$\frac{\mathrm{d}^2V(z)}{\mathrm{d}z^2} + \frac{1}{L}\frac{\mathrm{d}L}{\mathrm{d}z}\frac{\mathrm{d}V(z)}{\mathrm{d}z} + \frac{k^2}{L}(c^2\rho - N)V(z) = 0 \tag{3}$$

On putting, $N = N_1\alpha^z$, $L = L_1\alpha^z$, and $\rho = \rho_1\alpha^z$ in Eq. (3), we get

$$\frac{\mathrm{d}^2V(z)}{\mathrm{d}z^2} + \log(\alpha)\frac{\mathrm{d}V(z)}{\mathrm{d}z} + \frac{k^2}{L_1}(c^2\rho_1 - N_1)V(z) = 0 \tag{4}$$

On solving Eq. (4) and using the result in the Eq. (2), we get

$$v_1 = (c_1\mathrm{e}^{z(-\log(\alpha)-\sqrt{-4A+(\log(\alpha))^2})} + c_2\mathrm{e}^{z(-\log(\alpha)+\sqrt{-4A+(\log(\alpha))^2})})e^{ik(x-ct)} \tag{5}$$

where $A = k^2(c^2/\beta - N_1/L_1)$, $\beta = \sqrt{L_1/\rho_1}$.

4 Solution for Isotropic Layer (Medium 2)

The equation of motion is given by (Biot [21])

$$\mu_2\frac{\partial^2 v_2}{\partial x^2} + \mu_2\frac{\partial^2 v_2}{\partial z^2} = \rho_2\frac{\partial^2 v_2}{\partial t^2} \tag{6}$$

Let us assume the solution be

$$v_2 = V(z)\mathrm{e}^{\{ik(x-ct)\}} \tag{7}$$

Using Eq. (7) in Eq. (6), we get

$$\frac{d^2V(z)}{dz^2} + k^2\left(\frac{c^2\rho_2}{\mu_2} - 1\right)V(z) = 0 \tag{8}$$

On solving Eq. (8), we get

$$v_2(z) = \left[c_3\mathrm{e}^{-ik\nu_1 z} + c_4\mathrm{e}^{ik\nu_1 z}\right]\mathrm{e}^{ik(x-ct)} \tag{9}$$

where, $\nu_1 = \sqrt{k^2\left(\frac{c^2}{\beta_2^2} - 1\right)}$ and $\beta_2^2 = \frac{\mu_2}{\rho_2}$

5 Solution for Non-homogeneous Half-Space (Medium 3)

The third layer has been taken as non-homogeneous isotropic half-space under initial horizontal compressive stress P along x direction. The equation of motion due to the Love wave can be written as (Biot [21])

$$\frac{\partial s_{21}}{\partial x}+\frac{\partial s_{23}}{\partial z}-\frac{P}{2}\left(\frac{\partial^2 v}{\partial x^2}\right)=\frac{\partial^2(\rho v)}{\partial x^2} \tag{10}$$

where s_{ij} are the incremental stress components in the half-space, P is the initial compressive stress along x direction, and ρ is the density of the material of the half-space.

The stress–strain relations for medium 3 are

$$s_{21}=2\mu e_{xy},\ \ s_{23}=2\mu e_{yz}$$

Also,

$$\mu=\mu_3\,(1+az)\,\mathrm{e}^{bz},\ \rho=\rho_3\,(1+az)\,\mathrm{e}^{bz},\ P=P_3\,(1+az)\,\mathrm{e}^{bz} \tag{11}$$

Substituting the above relation in Eq. (10), we get

$$\begin{aligned}&\mu_3\,(1+az)\frac{\partial^2 v_3}{\partial x^2}+\mu_3\,(a+b+abz)\frac{\partial v_3}{\partial z}\\&+\mu_3\,(1+az)\frac{\partial^2 v_3}{\partial z^2}\\&-\frac{P_3}{2}\,(1+az)\frac{\partial^2 v_3}{\partial x^2}\\&=\rho_3\,(1+az)\frac{\partial^2 v_3}{\partial t^2}\end{aligned} \tag{12}$$

Taking,

$$v_3=V(z)\mathrm{e}^{ik(x-ct)} \tag{13}$$

Using Eq. (13) in Eq. (12), we get

$$\frac{\mathrm{d}^2V}{\mathrm{d}z^2}+\frac{(a+b+abz)}{(1+az)}\frac{\mathrm{d}V}{\mathrm{d}z}+k^2\left[\frac{P_3}{2\mu_3}+\frac{c^2}{\beta_3^2}-1\right]V=0 \tag{14}$$

where $\beta_3^2=\frac{\mu_3}{\rho_3}$.

On solving Eq. (14) with the help of Eq. (13), we find

$$v_3 = c_5 e^{-\frac{1}{2}\left(b+\sqrt{b^2-4A_1}\right)z}$$
$$\text{Hypergeometric } U\left[\frac{b+\sqrt{b^2-4A_1}}{2\sqrt{b^2-4A_1}}, 1, \frac{\sqrt{b^2-4A_1}}{a} + \sqrt{b^2-4A_1}z\right] e^{ik(x-ct)}$$
$$+ c_6 e^{-\frac{1}{2}\left(b+\sqrt{b^2-4A_1}\right)z}$$
$$\text{Laguerre } L\left[\frac{-b-\sqrt{b^2-4A_1}}{2\sqrt{b^2-4A_1}}, 1, \frac{\sqrt{b^2-4A_1}}{a} + \sqrt{b^2-4A_1}z\right] e^{ik(x-ct)} \tag{15}$$

Imposing the condition that as $z \to \infty,\ v_3 \to 0$, Eq. (15) reduces to

$$v_3 = c_5 e^{-\frac{1}{2}\left(b+\sqrt{b^2-4A_1}\right)z}$$
$$\text{Hypergeometric } U\left[\frac{b+\sqrt{b^2-4A_1}}{2\sqrt{b^2-4A_1}}, 1, \frac{\sqrt{b^2-4A_1}}{a} + \sqrt{b^2-4A_1}z\right] e^{ik(x-ct)} \tag{16}$$

where $A_1 = k^2\left[\zeta + \frac{c^2}{\beta_3^2} - 1\right]$ and $\zeta = \frac{P_3}{2\mu_3}$.

6 Boundary Conditions

1. At $z = \xi_1(x) - 2h$, as the surface is free

$$\left[\tau_{zy} - \xi_1' \tau_{xy}\right]_{M_1} = 0 \tag{17}$$

2. At $z = \xi_2(x) - h$

$$\left[\tau_{zy} - \xi_2' \tau_{xy}\right]_{M_1} = \left[\tau_{zy} - \xi_2' \tau_{xy}\right]_{M_2} \tag{18}$$

$$v_1 = v_2 \tag{19}$$

3. At $z = \xi_3(x)$

$$\left[\tau_{zy} - \xi_2' \tau_{xy}\right]_{M_2} = \left[\tau_{zy} - \xi_2' \tau_{xy}\right]_{M_3} \tag{20}$$

$$v_2 = v_3 \tag{21}$$

where

$$\xi_1' = \frac{\partial \xi_1}{\partial x},\ \xi_2' = \frac{\partial \xi_2}{\partial x}$$

τ_{yz} and τ_{xy} are stress tensors and their expressions are given by

$$\left[\tau_{yz}\right]_{M_1} = L_1\frac{\partial v_1}{\partial z},\ \left[\tau_{xy}\right]_{M_1} = L_1\frac{\partial v_1}{\partial x},\ \left[\tau_{yz}\right]_{M_2} = \mu_2\frac{\partial v_2}{\partial z},\ \left[\tau_{xy}\right]_{M_2} = \mu_2\frac{\partial v_2}{\partial x},$$

$$\left[\tau_{yz}\right]_{M_3} = \mu_3\frac{\partial v_3}{\partial z},\ \left[\tau_{xy}\right]_{M_3} = \mu_3\frac{\partial v_3}{\partial x}$$

where M_1, M_2 and M_3 are subscript for medium 1, medium 2 and medium 3, respectively.

Applying the above boundary conditions, we get

$$\left(\frac{1}{2}\left(-\log\alpha - \sqrt{-4A+(\log\alpha)^2}\right) - ik\xi_1'\right)\mathrm{e}^{\frac{\xi_1(x)-2h}{2}\left(-\log\alpha-\sqrt{-4A+(\log\alpha)^2}\right)}c_1$$
$$+\left(\frac{1}{2}\left(-\log\alpha + \sqrt{-4A+(\log\alpha)^2}\right) - ik\xi_1'\right)\mathrm{e}^{\frac{\xi_1(x)-2h}{2}\left(-\log\alpha+\sqrt{-4A+(\log\alpha)^2}\right)}c_2 = 0 \tag{22}$$

$$\frac{L_1}{2}(-\log\alpha - \sqrt{-4A+(\log\alpha)^2} - ik\xi_1')\mathrm{e}^{(\xi_2(x)-h)}(-\log\alpha - \sqrt{-4A+(\log\alpha)^2})c_1$$
$$+\frac{L_1}{2}(-\log\alpha + \sqrt{-4A+(\log\alpha)^2} - ik\zeta_1')\mathrm{e}^{\xi_2(x)-h})(-\log\alpha - \sqrt{-4A+(\log\alpha)^2})c_2$$
$$-\mu_2(\nu_1 - \xi_2')ik\mathrm{e}^{ik\nu_1(\xi_2(x)-h)}c_3 + \mu_2(\nu_1 + \xi_2')ik\mathrm{e}^{-ik\nu_1(\xi_2(x)-h)}c_4 = 0 \tag{23}$$

$$\mathrm{e}^{\frac{\xi_2(x)-h}{2}\left(-\log\alpha-\sqrt{-4A+(\log\alpha)^2}\right)}c_1 + \mathrm{e}^{\frac{\xi_2(x)-h}{2}\left(-\log\alpha+\sqrt{-4A+(\log\alpha)^2}\right)}c_2$$
$$-\mathrm{e}^{ik\nu_1(\xi_2(x)-h)}c_3 - \mathrm{e}^{-ik\nu_1(\xi_2(x)-h)}c_4 = 0 \tag{24}$$

$$ik\mu_2(\nu_1 - \xi_2')\mathrm{e}^{ik\nu_1\zeta_3(x)}c_3 - ik\mu_2(\nu_1 + \xi_2')\mathrm{e}^{-ik\nu_1\xi_3(x)}c_4$$
$$-\{\mu_3\left(-b - \sqrt{-4A_1+b^2}\right)$$
$$\mathrm{e}^{\frac{\xi_3(x)}{2}\left(-b-\sqrt{-4A_1+b^2}\right)}$$
$$\text{Hypergeometric}\,U\left(\frac{b+\sqrt{-4A_1+b^2}}{2\sqrt{-4A_1+b^2}}, 1, \frac{\sqrt{-4A_1+b^2}}{a} + \sqrt{-4A_1+b^2}\xi_3(x)\right)$$
$$-\mu_3\left(-b - \sqrt{-4A_1+b^2}\right)\mathrm{e}^{\frac{\xi_3(x)}{2}\left(-b-\sqrt{-4A_1+b^2}\right)}$$
$$\text{Hypergeometric}\,U\left(1+\frac{b+\sqrt{-4A_1+b^2}}{2\sqrt{-4A_1+b^2}}, 2, \frac{\sqrt{-4A_1+b^2}}{a} + \sqrt{-4A_1+b^2}\xi_3(x)\right) \tag{25}$$
$$+\frac{ik\xi_3'}{\frac{1}{2}\left(-b - \sqrt{-4A_1+b^2}\right)}$$
$$\text{Hypergeometric}\,U\left(\frac{b+\sqrt{-4A_1+b^2}}{2\sqrt{-4A_1+b^2}}, 1, \frac{\sqrt{-4A_1+b^2}}{a} + \sqrt{-4A_1+b^2}\xi_3(x)\right)\}$$
$$c_5 = 0$$

$$e^{ik\nu_1\xi_3(x)}c_3 + e^{-ik\nu_1\xi_3(x)}c_4 - e^{\frac{1}{2}\left(-b-\sqrt{-4A_1+b^2}\right)\xi_3(x)}$$

$$\text{Hypergeometric } U\left(\frac{b+\sqrt{-4A_1+b^2}}{2\sqrt{-4A_1+b^2}}, 1, \frac{\sqrt{-4A_1+b^2}}{a} + \sqrt{-4A_1+b^2}\xi_3(x)\right)$$

$$c_5 = 0 \tag{26}$$

For nonzero solution of c_1, c_2, c_3, c_4 and c_5, we must have

$$[D_{ij}]_{5\times5} = 0, \quad i,j = 1, 2...5. \tag{27}$$

Equation (27) gives the required frequency equation for Love-type waves in a three-layered model (see Appendix-I for D_{ij}).

7 Particular Cases

Case-I: When $\xi_1 = h$ and $\xi_2 = 0$, medium 1 and medium 2 coincide with flat upper surface at flat surface $z = -h$ and corrugation only stays at the interface of the layer and the half-space when $\xi_3 = d\cos px$.

$$\tan\{k\sqrt{\frac{c^2}{\beta_2^2} - 1}\,(h + d\cos px)\} = \frac{\left(\frac{\mu_3}{\mu_2}\right)\sqrt{\frac{c^2}{\beta_3^2} - 1}}{\sqrt{\frac{c^2}{\beta_2^2} - 1}} + i\frac{\left(\frac{\mu_3}{\mu_2}\right) - 1}{\sqrt{\frac{c^2}{\beta_2^2} - 1}}pd\sin px \tag{28}$$

Subcase: If the amplitude of the corrugated surface is zero, i.e. $d = 0$ then we get

$$\tan\{k\left(\sqrt{\frac{c^2}{\beta_2^2}} - 1\right)h\} = \frac{\left(\frac{\mu_3}{\mu_2}\right)\left(\sqrt{\frac{c^2}{\beta_3^2}} - 1\right)}{\sqrt{\frac{c^2}{\beta_2^2}} - 1} \tag{29}$$

Equation (29) is the classical equation as derived by Love [1].

Case-II: When $\xi_1 = h + d\cos px$ and $\xi_2 = d\cos px$, medium 1 and medium 2 coincide with upper corrugated surface, whereas the interface of the layer and the half-space stays flat when $\xi_3 = 0$

$$\tan\left\{k\sqrt{\frac{c^2}{\beta_2^2}-1}\,(h-d\cos px)\right\}=\left(\frac{\mu_3}{\mu_2}\right)\frac{\sqrt{\frac{c^2}{\beta_2^2}-1}\sqrt{\frac{c^2}{\beta_3^2}-1}\left(\left(\frac{c^2}{\beta_2^2}-1\right)+(pd\sin px)^2\right)}{\left(\frac{c^2}{\beta_2^2}-1\right)^2+\left(\frac{\mu_3}{\mu_2}\sqrt{\frac{c^2}{\beta_3^2}-1}pd\sin px\right)^2}+i\frac{\sqrt{\frac{c^2}{\beta_2^2}-1}pd\sin px\left(\frac{c^2}{\beta_2^2}-1\right)\left(\frac{\mu_3}{\mu_2}\right)^2\left(\frac{c^2}{\beta_3^2}-1\right)}{\left(\frac{c^2}{\beta_2^2}-1\right)^2+\left(\frac{\mu_3}{\mu_2}\sqrt{\frac{c^2}{\beta_3^2}-1}pd\sin px\right)^2} \tag{30}$$

Case-III: When $\xi_1 = d\cos px + h$, $\xi_2 = d\cos px$ and $\xi_3 = d\cos px$. This is the case when medium 1 and medium 2 coincide each other with corrugated upper boundary plane and corrugated interface

$$\tan\{k\sqrt{\frac{c^2}{\beta_2^2}-1}\,((d_2-d_1)\cos px+h)\}=\left(\frac{\mu_3}{\mu_2}\right)\sqrt{\frac{c^2}{\beta_3^2}-1}\frac{\left(\sqrt{\frac{c^2}{\beta_2^2}-1}\right)^3+T_1}{q_1^2+q_2^2}+i\frac{\left(\sqrt{\frac{c^2}{\beta_2^2}-1}\right)T_2}{q_1^2+q_2^2} \tag{31}$$

where
$q_1=\left(\frac{c^2}{\beta_2^2}-1\right)+\left(\frac{\mu_3}{\mu_2}-1\right)p^2d_1d_2\sin^2 px,\ q_2=\left(\frac{\mu_3}{\mu_2}\right)\left(\frac{c^2}{\beta_3^2}-1\right)pd_1\sin px$
$T_1=\sqrt{\frac{c^2}{\beta_2^2}-1}p^2d_1d_2\left(\frac{\mu_3}{\mu_2}-1\right)\sin^2 px-p^2d_1\{\left(\frac{\mu_3}{\mu_2}-1\right)d_2+d_1\}\sin^2 px,$
$T_2=\{\left(\frac{\mu_3}{\mu_2}-1\right)d_2+d_1\}p_1-\left(\frac{\mu_3}{\mu_2}\right)^2\left(\frac{c^2}{\beta_3^2}-1\right)d_1.$

The result given by case 1, case 2 and case 3 is in agreement with the result of Singh [7] and Noyer [12].

8 Numerical Computation and Discussion

Numerical calculations have been performed for Eq. (27) for different cases as discussed in Tables 1 and 2. The graphs have been plotted for dimensionless phase velocity versus dimensionless wave number to study the impact of inhomogeneity along with the corrugated surfaces. Considering $d = 0.1, pd = 0.05,\ ph = 1.40$ and $\frac{x}{h} = 0.04$, phase velocity decreases as the wave number increases for each of the cases as discussed in the table below.

Table 1 One periodic corrugated surface at a time

Cases	$\xi_1(x)$	$\xi_2(x)$	$\xi_3(x)$	Geometry of the problem
I	0	0	$d_1 \cos px$	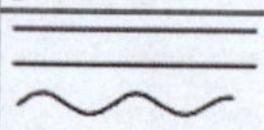
II	0	$d_1 \cos px$	0	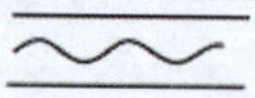
III	$d_1 \cos px$	0	0	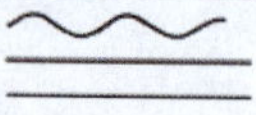

Table 2 One periodic corrugated surface at a time

Cases	$\xi_1(x)$	$\xi_2(x)$	$\xi_3(x)$	Geometry of the problem
IV	$d_1 \cos px$	$d_2 \cos px$	0	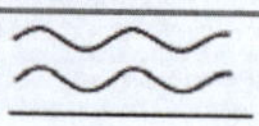
V	0	$d_1 \cos px$	$d_2 \cos px$	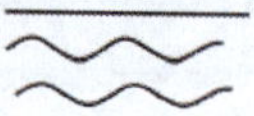
VI	$d_1 \cos px$	0	$d_2 \cos px$	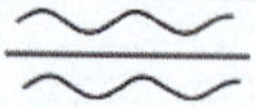

In Fig. 2, as the amplitude decays, the phase velocity decreases at a particular wave number in all the cases. However, when the corrugation lies in the top, the phase velocity is maximum at $d_1 = 1.0$ while it is least when the corrugated surface lies in the bottom.

In Fig. 3, as the inhomogeneity parameter α increases, the phase velocity also increases. The impacts are more visible in case I and II, and less in case III as curves are a bit overlapped.

Figures 4 and 5 show the fluctuation of phase velocity for the values of inhomogeneity parameter a and b. The effects are self-explanatory and reveals that it has great bearing on the velocity of Love wave.

Figure 6 reflects the interesting influence of initial stress associated with medium 3. It says, as the magnitude of the stress increases, the phase velocity decreases. However, when the corrugated surface lies in the middle, the phase velocity is highest and least when corrugated surface is at the bottom as per Table 1.

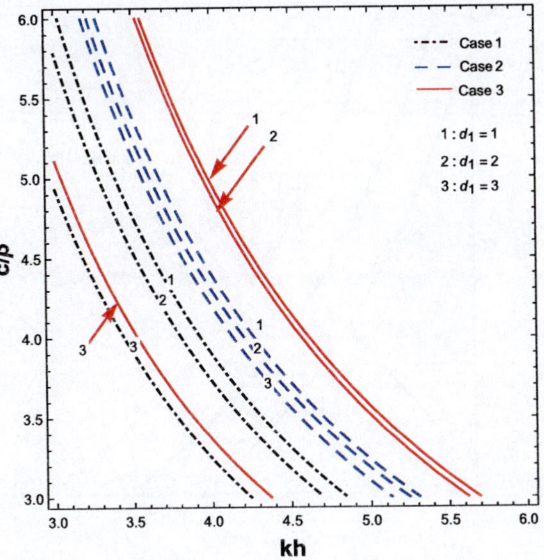

Fig. 2 Variation of dimensionless phase velocity against dimensionless wave number to study the impact of amplitude (d_1) of periodically corrugated surface

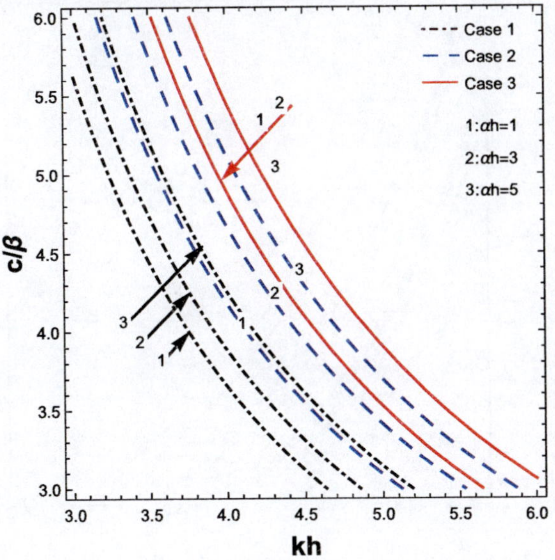

Fig. 3 Variation of dimensionless phase velocity against dimensionless wave number to study the inhomogeneity parameter (αh) associated with medium-1

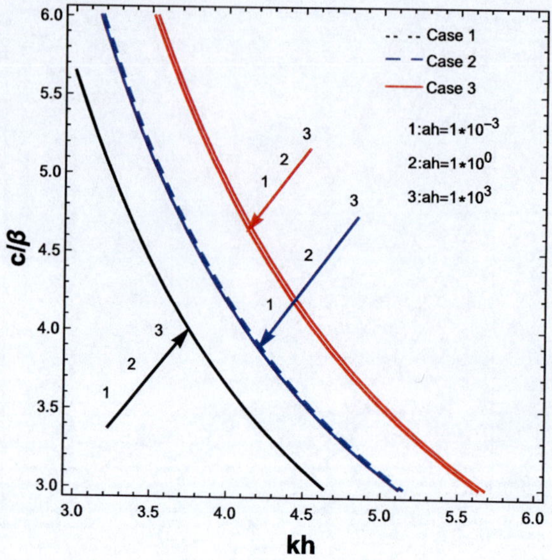

Fig. 4 Variation of dimensionless phase velocity against dimensionless wave number to study the inhomogeneity parameter (*ah*) associated with medium-3

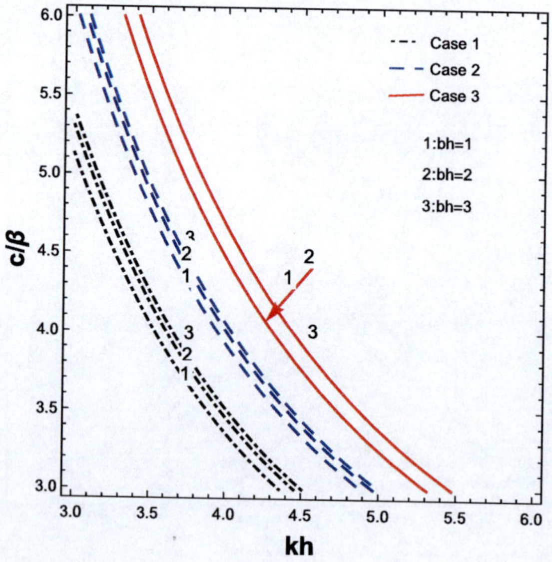

Fig. 5 Variation of dimensionless phase velocity against dimensionless wave number to study the inhomogeneity parameter(*bh*) associated with medium-3

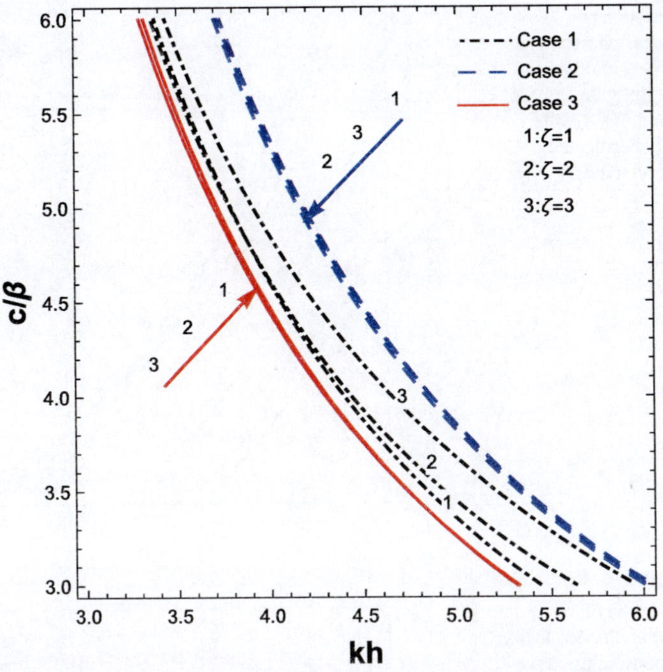

Fig. 6 Variation of dimensionless phase velocity against dimensionless wave number to study the dimensionless initial stress($\zeta = \frac{p_3}{2\mu_3}$) associated with medium-3

In Figs. 7 and 8, study has been made on the phase velocity for the different magnitude of the amplitude where the pattern of curves are opposite to each other. With the help of table II, it is found that as the amplitude d_1 increases (or constant), keeping the amplitude d_2 constant (or increases), the phase velocity decreases (or increases). It reveals the fact that velocity of wave is greatly moved by the degree of corrugation. Moreover, velocity almost coincides for the higher range of wave number, thereby providing a upper bound of $kh = 5.5$ and $kh = 5.0$ (approx) respectively.

9 Conclusion

In the present paper, analytic study has been performed to study the phase velocity of Love-type waves in a three-layered medium. The medium has been assumed with irregular surfaces along with inhomogeneities associated with the rigidities and densities of the medium. The variations in rigidities, densities, and initial stress have been assumed as the product of linear algebraic function and the exponential function of depth. Separate displacements have been found in all three media. Suitable

Fig. 7 Variation of dimensionless phase velocity against dimensionless wave number to study the impact of amplitude (d_1, d_2) of periodically corrugated surface (for fixed d_1)

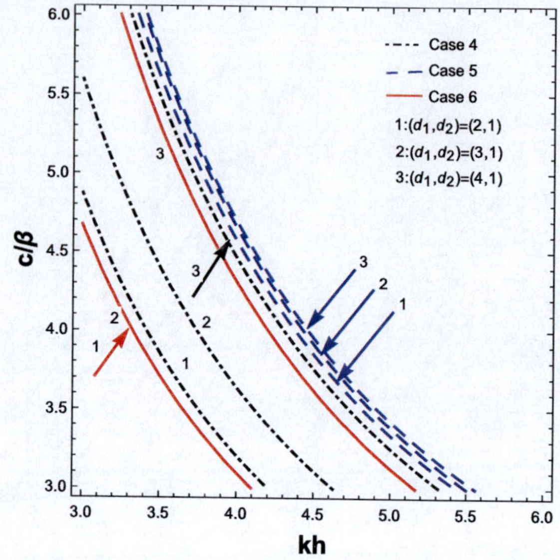

Fig. 8 Variation of dimensionless phase velocity against dimensionless wave number to study the impact of amplitude (d_1, d_2) of periodically corrugated surface (for fixed d_2)

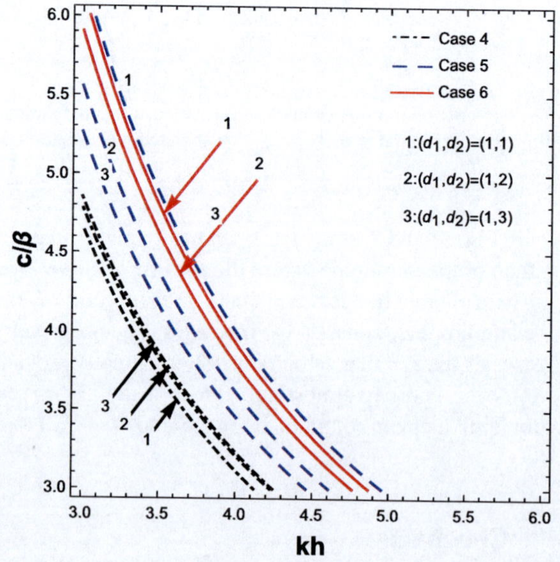

boundary conditions have been employed, and the dispersion equation has been derived in the closed-form. Case-wise investigations have been carried out which concludes the following results.

- The fluctuations of phase velocity are high when the amplitude of the periodic irregularities of the surfaces increases. However, the information about the level of decrement and increment can be achieved by comparing the graphs of various cases.
- It has been found that as the magnitude of the homogeneity increases, the phase velocity decreases or vice versa. In some cases, the increasing magnitude also increases the phase velocities. However, the trends are strictly followed in some cases and remain unfollowed in a few cases. It reflects the fact that inhomogeneity has a great bearing on the phase velocity of Love-type seismic waves.
- It has also been found that initial stress affects the velocity to a great extent, which is clearly visible in the graphs. Few critical points have been found where the curves are intersecting each other, thereby reflecting the fact that although the initial stress changes but the velocities remain same.

Acknowledgements Authors extend their sincere thanks to SERB-DST, New Delhi, for providing financial support under Early Career Research Award with Ref. No. ECR/2017/001185. Authors are also thankful to DST New Delhi, for providing DST-FIST grant with Ref. no. 337 to Department of Mathematics, BITS-Pilani, Hyderabad Campus.

10 Appendix I

$$D_{11} = \left(\frac{1}{2}\left(-\log\alpha - \sqrt{-4A + (\log\alpha)^2}\right) - ik\xi_1'\right) \mathrm{e}^{\frac{\xi_1(x)-2h}{2}\left(-\log\alpha - \sqrt{-4A+(\log\alpha)^2}\right)}$$
$$D_{12} = \left(\frac{1}{2}\left(-\log\alpha + \sqrt{-4A + (\log\alpha)^2}\right) - ik\xi_1'\right) \mathrm{e}^{\frac{\xi_1(x)-2h}{2}\left(-\log\alpha + \sqrt{-4A+(\log\alpha)^2}\right)}$$
$$D_{21} = \frac{L_1}{2}(-\log\alpha - \sqrt{-4A + (\log\alpha)^2} - ik\xi_1')\mathrm{e}^{(\xi_2(x)-h)}(-\log\alpha - \sqrt{-4A + (\log\alpha)^2})$$
$$D_{22} = \frac{L_1}{2}(-\log\alpha + \sqrt{-4A + (\log\alpha)^2} - ik\xi_1')\mathrm{e}^{\xi_2(x)-h)}(-\log\alpha - \sqrt{-4A + (\log\alpha)^2})$$
$$D_{23} = -\mu_2(\nu_1 - \xi_2')ik\mathrm{e}^{ik\nu_1(\xi_2(x)-h)}$$
$$D_{24} = -\mu_2(\nu_1 + \xi_2')ik\mathrm{e}^{-ik\nu_1(\xi_2(x)-h)}$$
$$D_{31} = \mathrm{e}^{\frac{\xi_2(x)-h}{2}\left(-\log\alpha - \sqrt{-4A+(\log\alpha)^2}\right)}$$
$$D_{32} = \mathrm{e}^{\frac{\xi_2(x)-h}{2}\left(-\log\alpha + \sqrt{-4A+(\log\alpha)^2}\right)}$$
$$D_{33} = -\mathrm{e}^{ik\nu_1(\xi_2(x)-h)}$$
$$D_{34} = -\mathrm{e}^{-ik\nu_1(\xi_2(x)-h)}$$
$$D_{43} = ik\mu_2(\nu_1 - \xi_2')\mathrm{e}^{ik\nu_1\xi_3(x)}$$
$$D_{44} = -ik\mu_2(\nu_1 + \xi_2')\mathrm{e}^{-ik\nu_1\xi_3(x)}$$
$$D_{45} = \mu_3\left(-b - \sqrt{-4A_1 + b^2}\right)\mathrm{e}^{\frac{\xi_3(x)}{2}\left(-b - \sqrt{-4A_1+b^2}\right)}$$
$$\text{Hypergeometric}\, U\left(\frac{b+\sqrt{-4A_1+b^2}}{2\sqrt{-4A_1+b^2}}, 1, \frac{\sqrt{-4A_1+b^2}}{a} + \sqrt{-4A_1 + b^2}\xi_3(x)\right)$$

$$-\mu_3\left(-b-\sqrt{-4A_1+b^2}\right)\mathrm{e}^{\frac{\xi_3(x)}{2}\left(-b-\sqrt{-4A_1+b^2}\right)}$$

$$\text{Hypergeometric}U\left(1+\frac{b+\sqrt{-4A_1+b^2}}{2\sqrt{-4A_1+b^2}},2,\frac{\sqrt{-4A_1+b^2}}{a}+\sqrt{-4A_1+b^2}\xi_3(x)\right)$$

$$\frac{ik\xi_3'}{\frac{1}{2}\left(-b-\sqrt{-4A_1+b^2}\right)}$$

$$\text{Hypergeometric}U\left(\frac{b+\sqrt{-4A_1+b^2}}{2\sqrt{-4A_1+b^2}},1,\frac{\sqrt{-4A_1+b^2}}{a}+\sqrt{-4A_1+b^2}\xi_3(x)\right)$$

$$D_{53}=\mathrm{e}^{ik\nu_1\xi_3(x)}$$

$$D_{54}=\mathrm{e}^{-ik\nu_1\xi_3(x)}$$

$$D_{55}=\mathrm{e}^{\frac{\xi_3(x)}{2}\left(-b-\sqrt{-4A_1+b^2}\right)}$$

$$\text{Hypergeometric}U\left(\frac{b+\sqrt{-4A_1+b^2}}{2\sqrt{-4A_1+b^2}},1,\frac{\sqrt{-4A_1+b^2}}{a}+\sqrt{-4A_1+b^2}\xi_3(x)\right)$$

$$D_{13}=D_{14}=D_{15}=D_{25}=D_{35}=D_{41}=D_{42}=D_{51}=D_{52}=0$$

References

1. Love AEH (1911) Some problems of geodynamics. Cambridge University Press, London
2. Bullen KE (1940) The problem of Earth's density variation. Bull Seismol Soc Am 30:235–250
3. Sari C, Salk M (2002) Analysis of gravity anomalies with hyperbolic density contrast: an application to the gravity data of Western Anatolia. J Balkan Geophys Soc 5:87–96
4. Wolf B (1970) Propagation of Love waves in layers with irregular boundaries. Pure Appl Geophys 78:48–57
5. Santos P, Tadeu A (2004) Scattering of seismic waves generated by an irregular seabed. Comput Struct 82:1793–1804
6. Zhang W, Chen X (2006) Traction image method for irregular free surface boundaries in finite difference seismic wave simulation. Geophys J Int 167:337–353
7. Singh S (2010) Love wave at a layer medium bounded by irregular boundary surfaces. J Vib Control 17:789–795
8. Poonia R, Madan DK (2014) Love wave propagation in an irregular fluid saturated porous anisotropic layer with rigid boundary. J Appl Sci Res 10:8
9. Mistri KC, Singh AK, Das A (2018) Attenuation and dispersion of SH-waves in a loosely bonded sandwiched fluid saturated porous layer. Soil Dyn Earthq Eng 107:350–362
10. Liu X, Chen J, Zhao Z et al (2018) Simulating seismic wave propagation in viscoelastic media with an irregular free surface. Pure Appl Geophys 175:3419–3439
11. Aki K (1960) The use of Love waves for the study of earthquake mechanism. J Geophys Res 65:323–331
12. Noyer JD (1961) The effect of variations in layer thickness on Love waves. Bull Seismol Soc Am 51:227–235
13. Wang T, Tang X (2003) Finite-difference modeling of elastic wave propagation: A nonsplitting perfect matched layer approach. Geophysics 68:1749–1755
14. Du J, Jin X, Wang J (2007) Love wave propagation in layered magneto-electo-elastic structure with initial stress. Acta Mechanica 192:169–189
15. Du J, Xian K, Wang J et al (2008) Propagation of Love waves in prestressed Piezoelectric layered structures loaded with viscous liquid. Acta Mechanica Solida Sinica 21:542–548
16. Gupta S, Majhi DK, Kundu S et al (2013) Propagation of Love waves in non-homogeneous substratum over initially stressed heterogeneous half-space. Appl Math Mech 34:249–258

17. Vishwakarma SK, Panigrahi TR, Kaur R (2018) Love wave frequency under the influence of linearly varying Shear moduli, initial stress and density of orthotropic half space. In: International conference on mathematics and computing, pp 209–223
18. Vishwakarma SK, Kaur R, Panigrahi TR (2018) Love wave frequency in an orthotropic crust over a doubled-layered anisotropic mantle. Soil Dyn Earthq Eng 110:86–92
19. Vishwakarma SK, Panigrahi TR, Kaur R (2019) SH-wave propagation in linearly varying fiber-reinforced viscoelastic composite structure uninitial stress. Arab J Geosci 12(2):59
20. Gubbins D (1990) Seismology and plate tectonics. Cambridge University Press, Cambridge
21. Biot MA (1940) The influence of initial stress on elastic waves. J Appl Phys 11:522–530

Numerical Study of Dynamic Loading on Ballastless Railway Tracks

Pranjal Mandhaniya, J. T. Shahu, and Mayuresh Bakare

Abstract In this study, an attempt is made to analyze the ballast-less track by modelling it as a 3D problem, and the solution is obtained using the ABAQUS CAE FEM package. Designs similar to RHEDA slab track has been modelled, which consists of three layers—concrete/asphalt bearing layer (CBL/ABL), hydraulically bonded layer (HBL) and frost protection layer (FPL, used for freezing temperatures) over subsoil base. The load is applied in static/linear step and motion is incorporated in dynamic/implicit step, point loads of 100 kN at the centre of each wheel are applied, and results for different speeds are obtained. Time increment is decreased to make the loading rate consistent. Indian broad gauge and UIC60 (approx.) rails are used. Mesh sizes are kept optimum to balance result precision and CPU time. For this study, speed of load is proportional to the difference between maximum downward deflection and maximum uplift. Maximum deflection also shows a linear correlation with both subcritical and supercritical speeds. The area under the curve for deflection versus normalized distance along path shows good proportionality at subcritical speeds.

Keywords Finite element analysis · Abaqus · Dynamic load · Traffic load · Ballastless rail track

1 Introduction

In India, construction of rail track is built according to the standards provided by RDSO [6, 7]. Generally, it is constructed in cut or as embankment, which is similar to road/highway but asphalt layer is replaced by superstructure. Ballasted track consists

P. Mandhaniya (✉) · J. T. Shahu · M. Bakare
IIT Delhi, New Delhi 110016, India
e-mail: pranjalmandhaniya@gmail.com

J. T. Shahu
e-mail: shahu@civil.iitd.ac.in

M. Bakare
e-mail: bakaremayuresh@gmail.com

T. G. Sitharam et al. (eds.), *Seismic Hazards and Risk*, Lecture Notes in Civil Engineering 116, https://doi.org/10.1007/978-981-15-9976-7_29

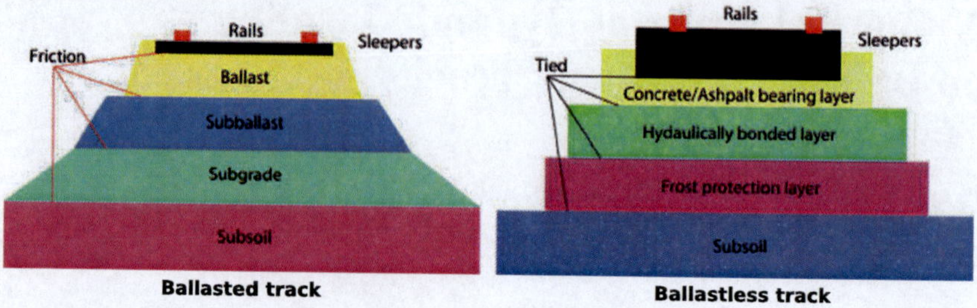

Fig. 1 Typical configurations of ballasted and ballastless rail tracks

of three layers: subgrade, subballast and ballast. All these are aggregate materials but with different properties i.e. grain size distribution, rock strength, CBR and cohesion. The role of ballast aggregates is to transfer the load to the ground. They not only absorb vibrations from moving load but also help in drainage. This vibration absorption can cause ballast fouling. Hence, it requires high maintenance. With time, fouled particles render the soil base impermeable, causing drainage problems. This is a major disadvantage of ballasted track systems. Such shortcomings motivates to think about its alternative, such as Ballastless track systems. Use of ballastless track is very limited in India. It is mainly used in metro rail projects, which uses BLT system similar to German slab track systems. BLT design similar to this are being employed in Delhi metro and recently built Pir Panjal tunnel in Jammu and Kashmir. Its design consists of sleeper resting on concrete/asphalt base layer as shown in Fig. 1, which a substructure is present. Substructure is also divided into several parts/layers of different material. Reason of it being called ballastless is the replacement of ballast with a combination of CBL/ABL and elastic pads. Concrete can not handle vibration like ballast. Thus, its design consists of anti-noise mechanisms and elastic pads to damp the vibrations.

Though its initial cost is high, maintenance cost is negligible and it provides long term benefits like reduction in life cycle cost and passenger comfort. There is an uncertainty in calculating its overall behavior because any existing ballastless line has not yet completed design period. This study will address the question of ballastless rail track design for high speed trains.

2 Major Literature Study

Feng [1] studied the influence of design parameters on dynamic response of the railway track structure by implementing Finite Element Method (FEM). Different railway track systems have been simulated, including: beam on discrete support model, discretely support track including ballast mass model and rail plus sleeper on continuum model. He modeled rail and sleeper as Euler-Bernoulli beam element. Railpads and the connection between sleeper and ballast are modelled using Spring and dashpot elements. FEM analysis is divided into three categories: eigenvalue, dynamic and general static. The eigen frequencies and corresponding vibration modes were extracted from all the models. Main part of the finite element modeling involves the steady-state dynamic analysis, in which receptance functions were obtained and used as the criterion for evaluating the properties of track components. Dynamic explicit analysis has been used for the simulation of a moving load, and the train speed effect has been studied. The displacement of trackbed has been evaluated and compared to the measurement taken by Kjorling [2].

Michas [5] also used ABAQUS CAE FEM software to model a 3D slab track design under static traffic loading. The results suggest that slab track have profoundly better stability and durability comparing to ballasted track mainly due to its higher stiffness and strength. 3D FEM modeling of rail track was also attempted by Sowmiya et al. [8]. In their model, ballast and a subballast layer placed over a compacted subgrade (formation soil) like conventional structure of a railway track system. MIDAS (GTS) software was used to create three dimensional finite element models.

3 Numerical Study in ABAQUS

ABAQUS model is shown in Fig. 2. The geometrical and material properties of different components used in ABAQUS modelling are shown in Table 1. The dimensions given in the table are formatted as width × thickness × length.

Moving load is modeled as a rigid wheel-axle assembly (half-bogie) of negligible mass and 200 kN load at its center with general contact condition including zero friction (smooth). This will be convenient for load transfer while performing motion. The center to center spacing between sleepers is 0.5 m. Sleeper dimensions are adapted from RDSO Indian railway standard [7].

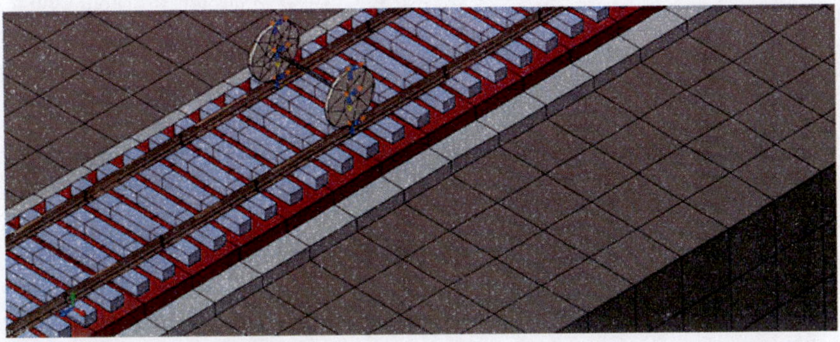

Fig. 2 ABAQUS snapshot of ballastless track model

Table 1 Component properties for ABAQUS

Component	Dimensions (m^3)	Mesh size	*E* (GPa)	Density (kg/m^3)	Poisson's ratio
Rail	UIC60 (approx.)	2	207	7850	0.28
Sleeper	0.25 × 0.21 × 2.6	1	70	2400	0.2
CBL	3 × 0.25 × 300	1	35	2400	0.2
HBL	4 × 0.25 × 300	1	5	2400	0.28
Subsoil	12 × 12 × 300	1	0.01	2000	0.4

A track length of 300 m is chosen for dynamic analysis which is appropriate for the values of speeds considered. The initial position of load is 100 m from any end, then the motion is performed for 100 m at various speeds.

3.1 Mesh, Interfaces and Boundary Conditions

Mesh sizes and types of components are defined using default setting in ABAQUS GUI. A structured C3D8R mesh is used for all the parts based on a mesh convergence study as shown in Fig. 3. According to the mesh convergence study, the mesh sizes are changed to check the computational time with precision and the mesh sizes which are used in the final analysis provides adequate precision without much computational efforts.

All layers are tied at their interfaces to allow for load transfer from top to bottom. Tied constraint allows no slip and no separation which is similar to rough contact in both tangential and normal directions. All vertical boundaries are applied with roller boundary i.e. allowing settlements without horizontal displacements. Bottom boundary of soil is fixed allowing no displacement and rotations (ENCASTRE).

3.2 Analytical Steps

Analysis is performed using two steps.

1. Static/general step of 0.1 s is used to apply load from 0 to 100 kN. Loads of 100 kN at each contact is applied i.e. gross moving load = 200 kN.

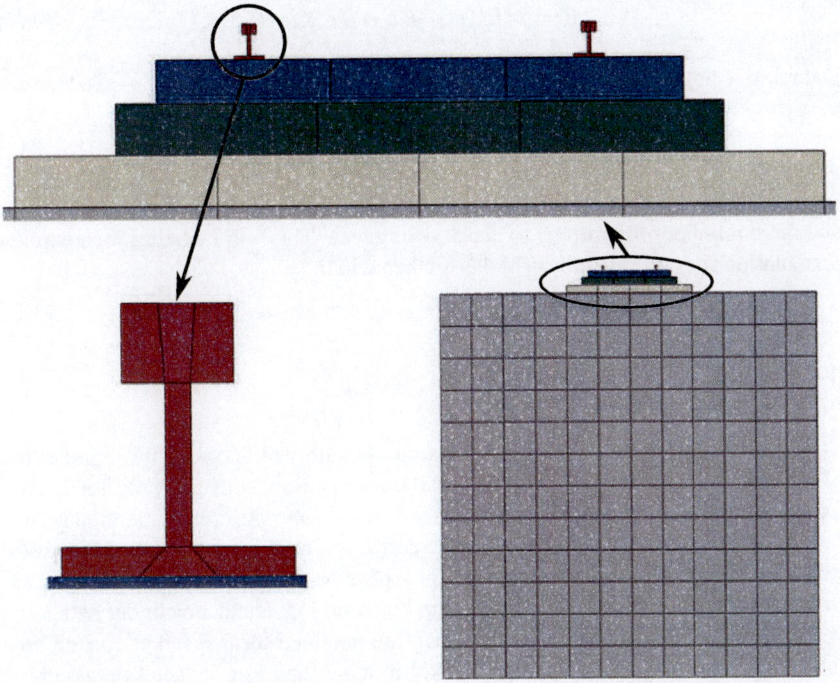

Fig. 3 Mesh view of BLT cross-section

Table 2 Speed variation in dynamic/implicit step

Speed (km/h)	50	100	150	200	250	300
Step time (s)	7.2	3.6	2.4	1.8	1.44	1.2

2. Dynamic/implicit step used after that for moving 100 kN load 100 m distance with initial increment 0.00625 s interval. Table 2 shows variation of step time to change speed of moving load. Results are obtained at 10 m of the motion for all speeds.

Static loading step is used as it will not interfere with the mass properties of the model based on its formulation given in Eq. (1). Static analysis will give stress based on the material stiffness only.

$$F = Ku(t) \tag{1}$$

Moving load analysis, on the other hand, will require addressing the dynamic effect and overall response of the structure. Dynamic/implicit formulation does that by solving the Eq. (2), where is calculates the stress from mass, damping and stiffness matrices.

$$F = Mu''(t) + Cu'(t) + Ku(t) \tag{2}$$

Damping is not used in current analysis so Eq. (2) becomes Eq. (3).

$$F = Mu''(t) + Ku(t) \tag{3}$$

Implicit formulation is intrinsically stable due to the incorporation of tolerances and large number of iterations to check divergence. This will be better then explicit formulation for current study with elastic behaviour.

4 Results

The results are recorded along the nodes on the path which passes through the top of the subsoil layer at the center of the rail track. These results are valid for 200 kN axle moving load at the conditions described above. Four different deflection related factors are analyzed to establish a speed-deflection relation. Maximum downward deflection in Fig. 4, maximum uplift in Fig. 5, peak-to-peak (difference between maximum downward deflection and maximum uplift) in Fig. 6 and area under deflection versus normalized distance curve in Fig. 7, are obtained for 6 speed at 10 step time intervals. Load magnitude used is 200 kN moving load as a wheel-axle assembly (100 kN per wheel) for all speeds.

All four quantities suggest that deflection increases till speed equals critical speed then deflection starts to decrease. For this track length i.e. 300 m, It can be inferred that the critical speed is near the value of 150 km/h as shown in Fig. 4.

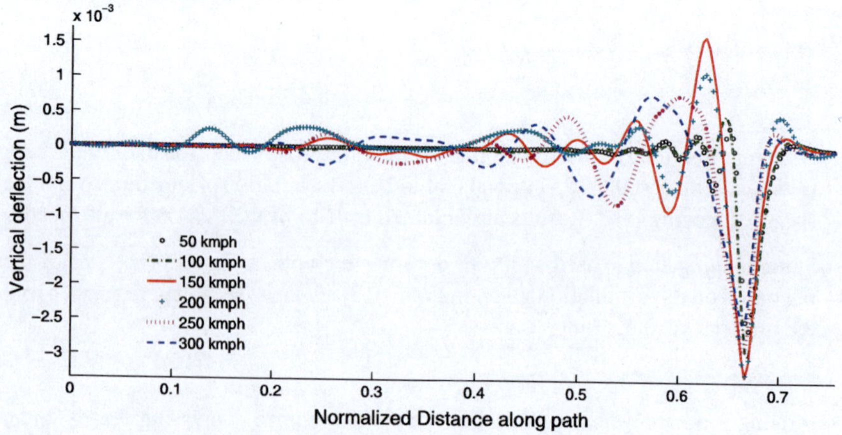

Fig. 4 Vertical deflection versus normalized distance along path curves when motion finishes

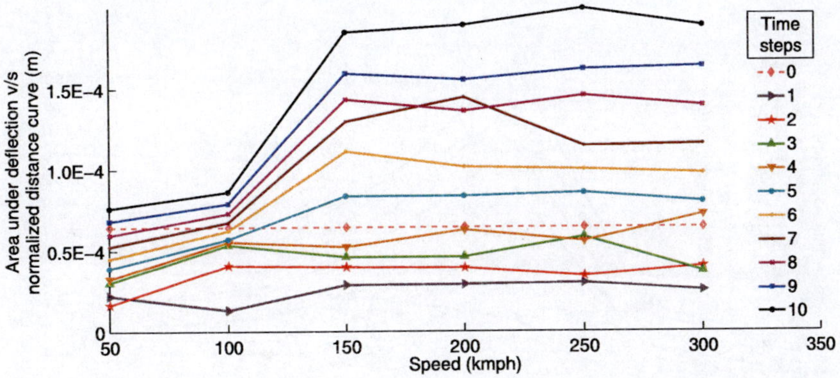

Fig. 5 Area under vertical deflection versus normalized distance along path curves when motion finishes

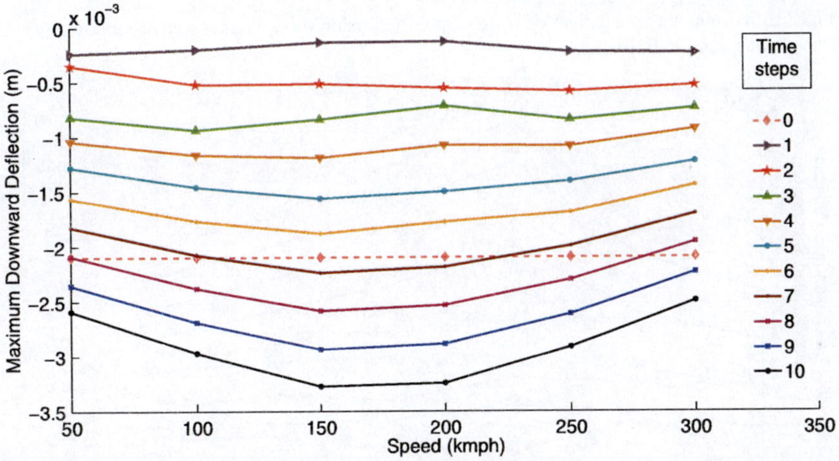

Fig. 6 Maximum downward deflection from vertical deflection versus normalized distance along path curves when motion finishes

Figure 5 shows that area under deflection versus normalized distance curve is proportional to speed for speed values of 50, 100 and 150 kmph, with adjusted R-squared 0.64, but after that it remains almost constant.

It can be inferred from Fig. 6 that maximum downward deflection shows linear correlation with speed. This linear pattern is increasing for subcritical and decreasing for supercritical speeds. Adjusted R-squared value for this fit is 0.99 for both subcritical and supercritical speeds. From maximum deflection, it can be inferred from Fig. 6 that critical speed falls between 150 and 200 kmph.

It can be inferred from Fig. 7 that critical speed for this model is very near to 150 kmph as uplift deflection took steep fall after 150 kmph. For subcritical speed

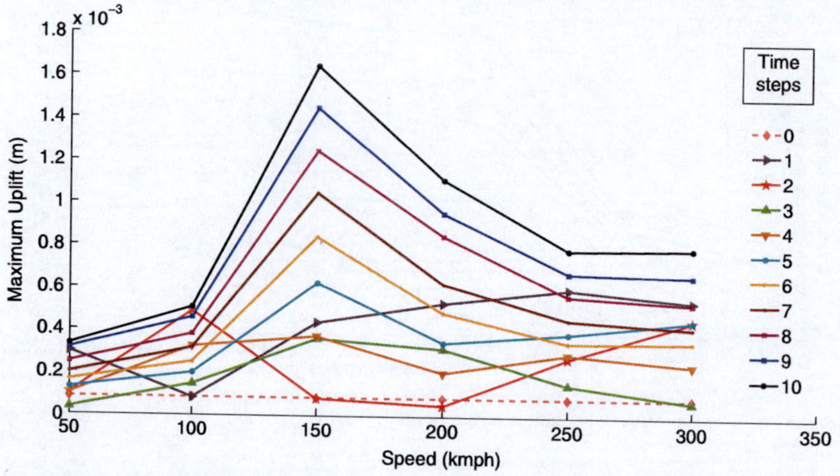

Fig. 7 Maximum upward deflection from vertical deflection versus normalized distance along path curves when motion finishes

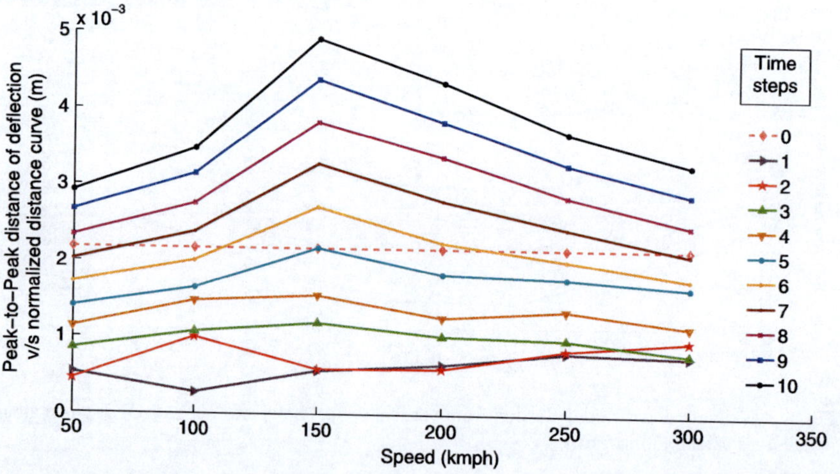

Fig. 8 Peak to peak deflection from vertical deflection versus normalized distance along path curves when motion finishes

(50, 100 and 150 kmph), linear correlation between maximum uplift and deflection is weak with adjusted R-squared of 0.7 and for supercritical speeds of 150, 200 and 250 kmph, correlation is almost linear with adjusted R-squared equaling 0.96. This shows maximum uplift is almost linearly correlates for supercritical speed.

Peak-to-peak deflection can be directly correlated with speed as shown in Fig. 8. Adjusted R-squared value for linear relation between deflection and speed for 50, 100 and 150 kmph comes out to be 0.878, whereas for 150, 200, 250 and 300 kmph

Fig. 9 Vertical deflection contours at 50, 100, 150 and 200 kmph

it comes out to be 0.989. It is evident that peak-to-peak deflection is most eligible for forming relation between deflection and speed.

Figure 9 shows stress field for four speeds 200 m when the motion of the wheel finishes. Blue color shows zero stress and yellowish-red shows maximum stress. Intermediate colors shows variation of stress in track strata. It can be seen that at supercritical speeds (150 and 200 kmph), significant deflection occurs behind the load. It can be inferred that the point of maximum deflection shifts behind the load with increasing speed [4]. Both results have been verified from Mallik et al. [3].

5 Conclusions

Based in the results shown above some conclusions can be drawn from the present study.

- Critical speed for moving load model is found to be between 150 and 200 kmph on the basis of maximum downward deflection and maximum uplift, in this hypothesis.
- It can be seen that speed is linearly correlated to maximum downward deflection for both subcritical and supercritical part having R-squared value 0.99. Peak-to-peak distance (difference between maximum downward deflection and maximum uplift) also shows linear correlation with adjusted R-squared of 0.989 for super critical speeds.

- Moving load produces impact due to deflection waves i.e. its effect remains at a point even though it might have moved away from the point of action, this is the reason for its higher intensity then stationary load.
- The point of maximum deflection shifts behind the load with increasing speed. At supercritical speeds of the load, significant deflection occurs behind the load. This also clarifies the definition of critical speed i.e. where maximum deflection and loading point moves with zero phase difference.

References

1. Feng H (2011) 3D-models of railway track for dynamic analysis. Master's thesis, KTH Sweden
2. Kjorling M (1995) Dynamic response of railway track components. Master's thesis, KTH Sweden
3. Mallik A, Chandra S, Singh AB (2006) Steady-state response of an elastically supported infinite beam to a moving load. J Sound Vib 291(3):1148–1169
4. Mandhaniya P, Chandra S (2015) Analysis of a moving load on ballastless rail track, vol 50. Indian Geotechnical Society, Indian Geotechnical Conference
5. Michas G (2012) Slab track systems for high-speed railways. Master's thesis, KTH Sweden
6. RDSO (2008) The technical diary of IPWE (India). Technical report, Institution of Permanent Way Engineers, India
7. RDSO L (2016) Indian railway standard specification for pre-tensioned prestressed concrete sleepers for broad gauge and metre gauge
8. Sowmiya L, Shahu J, Gupta K (2010) Three dimensional finite element analysis of railway track. In: Indian geotechnical conference. Indian Geotechnical Conference

Ultrasonic P Wave Velocity Measurement of Sedimentary Rocks in Laboratory and Its Dependency on Different Parameters by Using Multivariate Regression Analysis

Siddharth Garia, Arnab Kumar Pal, Archana M. Nair, and K. Ravi

Abstract Ultrasonic Pulse Velocity (UPV) is one of the non-destructive geophysical method commonly used in different fields such as civil, mining, geotechnical, and rock engineering. Ultrasonic technique is generally used to reproduce seismic waves in the laboratory. By doing so, different petrophysical and mechanical properties of rocks can be investigated by correlating it with the ultrasonic velocity. However, the range of frequency of the ultrasonic technique, to be used in laboratory or field, may also influence the velocity measurement, although it is not yet highlighted in the literature. The standards (ASTM/ISRM codes) only stipulate that as long as the wavelength is significantly greater than the grain size, the velocity values determined do not appear to be frequency dependent. In order to explore the influence of frequency, UPV measurements were conducted at 54 and 500 kHz frequency on dry rock cores of varying dimensions at ambient conditions in the laboratory. It was observed that 500 kHz transducers gave higher velocity measurement values than 54 kHz transducers. UPV_{54kHz} and UPV_{500kHz} were correlated using regression analysis, and R^2 obtained was 0.76. Moreover, a method based on multivariate analysis was developed to determine the influence of different parameters, namely diameter, length, aspect ratio, bulk density of sedimentary rocks on UPV measurements, and subsequently, R^2 increased to 0.91.

Keywords Ultrasonic pulse velocity · P wave velocity · Multivariate analysis

The original version of this chapter was revised: The source details of Figure 3 has been updated in its caption. The correction to this chapter is available at https://doi.org/10.1007/978-981-15-9976-7_34

S. Garia (✉) · A. K. Pal · A. M. Nair · K. Ravi
Department of Civil Engineering, IIT Guwahati, Guwahati 781039, India
e-mail: sidd_41@iitg.ac.in

A. K. Pal
e-mail: arnab.pal@iitg.ac.in

A. M. Nair
e-mail: nair.archana@iitg.ac.in

K. Ravi
e-mail: ravi.civil@iitg.ac.in

T. G. Sitharam et al. (eds.), *Seismic Hazards and Risk*, Lecture Notes in Civil Engineering 116, https://doi.org/10.1007/978-981-15-9976-7_30

1 Introduction

1.1 Overview

Ultrasonic P wave velocity test, commonly recognized as UPV test is a non-destructive, cost effective, reliable technique for both on-site and laboratory measurements and is extensively used for applications pertaining to the field of mining, geotechnical, and civil engineering practices [7, 10, 11, 14, 15, 18, 19, 21, 22, 28, 29, 32]. For rock engineering applications, this technique is used for accessing the dynamic properties of rock as the velocity which investigates the structure and texture of the rock [32]. Most of the literature correlated this ultrasonic P wave velocity to different petrophysical and mechanical properties [4–6, 9–11, 13–16, 18–32] and established this UPV technique for classification and characterization of different types of rocks. UPV testing is one type of laboratory-based techniques to reproduce and replicate seismic wave velocity through ultrasonic technique. There are several factors such as porosity, confining pressure, unit weight, lithology, grain size and shape, anisotropy, temperature, and weathering that affect P wave measurement [4–6, 14–23, 32]. Apart from these petrophysical and geomechanical factors that affect Vp measurement, exponential, and polynomial relationships were found while investigating sample diameter effect on ultrasonic velocity for fixed length of specimen [8].

1.2 Ultrasonic Velocity Measurement

Portable ultrasonic non-destructive digital indicating tester, also widely known as PUNDIT, is generally used to estimate the velocity of wave through the rock cores [3, 16, 18, 24, 32]. It consists of a pair of transducers (one acting as a transmitter, the other one as a receiver) of varying frequency, 54 kHz to 1 MHz, pulse generator, and an electronic counter for propagation time measurements. Apart from this, the velocity of the rock samples can also be measured by using an arrangement of transducers (transmitter and receiver) of varying frequency, 54 kHz to 1 MHz, pulse generator, and an oscilloscope [15, 25, 31]. The different ultrasonic techniques, namely pulse echo [1, 10, 19] and through transmission [11, 15, 18, 21, 24, 28, 32], have been used to measure Vp, Vs (compressional shear wave velocity) in different rock specimens by various researchers [13–32] in their respective studies. Generally, when only one transducer that acts both as a transmitter and receiver is used for measuring Vp, Vs, it is known as pulse-echo technique; while when a pair of transducers is used for measuring velocity, it is known as through transmission technique. Indirect measurement (pulse echo) may employ two transducers. However, the range of frequency of the ultrasonic technique, to be used in laborarory or field, may also influence the velocity measurement, although it is not yet highlighted in the literature. As per ISRM/ASTM standards, the velocity values determined do not appear to be dependent on frequency as long as the associated wavelength is significantly greater than the sample grain size [2, 12]. This study investigates the effect of transducer frequency on the UPV measurement in sedimentary rocks, especially sandstone and carbonates taking into consideration other physical parameters. UPV measurements were conducted in the laboratory by using 54 and 500 kHz frequency of transducers at through transmission mode on dry rock cores at ambient conditions. The test results were statistically analyzed, and a method based on multivariate analysis was developed to determine the influence of different parameters such as diameter, length, aspect ratio, and bulk density of sedimentary rocks on UPV measurements based at laboratory scale.

2 Materials and Methodology

2.1 *Sampling*

For this study, total eleven number of samples of varying lengths and diameter (Fig. 2) was taken from different parts of India; out of which nine sandstone samples were collected from upper Assam region (mainly Mizoram, Margherita), and the rest two-carbonate samples were collected from Bombay offshore region as shown in Fig. 1. These regions are considered as prolific hydrocarbon producing region in India, especially Upper Assam and Bombay offshore basin.

Rock cores, free of any visible cracks/joints, and unweathered were analyzed in this study. As per ASTM/ISRM code, the travel distance through the rock samples should be ten times the average grain size of the rock sample. All the samples follow these standard guidelines for performing UPV measurements.

A variety of samples with different lengths and diameter were taken for analysis. The description of the analyzed rock core samples that include lithology, length, and diameter of all the core plugs was measured and tabulated in Table 1. The range of grain size varies from 0.1 to 0.25 mm.

2.2 *Experimental Procedure*

Laboratory-based coring and sawing machine were used to cut uneven rock samples collected from the site to make it suitable for use in the ultrasonic pulse velocimetry test. UPV tests were conducted on all the samples of different diameter and of varying length as shown in Fig. 2. Before performing the UPV test, the samples were oven dried (temperature 105 °C, for 24 h). UPV tests were performed by PUNDIT based on the measurement of the time of propagation of ultrasound pulses by a pair of transducers of 54 and 500 kHz frequency according to ISRM 2007 (Fig. 3). There are commercially available transducers of different frequencies such as 54, 150, 250,

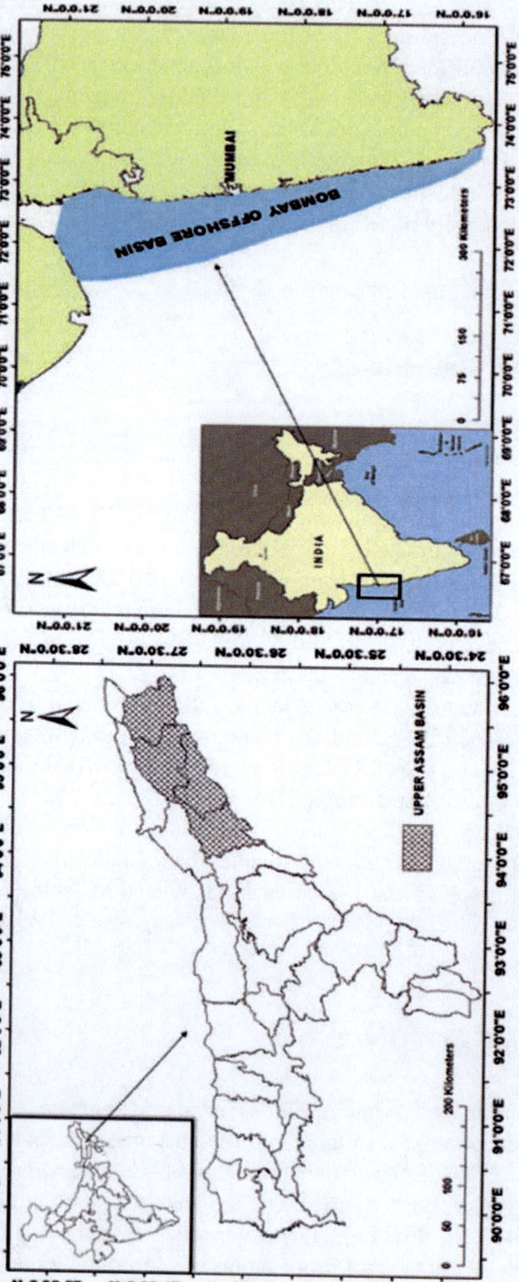

Fig. 1 Map showing sampling locations across different parts of India

Table 1 Description of the rock cores used for the UPV test

S. no.	Rock code	Lithology	Length (mm)	Diameter (mm)	Length to diameter ratio (Aspect ratio, AR)
1	A1	Sandstone	51	54.6	0.93
2	A2	Sandstone	32	54	0.59
3	A3	Sandstone	33.76	54.6	0.62
4	A4	Sandstone	30.4	54.9	0.55
5	A5	Sandstone	55.7	56	0.99
6	A6	Sandstone	30	54.5	0.55
7	A7	Sandstone	70.4	48.3	0.77
8	A8	Sandstone	37.45	25.14	1.49
9	A9	Sandstone	56.11	25.45	2.20
10	A10	Carbonate	73.05	25.29	2.89
11	A11	Carbonate	47.33	25.41	1.86

Fig. 2 Core samples collected for performing ultrasonic pulse velocimetry measurement (A1 to A11 from left to right)

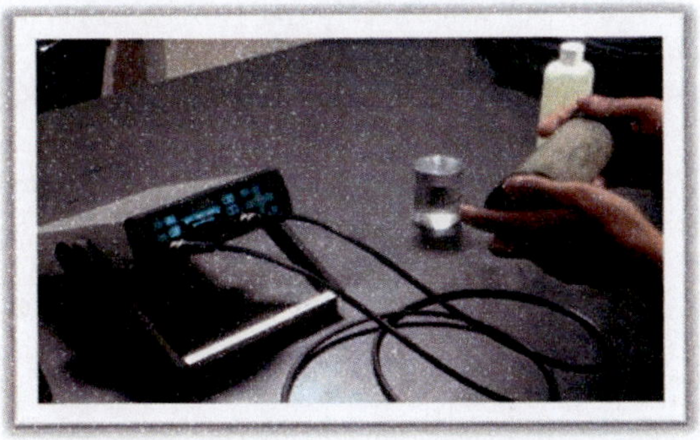

Fig. 3 Determination of P wave velocity in laboratory for rock core plugs by PUNDIT setup (Garia et al.., 2021 in the Journal of Petroleum Science and Engineering- ELSEVIER)

500 kHz that have been used in the literature. These two transducers, 54 and 500 kHz, were used in this study to incorporate the maximum and minimum set of transducers range that may be used for carrying out UPV measurements in the laboratory.

Length of the samples was measured with manual vernier caliper. The flat end surfaces of core samples were smoothened, which will provide contact between the transducer/receiver and the rock sample. To reduce air pocket occurrence and ensure full contact between transducers and the rock core, coupling gel was applied on the side faces of the rock sample. After calibrating the device through calibrated core made of acrylic glass, the transducers of both frequencies were placed on the opposite end parallel surfaces of the rock specimens. UPV values were obtained based on division of the length of the sample core and the transit travel time. The measurements were done three times on each rock core samples, and mean UPV values were evaluated. For correct measurement, three observations have been taken into consideration, and average of them has been used for further analysis. For the average, 5% tolerance has been taken into consideration. Core samples used for UPV measurements were also used for the measurement of bulk density. For the bulk density measurement, dimension taken through vernier calipers was used. The rock sample volume was thus determined from an average of three caliper readings while dry weight was defined after drying for 24 h at 105 °C temperature. Obtained data through both 54 and 500 kHz frequency transducer were statistically analyzed using least square regression method. R software was used to fit the regression equations by the method of least squares.

3 Results and Discussion

The range of Vp values obtained for different samples by using transducers of two different frequencies was plotted in the form of histogram as shown in Fig. 4. The highest and lowest recorded Vp (value ± standard deviation) were (3174 ± 30.55) m/s (for sample A11) and 1920 ± 10.11) m/s (for sample A5) for 54 kHz transducer, while 3371 ± 34.60) m/s (for sample A11) and 2126 ± 11.54) m/s (for sample A5) for 500 kHz transducer, respectively.Obtained data through both 54 and 500 kHz frequency transducer were statistically analyzed using least square regression method. UPV_{54kHz} and UPV_{500kHz} were correlated using regression analysis, and it was observed that R^2 obtained was 0.76 (Fig. 5). Bulk density was also measured for all the samples as shown in Fig. 6, and it ranges from 1.89 to 2.54 gm/cc.

The measurements done by using 500 kHz transducers gave higher velocity measurement values than 54 kHz transducers and the percentage increase varied from 3 to 36%. To assess the influence of sample dimensions on UPV measurements, different rock cores of varying length and diameter were taken for analysis. P wave velocities were correlated with the sample dimension for each rock core samples. Upon introducing length and diameter of the rock core plug, the coefficient of determination R^2 relating 500 and 54 kHz transducer velocity is increased from 0.792 to 0.80. While introducing parameters like aspect ratio (length/diameter ratio),

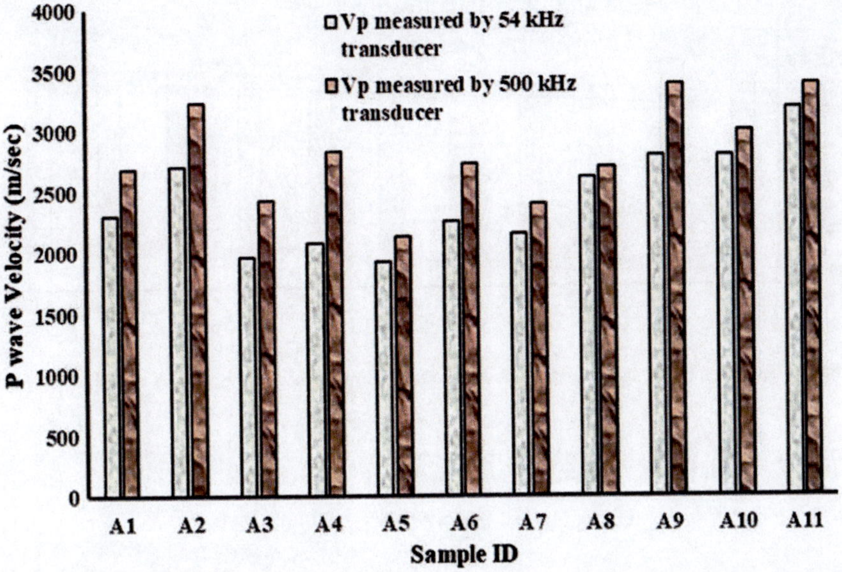

Fig. 4 Ultrasonic pulse velocity values of the sandstone and carbonate samples used in the study measured by 54 and 500 kHz transducers

Fig. 5 Correlated ultrasonic pulse velocity values of the sandstone and carbonate samples used in the study measured through 54 and 500 kHz transducer

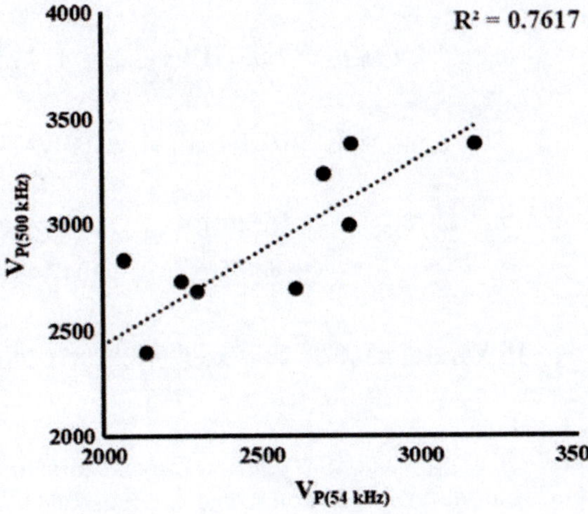

R^2 obtained was 0.87 and introducing the effect of bulk density, R^2 increased up to 0.91. The multivariate regression analysis was performed for each rock specimen taking into account UPV results obtained through 54 and 500 kHz transducer, and other properties like length, diameter of rock core (dimensions), aspect ratio, and bulk density and the corresponding statistically equations are listed as Eqs. 1–7. The

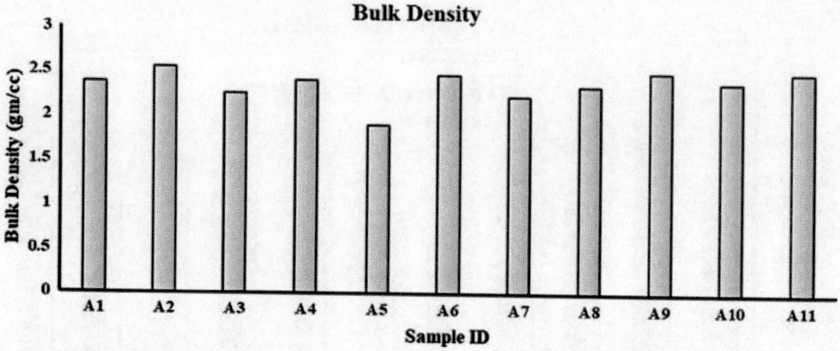

Fig. 6 Bulk density of the sandstone and carbonate samples used in the study

measured and predicted Vp were thus plotted as shown in Fig. 7.

$$\mathrm{UPV}_{500\mathrm{kHz}} = 0.8788\mathrm{UPV}_{54\mathrm{kHz}} + 670.264 \tag{1}$$

$$\mathrm{UPV}_{500\mathrm{kHz}} = 0.9439\mathrm{UPV}_{54\mathrm{kHz}} - 5.6046L + 760.259 \tag{2}$$

$$\mathrm{UPV}_{500\mathrm{kHz}} = 1.1533\mathrm{UPV}_{54\mathrm{kHz}} - 9.4083D - 407.484 \tag{3}$$

$$\mathrm{UPV}_{500\mathrm{kHz}} = 1.0322\mathrm{UPV}_{54\mathrm{kHz}} - 114.225\mathrm{AR} + 437.87 \tag{4}$$

$$\mathrm{UPV}_{500\mathrm{kHz}} = 1.1267\mathrm{UPV}_{54\mathrm{kHz}} + 7.1021D - 3.5033L - 87.0477 \tag{5}$$

$$\begin{aligned}\mathrm{UPV}_{500\mathrm{kHz}} = {} & 1.0549\mathrm{UPV}_{54\mathrm{kHz}} + 34.7837D - 29.6106L \\ & + 880.5092\mathrm{AR} - 1047.4811\end{aligned} \tag{6}$$

$$\begin{aligned}\mathrm{UPV}_{500\mathrm{kHz}} = {} & 0.6203\mathrm{UPV}_{54\mathrm{kHz}} + 12.3283D - 6.0407L + 279.93\mathrm{AR} \\ & + 1054.18\rho_b - 1795.22\end{aligned} \tag{7}$$

where $\mathrm{UPV}_{500\mathrm{kHz}}$ and $\mathrm{UPV}_{54\mathrm{kHz}}$ correspond to velocity measured using 500 and 54 kHz transducer, and D, L, AR, and ρ_b correspond to diameter, length, aspect ratio, and bulk density of the samples, respectively.

On the basis of 11 samples, two classes of aspect ratio (AR) were categorized. One with AR < 1 and the other with AR > 1. With that idea, equations were developed, and it was observed that R^2 increased from 0.81 (Eq. 5) to 0.87 (Eq. 6) on inclusion of L, D, and AR. This may be useful to understand the effect of geometry on the velocity measurements, since proper core samples may not be always available for testing.

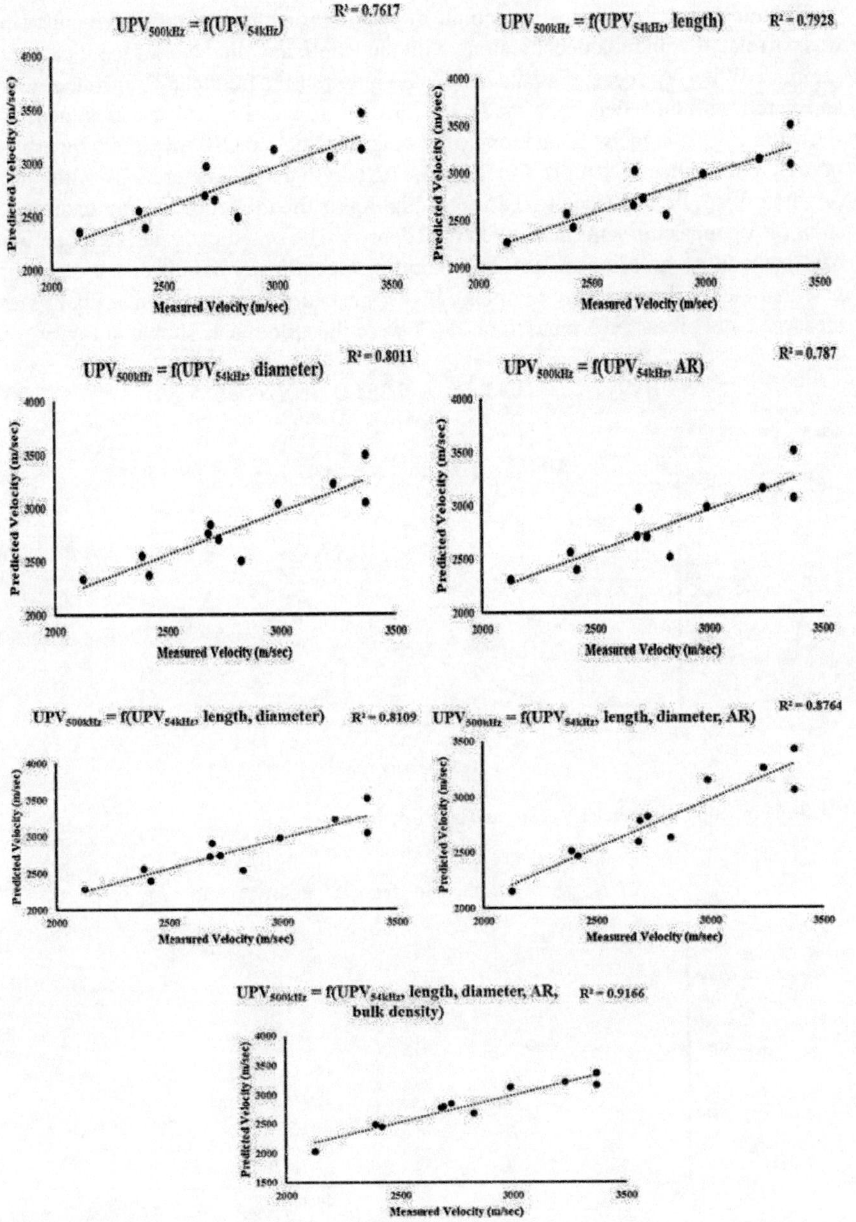

Fig. 7 Multivariate regression analysis performed to correlate UPV_{500kHz} and UPV_{54kHz} data measured

To understand the influence of sample dimensions on the UPV measurements, Vp was correlated with bulk density along with the length and diameter of the rock core samples. When Vp measurements conducted by 500 kHz frequency transducer was correlated with bulk density, a moderate correlation with R^2 0.74 was obtained, as shown in Fig. 8, since with an increase in bulk density, Vp also increases. Incorporating the influence of sample dimensions, i.e., length and diameter along with bulk density on Vp, R^2 increased to 0.86. This highlights the influence of sample dimensions on Vp measurements along with bulk density. The statistically derived equation based on multi-regression analysis incorporating the influence of bulk density, length, and diameter of the rock core samples on UPV measurements is shown in Eq. 8. The measured and predicted Vp based on Eq. 8 were thus plotted as shown in Fig. 9.

$$UPV_{500kHz} = 1890.8\rho_b - 5.521D + 6L - 1665.5 \tag{8}$$

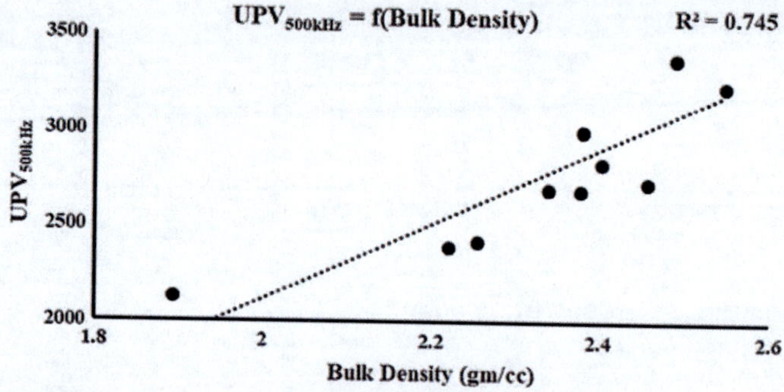

Fig. 8 Correlation between UPV_{500kHz} and bulk density

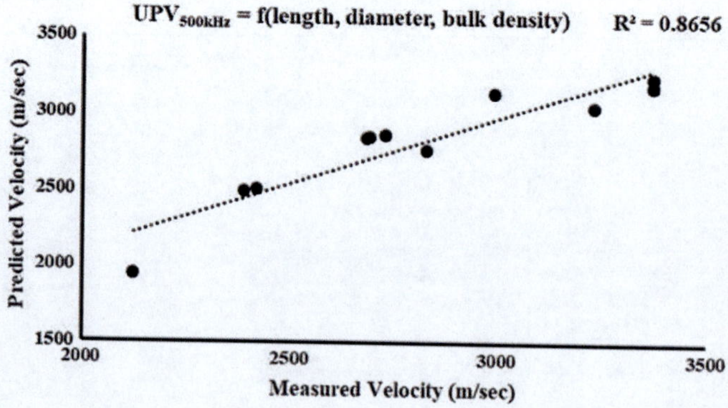

Fig. 9 Multivariate regression analysis performed to correlate UPV_{500kHz} with bulk density and sample dimensions

R^2 reported with the correlation reported for the various regression equation itself shows the standard error as R^2 value signifies the deviation of the values from the mean value considered.

The statistically developed relationships (Eqs. 1–7) highlight the effect of different frequency transducer, sample dimensions along with the effect of bulk density on the measurement of P wave velocity based on laboratory-scaled based measurements. Laboratory measurements of velocity of elastic waves in rocks are needed for the interpretation of seismic velocities, which in turn can be used for rock physics modeling. Equation 8 can be used to predict UPV based on sample dimensions and bulk density for different samples which would be helpful in characterizing different rock masses on the basis of UPV values. These equations are simple, practical, and accurate enough to apply and are recommended in practice in various fields like mining, geotechnical, and engineering purposes.

4 Conclusions

This study was aimed to investigate the effect of transducer frequency on the P wave ultrasonic velocity measurement in sedimentary rock samples of varying length and diameter. UPV_{500kHz} and UPV_{54kHz} data pairs were statistically analyzed and correlated using multivariate regression analysis. Based on the findings presented, it was noted that measurements done by using 500 kHz transducers gave higher velocity measurement values than the 54 kHz transducers. UPV_{54kHz} and UPV_{500kHz} were correlated using regression analysis. A method based on multivariate analysis was developed to determine the influence of different parameters such as diameter, length, aspect ratio, and bulk density of sedimentary rocks on UPV measurements by transducers of different frequencies. Moreover, incorporating the influence of sample dimensions, i.e., length and diameter along with bulk density on Vp, a better correlation was obtained which highlights the influence of sample dimensions on Vp measurements. Hence, apart from petrophysical and mechanical properties of rocks, sample dimensions also have a significant role to play in influencing Vp measurements at ambient conditions in the laboratory.

Acknowledgments The authors would like to acknowledge the valuable contributions of KDMIPE, ONGC, and North Eastern Coalfield, Margherita.

References

1. Assefa S, Mccann C, Sothcott J (2003) Velocities of compressional and shear waves in limestones. Geophys Prospect 51:1–13
2. ASTM D 2845-08 (2005) Standard test method for laboratory determination of pulse velocities and ultrasonic elastic constants of rock

3. Brotons V, Tomas R, Ivorra S, Grediaga A, Martınez J, Benavente D, Gomez M (2016) Improved correlation between the static and dynamic elastic modulus of different types of rocks. Mater Struct 49:3021–3037
4. Chang C, Zoback MD, Khaksar A (2006) Empirical relations between rock strength and physical properties in sedimentary rocks. J Petrol Sci Eng 51:223–237
5. Çobanoğlu İ, Çelik SB (2008) Estimation of uniaxial compressive strength from point load strength, Schmidt hardness and P-wave velocity. Bull Eng Geol Env 67(4):491–498
6. D'Andrea DV, Fisher RL, Fogelson DE (1964) Prediction of compression strength from other rock properties. Colo Sch Mines Q 59(4B):623–640
7. Ercikdi B, Karaman K, Cihangir F, Yılmaz T, Aliyazıcıoğlu Ş, Kesimal A (2016) Core size effect on the dry and saturated ultrasonic pulse velocity of limestone samples. Ultrasonics 72:143–149
8. Fener M (2011) The effect of rock sample dimension on the P-wave velocity. J Nondestr Eval 30(2):99–105
9. Fjaer E (1999) Static and dynamic moduli of weak sandstone, Rock Mechanics for industry. Balkema, pp 675–681
10. Gaviglio P (1989) Longitudinal waves propagation in a limestone: the relationship between velocity and density. Rock Mech Rock Eng 22(4):299–306
11. Han D, Nur A, Morgan D (1986) Effects of porosity and clay content on wave velocities in sandstones. Geophysics 51:2093–2107
12. ISRM (1978) Suggested method for determining sound velocity. Int J Rock Mech Min Sci Geomech Abstr 15:53–58
13. Kahraman S (2001) A correlation between P-wave velocity, number of joints and Schmidt hammer rebound number bit. J Rock Mech Min Sei 38:729–773
14. Kahraman S (2007) The correlations between the saturated and dry P-wave velocity of rocks. Ultrasonics 46(4):341–348
15. Kahraman S, Yeken T (2008) Determination of physical properties of carbonate rocks from P-wave velocity. Bull Eng Geol Environ 67:277–281
16. Karakul H, Ulusay R (2013) Empirical correlations for predicting strength properties of rocks from p-wave velocity under different degrees of saturation. Rock Mech Rock Eng 46:981–999
17. Karaman K, Kaya A, Kesimal A (2015) Effect of the specimen length on ultrasonic P-wave velocity in some volcanic rocks and limestones. J Afr Earth Sci 112:142–149
18. Kilic A, Teymen A (2008) Determination of mechanical properties of rocks using simple methods. Bull Eng Geol Environ 67(2):237–244
19. Klimentos T (1991) The effects of porosity-permeability-clay content on the velocity of compressional waves. Geophysics 56:1930–1939
20. Kurtulus C, Cakir S, Yogurtcuoglu AC (2016) Ultrasound study of limestone rock physical and mechanical properties. Soil Mech Found Eng 52(6):27–31
21. Madhubabu N, Singh PK, Kainthola A, Mahanta B, Tripathy A, Singh TN (2016) Prediction of compressive strength and elastic modulus of carbonate rocks. Measurement 88:202–213
22. McDowell PW, Millett N (1984) Surface ultrasonic measurement of transverse velocities through rock sample. Intl J of Rock Mech & Mining Sci & Geomechanic Abs 21(4)
23. Moos D, Zoback M, Bailey L (1999) Feasibility study of the stability of open hole multilaterals, Cook Inlet, Alaska. In: SPE mid-continent operations symposium 1999, Society of petroleum engineers: Oklahoma City, OK, USA, 2001
24. Moradian ZA, Behnia M (2009) Predicting the uniaxial compressive strength and static Young's modulus of intact sedimentary rocks using the ultrasonic test. Int J Geomech 9(1):14–19
25. Mueller HJ (2013) Measuring the elastic properties of natural rocks and mineral assemblages under Earth's deep crustal and mantle conditions. J Geodyn 71:25–42
26. Pappalardo G, Mineo S (2016) Geotechnical characterization of limestones employed for the reconstruction of a UNESCO world heritage Baroque monument in southeastern Sicily (Italy). Eng Geol 212:86–97
27. Parent T, Domede N, Sellier A, Mouatt L (2015) Mechanical characterization of limestone from sound velocity measurement. Int J Rock Mech Min Sci 79:149–156

28. Sarkar K, Vishal V, Singh TN (2012) An empirical correlation of index geomechanical parameters with the compressional wave velocity. Geotech Geol Eng 30:469–479
29. Sharma PK, Singh TN (2008) A correlation between P-wave velocity, impact strength index, slake durability index and uniaxial compressive strength. Bull Eng Geol Environ 67:17–22
30. Soroush H, Qutob H, Oil W, Me T (2011) Evaluation of rock properties using ultrasonic pulse technique and correlating static to dynamic elastic constants. In: 2nd south Asian geoscience conference and exhibition, GEOIndia 2011, Greater Noida, New Delhi, India
31. Vilhelm J, Ivankina T, Lokají T, Rudajev V (2016) Comparison of laboratory and field measurements of P and S wave velocities of a periodite rock. Int J Rock Mech Min Sci 88:235–241
32. Yasar E, Erdogan Y (2004) Correlating sound velocity with the density, compressive strength and Young's modulus of carbonate rocks. Int J Rock Mech Min Sci 41:871–875

Tsunamigenic Seismic Activity (Earthquakes) Prediction from III-Component Seismic Data

Tammineni Gnananandarao, Rakesh Kumar Dutta, and Vishwas Nandkishor Khatri

Abstract In this study, two classification methods such as the artificial neural network with feed-forward backpropagation (ANN) and random forest regression (RFR) were used to identify tsunamigenic earthquakes at a regional distance with shallow focus depth (less than 70 km). To attain the tsunamigenic earthquake classification, depth (H), distance (D), azimuth (A), moment magnitude (Mw), and fault parameter (ϕ_s, δ, λ) were used as a input parameters. The total collected data was divided into two parts such as for training (70% of total data) and testing (remaining 30% data). The trained ANN and RFR were found to successfully categorize as tsunamigenic or non-tsunamigenic 100% and 89%, respectively. The ANN and RFR methods may be useful where insufficient weak tsunami alert systems (TAS) are available in the vicinity.

Keywords Potentially tsunamigenic earthquake · Tsunami · Artificial neural networks · Random forest regression

1 Introduction

For the coastal areas, one of the greatest natural threats worldwide is a tsunami, which in a short span of time causes serious damage to life, and property. As evidence from the 2004 tsunami 2.3 lack people were killed in 14 difference countries, where 18,000 above were alone killed in India [1] and an average tsunami loss is estimated as $19.14 million [2]. Tsunamigenic earthquakes landsides and volcanic eruptions in the sea are causes for large-scale displacement of seawater toward offshore. However, the

T. Gnananandarao (✉) · R. Kumar Dutta
NIT Hamirpur, Hamirpur, Himachal Pradesh, India
e-mail: anandrcwing@gmail.com

R. Kumar Dutta
e-mail: rakeshkdutta@yahoo.com

V. N. Khatri
IIT Dhanbad, Dhanbad, Jharkand, India
e-mail: vishwas@iitism.ac.in

T. G. Sitharam et al. (eds.), *Seismic Hazards and Risk*, Lecture Notes in Civil Engineering 116, https://doi.org/10.1007/978-981-15-9976-7_31

key reason for generating tsunamis in different regions of the globe is tsunamigenic earthquakes. Sumatra geographical region was the most influential area in Indian Ocean for tsunami. It is situated near the eastern coast of India, Sri Lanka, and various other islands including Singapore, Malaysia, and Indonesia.

2 Background

The Okada model has been used widely accepted method to predict tsunami generation by terrible earthquakes [3], that models sea surface deformations in half-space due to shear and tensile failure. Nevertheless, solitary parameters of magnitude and hypocenter are commonly available in the earthquake originating point for initial few minutes. Focal parameters and seismic moments will become readily accessible much later (nearly half an hour), generally requiring azimuthal-distributed multi-station data. For nearby coastal regions, this much time may be exorbitantly long. Hence, for the worst-case scenario such as 45° dip and 90° slip, fault parameters needed for the modeling the tsunami generation (boundary conditions). In addition, most tsunami warning centers generate and preserve a precomputed database under the worst circumstances at particular areas, based on the expected center of hypo and magnitude to overcome the time losses for simulation of the model. Empirical relationships are often applicable at certain local distances between the moment magnitude and the fault parameters [4]. Thus, most tsunami warning centers are primarily focused on simulation results on the first warning of an imminent tsunami. Those warnings will later be upgraded or downgraded if specific focal parameters come into being and the tsunami generation with bottom pressure recorders (BPR) was verified or declined and tide gauge data. This paper is expected to help current approaches by reducing false alarms, and focal parameters are integrated in the direction of seismic phase amplitudes, as well as signifying the use of ANN and RFR for tsunami warning. In this article, tsunamigenic early detection of shallow focus earthquakes at a regional distance (almost 2 thousand km) using ANN and RFR based on seismogram reported at a single three-component broadband station was addressed. It is worth noting that an earthquake is referred to in this article as tsunamigenic if it causes a tsunami [5] while a potential tsunamigenic earthquake (PTE) occurs at a fault just below the sea and can cause a tsunami depending on other circumstances. The aim is to find the PTE as a tsunamigenic, resource-limited earthquake is very useful in minimizing casualties due to local and global tsunamis from poorly controlled world regions. The inspiration for this activity comes from the fact that all the source information for the earthquake is included in a single three-component seismogram so that it can be used as a basic predictor of a tsunamigenic earthquake. In practice, the data used in this article might not be adequate to approximate the all earthquake source parameter numerical values, but it is certainly enough to recognize the tsunami causing capability.

3 Method

In this analysis, new methods such as the artificial neural network (ANN) with feed-forward backpropagation and random forest regression (RFR) were used to predict the tsunami generation potential of earthquakes from the Sumatra area using single three-component seismograms collected at PALK station to recognize the shallow focus (less than 70 km deep) of tsunamigenic earthquakes at a provincial distance.

4 Soft Computing Techniques

4.1 Artificial Neural Networks

Artificial neural network (ANN) is one of the branches of the artificial intelligence (AI). ANN means that the developers try to imitate human brains in systems working. These neural networks are having the capability to relate the specified input data with corresponding output data. The prediction capability that is not dependent on the number of input parameters is chosen to predict the output data. Generally, the artificial neural networks are consisting of input layer, hidden layer, and output layer, and each layer has their individual number of processing neurons. An artificial neural network architecture of the present study was shown in Fig. 1 for the reference.

The main aspect in developing the ANN is to generate the structure of the neural network and number of hidden nodes and hidden layers. Researchers left with no option to get the structure of the network except a trial-and-error procedure. Initial weight assignment, as well as some other related parameters, also impacts the prediction capability of the neural network. A complicated process for a feed-forward

Fig. 1 Architecture of the ANN model (7-5-1)

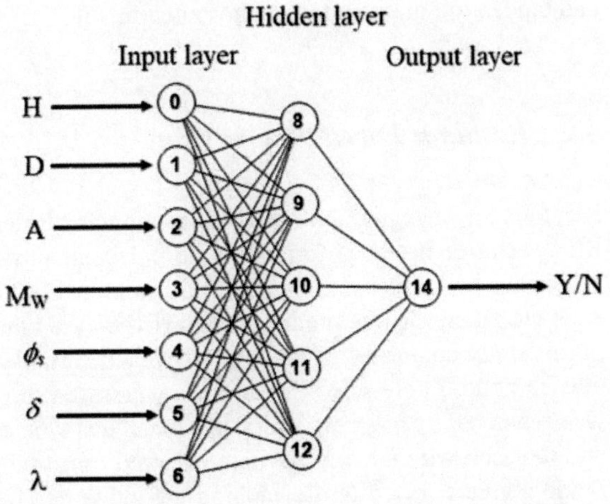

Table 1 Final weights between the input neuron and hidden neuron as well as hidden neuron and output neuron

	Weights (w_{ji})								Biases	
Hidden neurons	H	D	A	Mw	$Ø_s$	δ	Λ	Y/N	b_{hk}	b_0
Hidden 1	0.81	−0.25	−0.25	−3.26	0.85	0.41	−0.85	0.85	−3.20	5.01
Hidden 2	0.95	−0.32	−0.26	−3.48	0.83	0.37	−0.89	1.02	−3.47	–
Hidden 3	−0.20	−0.20	−1.99	0.28	0.01	−0.76	0.07	−0.20	−1.64	–
Hidden 4	−0.35	−0.34	−4.03	1.02	0.48	−0.99	1.35	−0.35	−4.20	–
Hidden 5	−0.25	−0.34	−3.35	0.85	0.39	−0.84	0.97	−0.25	−3.30	–

backpropagation algorithm with multilayer network was the choice of the number of hidden layers and the number of neurons in each layer. The number of nodes in the hidden layer shall be established first, before the number of the hidden layer selection. However, the number of neurons in the hidden layer is 2/3 or 70% of the input layer level as per the literature [6–9]. Further, if there is an insufficient number of neurons in the hidden layer, then the neurons could be added to the output layer later. The neurons number in the hidden layer must be less than twice as many as the input layer neurons as reported by [10]. It was reported by [11] that, in the hidden layer, the neuron size was kept in between the layer size of input layer and output layer. By the context of the above, using the law followed by [10], the number of neurons in the hidden layer was calculated. The final architecture of the neural network as 7-5-1, where 7 is the input parameters, 5 is the hidden layer neurons, and 1 is the output parameter. The weights and biases are generated between the input to hidden and hidden to output and were presented in Table 1.

4.2 *Random Forest Regression*

Random forest regression (RFR) is a method of classification and regression. In RFR each tree has been formed by an independent random vector and the random vector has been selected from the input vector. Unlike the class labels used by the RFR classifier, the tree predictor is carried on the numerical values. The RFR used in this study comprises of randomly selected variables or combination of variables at each node. These selected variables were used to grow a tree. The training data was generated by bagging, which is a technique where the data are randomly drawn and replaced with the original data reserved for training. The training data can also randomly be selected for developing the individual tree for each of the proceeding combination. In the present study, 70% of the input data were randomly selected for training and the remaining one-third data were left out from every tree grown. In order to design a tree predictor, there is a need to have a pruning method as well as a variable selection procedure. In this context, a large number of approaches [12, 13] were available in literature. These approaches were information gain ratio criterion and gini index which assign the quality measures directly to the variable. The RFR approach used in this study was based on the gini index variable selection procedure. This variable selection procedure measures the impurity of the variable with respect to the output. Using a combination of variables, the RFR design permits to grow a tree to the maximum depth of the new training data, where such fully grown trees cannot be pruned back.

5 Dataset

Total earthquake data 66 events were collected form the literature [14]. It is a list of events happened at Sumatra region between 2001 and 2016 with magnitude 6 and above, and it is shown in Fig. 2 with yellow color. The input parameters are depth (H), distance (D), azimuth (A), moment magnitude (M_w), and fault parameter ($Ø_s$, δ, λ). The range of collected data was shown in Table 2. The chosen output was binary decision classification that is either 0 (for χ) or 1 (for Y).

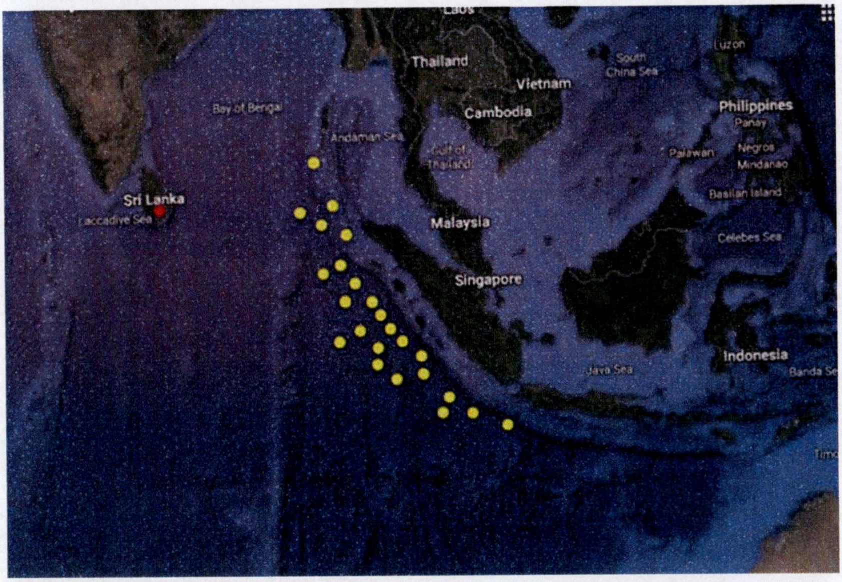

Fig. 2 Geographical image with potentially tsunamigenic earthquake events (in yellow dots) and seismogram recording station in Sri Lanka (in red dot. Courtesy Google map

Table 2 Range of data collected from literature

Input and output parameters	Dataset			
	Minimum	Maximum	Mean	Standard deviation
Depth *(H)*	2.5	46	26.25	8.63
Distance *(D)*	10.62	37.63	20.41	5.81
Azimuth *(A)*	97.1	131.57	112.49	6.75
Moment magnitude M_W	6	9.1	6.56	0.67
$\emptyset_s$	12	358	285.55	74.36
δ	5	89	32.06	27.38
λ	−170	165	68.69	71.21
Classification	No Tsunami = 58 (Events)		Tsunami = 8 (Events)	

6 Results and Discussions

ANN and RFR were trained with 70% of the data from the vicinity of the Sumatra region with the help of single three-component seismograms recorded at Palk station. The success of prediction of the tsunami was 100 and 94% form the training for ANN and RFR, respectively, and the same was tabulated in Table 3. After successful complication of the training, the remaining data (30%) was used to test the developed models. The results of the testing reveal that the prediction of tsunami was 100% successful with ANN and 89% for the RFR, and the same was tabulated in Table 4. The study of Tables 3 and 4 reveals that the prediction with ANN was more accurate than the RFR. However, The RFR is also having the capability in predicting the tsunami in the region of the Sumatra. Further study is recommended to improve the ANN model by incorporating more data from different region for the prediction of tsunamigenic earthquake. The authors believe that application of soft computing technique is cost effective in order to predict the tsunamigenic earthquake.

7 Conclusions

Artificial neural networks (with feed-forward backpropagation) and random forest regression are used to classify tsunamigenic earthquakes promptly using a single three-component recording station at a regional distance from the generated source of the tsunami. The study shows that ANN can effectively approximate the classification of a tsunamigenic earthquake in comparison to random forest regression (RFR) with depth (H), distance (D), azimuth (A), moment magnitude (Mw), and fault parameter (ϕ_s, δ, λ) as inputs. The successful estimation of potentially tsunamigenic earthquake with ANN is 100% for training and testing, whereas with RFR is 94 and 89% for training and testing, respectively.

Table 3 Comparison of the training data with ANN and RFR models

S. no	Class	ANN	RFR	S. no	Class	ANN	RFR
1	χ	χ	χ	25	χ	χ	χ
2	χ	χ	χ	26	χ	χ	χ
3	χ	χ	χ	27	χ	χ	χ
4	χ	χ	χ	28	χ	χ	Y
5	Y	Y	Y	29	χ	χ	χ
6	χ	χ	χ	30	χ	χ	χ
7	χ	χ	χ	31	χ	χ	χ
8	χ	χ	χ	32	χ	χ	χ
9	χ	χ	χ	33	χ	χ	χ
10	Y	Y	Y	34	χ	χ	χ
11	χ	χ	χ	35	χ	χ	χ
12	χ	χ	χ	36	χ	χ	χ
13	χ	χ	χ	37	Y	Y	Y
14	χ	χ	χ	38	χ	χ	χ
15	Y	Y	χ	39	χ	χ	χ
16	Y	Y	Y	40	χ	χ	χ
17	χ	χ	χ	41	χ	χ	χ
18	χ	χ	χ	42	χ	χ	Y
19	χ	χ	χ	43	χ	χ	χ
20	χ	χ	χ	44	χ	χ	χ
21	χ	χ	χ	45	χ	χ	χ

(continued)

Table 3 (continued)

S. no	Class	ANN	RFR	S. no	Class	ANN	RFR
22	χ	χ	χ	46	χ	χ	χ
23	χ	χ	χ	47	χ	χ	χ
24	χ	χ	χ	–	–	–	–

Table 4 Comparison of the testing data with ANN and RFR models

S. no.	Class	ANN	RFR	S. no	Class	ANN	RFR
1	χ	χ	χ	11	χ	χ	χ
2	χ	χ	χ	12	χ	χ	χ
3	Y	Y	Y	13	χ	χ	χ
4	χ	χ	χ	14	χ	χ	χ
5	χ	χ	Y	15	χ	χ	χ
6	χ	χ	χ	16	χ	χ	χ
7	Y	Y	Y	17	χ	χ	χ
8	χ	χ	χ	18	χ	χ	χ
9	χ	χ	χ	19	Y	Y	χ
10	χ	χ	χ	–	–	–	–

Symbols

H	Depth
D	Distance
A	Azimuth
M_W	Moment magnitude
ϕ_s, δ and λ	Fault parameter
χ	No tsunami
Y	Tsunami
ANN	Artificial neural network
RFR	Random forest regression
$	USD
AI	Artificial intelligence

References

1. https://timesofindia.indiatimes.com/event/2004-Indian-Ocean-tsunami/articleshow/5507 1172.cms
2. https://www.preventionweb.net/publications/view/42895
3. Okada Y (1985) Surface deformation due to shear and tensile faults in a half-space. Bull Seismol Soc Am 75 (4)
4. Papazachos BC, Scordilis EM, Panagiotopoulos D, Papazachos CB, Karakaisis GF (2004) Global relations between seismic fault parameters and momentmagnitude of earthquakes. Bull Geol Soc Greece 36
5. Satake K, Tanioka Y (1999) Sources of tsunami and tsunamigenic earthquakes in subduction zones. Pure Appl Geophys 154, 467–483
6. Boger Z, Guterman H (1997) Knowledge extraction from artificial neural network models. IEEE Int Conf Comput Cybern Simul 4:3030–3035
7. Gnananandarao T, Dutta RK, Khatri VN (2019) Application of artificial neural network to predict the settlement of shallow foundations on cohesionless soils. Geotechnical Applications, Lecture notes in Civil Engineering 13:51–58. https://doi.org/10.1007/978-981-13-0368-5_6
8. Gnananandarao T, Dutta RK, Khatri VN (2017) Artificial neural networks based bearing capacity prediction for square footing resting on confined sand. In: Indian geotechnical conference, GeoNEst, pp 14–16 December 2017, IIT Guwahati, India
9. Dutta RK, Rani R, Gnananandarao T (2018) Prediction of ultimate bearing capacity of skirted footing resting on sand using artificial neural networks. J Soft Comput Civil Eng 2(4):34–46. https://doi.org/10.22115/SCCE.2018.133742.1066
10. Berry MJA, Linoff G (1997) Data mining techniques. Wiley, New York
11. Blum A (1992) Neural netw in C++. Wiley, New York
12. Breiman L, Friedman JH, Olshen RA, Stone CJ (1984) Classifcation and regression trees. Wadsworth, Monterey
13. Quinlan JR (1992) Learning with continuous classes. In: Proceedings of Australian joint conference on Artifcial Intelligence., Singapore: World Scientifc Press, pp 343– 348
14. Kundu A, Kundu Bhadauria YS, Basu S, Mukhopadhyay S (2016) Prompt identification of tsunamigenic earthquakes from 3-component seismic data. Physics of the Earth and Planetary Interiors, 259:10–17

OLR Anomaly and Tidal Amplitude Irregularities (Tai) as Short-Term Earthquake Precursors—A Study of Doublet Earthquakes (1990–2010)

N. Venkatanathan, M. Senthil Kumar, and Ramya Jeyaraman

Abstract Pair of earthquakes occurred with similar focal mechanism is called doublet earthquakes. Both earthquakes have their focal region around 100 km diameter and the rupture zone intersect each other. The time interval between the two earthquakes is expressively small compared to the time span needed for the accumulation of strain released during the first event. Anomalous variations in OLR flux above + 2σ significant level were observed 3 to 30 days before the occurrence of big earthquakes. From the analysis of Kuril doublet earthquakes, the OLR flux index increases with the magnitude. For M8.3 earthquakes, the cumulative OLR index was 4.608, and for M8.1 earthquake, the cumulative OLR index was 4.297. Anomalous variations along the seismically vulnerable region may be due to the radon gas emanation from the interior of the Earth. Increase in tectonic activity might have been triggered by the irregularities observed in M2 component of solid Earth tides during a particular lunar phase. These tidal amplitude irregularities (TAI) affect the rotational speed of Earth in order to conserve the angular momentum of the Earth–Sun–Moon system, which in turn affects the movement of tectonic plates. From the analysis of doublet earthquakes occurred across the globe between 1990 and 2010, it was found that the TAI varies between 15 and 40%. Thus, the authors concluded that OLR anomaly and TAI can be used as a short-term earthquake forecasting tool.

Keywords OLR · TAI · Doublet earthquake · Remote sensing

1 Introduction

From the investigation of pair of all shallow earthquakes with the magnitude greater than 7.5, it was found that most of the pairs have similar focal mechanisms [1]. Deep event doublet earthquakes often occur in South America. Similar rupture process was observed between two earthquakes occurred on November 24, 2015 with the magnitude of Mw 7.5 and Mw 7.6 along the Peru–Brazil boundary [2]. From the analysis

N. Venkatanathan (✉) · M. Senthil Kumar · R. Jeyaraman
Department of Physics, SASTRA Deemed To be University, 613 401, Thanjavur, Tamil Nadu, India
e-mail: physics16972@gmail.com

T. G. Sitharam et al. (eds.), *Seismic Hazards and Risk*, Lecture Notes in Civil Engineering 116, https://doi.org/10.1007/978-981-15-9976-7_32

of fault geometry, rupture velocity, stress drop, and radiated energy, it is inferred that the ruptures are simple elliptical [2]. Doublet earthquakes have their focal region around 100 km diameter and the rupture zone intersect each other. The time interval between the two earthquakes is expressively small compared to the time span need for the accumulation of strain released during the first event [3]. Recent advances in remote sensing instruments have helped to advance the scientific understanding of earthquake signals in the atmosphere. Thermal imaging data from satellites show stationary thermal anomalies associated with large linear structures and fault systems in the Earth's crust [4], but also transient anomalies before major earthquakes [5, 6]. There have been many attempts for signs of tidal activation of earthquakes. The more rigorous studies in continental settings tend to find no correlation or only a very slight correlation with the exception of volcanically active areas [7]. The stress levels from the Earth's tides can be much higher than tectonic stress accumulation rates, postulating the tidal stresses to cause an earthquake [8]. Small stress change due to the Earth tide facilitates earthquake occurrence when the stress is near a critical condition in the potential focus field [9]. Temporal tidal triggering shifts have now also been observed in several subduction zones. Precursory activity of initiating signals before major earthquakes occur is of special interest [10]. Tidal attractions also generate elastic deformation of the solid Earth, exerted by the Moon and the Sun [11]. The Moon causes the fault to slip more—and faster—when it pulls in the same direction that the fault is slipping. Just like ocean tides, when the Sun and Moon are aligned, the strongest Earth tides occur, and the weakest occur when they are 90 degrees apart. The same gravitational forces reach out and compress the surface of the Earth. Some faults, such as offshore faults like the Cascadia subduction zone off the Pacific Northwest coast, are more susceptible to tidal triggering than others. The tidal response is also influenced by other aspects of the fault, such as its direction or how close it is to the surface of the Earth. On Earth, the Sun and Moon exert a gravitational tug which stretches and compresses crustal rocks. This cyclic stress can encourage or inhibit fault slip, particularly at the deep roots of faults. The amplitude of the Earth's solid tide varies over a period of fortnightly (2-week), as the Sun and Moon change their relative positions in the sky [12]. Clear evidence of earthquake triggering from Earth tides has been elusive for over a century. Using the largest global earthquake catalog available, we observe a clear correlation between the solid Earth tide phase and the timing of seismic events (with a confidence of some 99%). The detection of tidally triggered events suggests that tidal stresses, typically smaller than 0.1–1% of the magnitudes of tectonic stress, are sufficient to trigger up to about 0.2–0.3% of all earthquakes [13].Satellite thermal imaging data gives spatial extent and temporal development may depend on local geology and tectonics, focus mechanism, meteorological conditions, and other factors. Research on the relationship between data from satellite thermal infrared (TIR) and precursors to an earthquake is based on data from both single and multi-instruments. One of the principal parameters used to describe the Earth's radiation environment is the outgoing long-wave Earth radiation measurement [14]. While the Earth's surface absorbs solar energy from incident shortwave (SW) approximately, one-third is reflected back into space. Due to the low thermal conductivity of the surfaces, the heat absorbed at the surface does not reach any significant depth. Instead, the absorbed energy is re-emitted mostly during the night as long wavelength radiation (LWR) energy. The energy that is released directly into the atmosphere is called sensitive heat (SH) and is transferred as radiant thermal energy or heat [15]. OLR was associated with the top of the atmosphere and it integrates the emissions from ground, lower atmosphere and cloud, [16] is used primarily to study Earth's radiative budget and climate studies [17]. Daily OLR data have been used to study the variability of OLR in the earthquake activity zone [18–20,and21]. An increase in radiation and a transient change in OLR was proposed to be linked to atmospheric thermodynamic processes over seismically active regions and defined as thermal radiation anomaly (TRA) [15].

2 Study Area and Data

For more than eight years, the several polar-orbiting satellites have recorded these OLR data twice-daily, forming time series data over the various periods of time along with the spatial coverage of the entire Earth. National Oceanic and Atmosphere Administration (NOAA) satellites observe the experimental data OLR. The initial OLR data are processed using the interpolation method to minimize the distance over which a value is interpolated in space or time. Liebmann and Smith have given details on the interpolation technique [14]. The data used for this analysis are derived from NOAA satellite having the spatial resolution of 1 x 1 degree. In this analysis, doublet earthquake from Kuril Island and two Chile region same date occurrence earthquakes are selected for evaluation (Table 1). We have selected doublet earthquake since they have nearly same waveform from the same location. This would help us to analyze the behavior of precursory phenomena for same tectonic set up and physical behavior of earthquakes. The November 15, 2006 Mw = 8.3 (USGS) earthquake occurred on the southern slope of the Central Kuril Islands. The earthquake was a type of thrust fault which occurred at the boundary of the interplate [22]. A second earthquake of similar magnitude (Mw = 8.1; USGS) to the November 2006 earthquake occurred in the same area on January 13, 2007, but with a source located on the ocean slope of the Kuril–Kamchatka Trench about 100 km southeast of the earlier earthquake site the 2007 earthquake was a natural fault event that occurred in the inner portion of the subducting plate [22]. Six months prior to the occurrence of earthquakes in that region, the OLR study was made in order to search for short-term seismic anomalies.

Table 1 Earthquake data

Event	Origin time	Lat	Long	Magnitude Mw
15–11–2006	11:14:13.570Z	46.592	153.266	8.3
13–01–2007	04:23:21.160Z	46.243	154.524	8.1
27–02–2010	21:30:36.570Z	−36.122	−72.898	8.8
27–02–2010	21:08:02.360Z	−37.773	−75.048	7.4

3 Methodology

3.1 Tidal Amplitude Irregularities (TAI)

For more than a century, scientists are trying to correlate the solid Earth tides and earthquakes. Results of the earlier studies suggest that the big earthquakes are triggered by the "Solid Earth Tides". In this paper, the authors have found a strong relationship among the amplitude irregularity of solid Earth tides, the appearance of anomalous atmospheric precursors and earthquake occurrence. The normal range of solid Earth tide is ±20 cm and sometimes it reaches 30 cm or more than that. Solid Earth tides were calculated using a model developed by Dennis Milbert, NASA, and it is based on an edited version of the "dehanttideine lMJD.f" source code, that has adhered the recommendations of IERS conventions 2003.

As the Earth rotates under the tidal force due to the Sun and Moon, each point on the Earth would experience harmonic variations and displace the Earth's surface vertically up and down in a periodic manner (2 highs and 2 lows per day). On Syzygy days, the M2 constituent of solid Earth tides is having maximum amplitude and during Quadrature phase of the Moon has minimum amplitude. As the Moon orbiting from Syzygy to Quadrature position, the M2 component gradually decreases and it increases gradually when the Moon moves from Quadrature to Syzygy again. Any deviation from this harmonic variation is termed as "tidal amplitude irregularity" (TAI). Probably, the TAI caused due to the tidal friction produced in the "Earth–Moon–Sun" system. The rotation and orbital motion of the tidally locked celestial bodies are synchronized by the tidal forces. As the orbit of the celestial bodies is eccentric, the orbital speed varies with the distance. Hence, the tidal bulge produced on the surface of the Earth would not point toward the line joining centers of Moon and Sun, and the tidal bulge would be either leading or lagging in terms of space and time. To conserve the angular momentum of the system, the rotational speed of the Earth gets adjusted. A backward torque slows down the Earth's rotation, hence the maximum M2 component of the one half of the lunar cycle was observed prior to the Syzygy days. Similarly, if the maximum M2 component of the one half of the lunar cycle was observed after Syzygy days then the Earth's rotation gradually speeds up, i.e., the variations in the M2 component of the solid Earth tide leads to the change in the rotational speed of the Earth. Change in rotational speed of the Earth influences the rate of tectonic plates collision due to their differential inertia, just as the passengers collide with each other when the bus driver applies the brake suddenly. Also, the friction of the tidal bulge being pulled back and forth through the solid body of the Earth heats the interior of the Earth. Hence, these stresses slowly transformed into the seismic strain and when this strain goes above the critical level along the seismically vulnerable region, earthquake gets triggered.

Instead of observing the peak amplitude of M2 component of the tide on Syzygy days it is observed either prior to the Syzygy days or after it. The percentage of error in vertical shift of M2 component (Δy) can be calculated as follows,

$$\Delta y = \left(\frac{A_{M2}^{\uparrow} - A_{M2}^{\leftrightarrow}}{A_{M2}^{\leftrightarrow}}\right) x 100\% \quad (1)$$

where

$$\Delta y = \text{error percentage in vertical displacement of M2 component of solid earth tide}$$

$$A_{M2}^{\uparrow} = \text{highest Amplitude of M2 component either prior or after syzygy days}$$

$$A_{M2}^{\leftrightarrow} = \text{Actual amplitude of M2 component on syzygy days}$$

Note: The positive "Δy" indicates that the maximum tidal amplitude was observed prior to the Syzygy days and hence lesser M2 component on Syzygy days, the Earth would be slowing down its rotation. The observation of maximum tidal amplitude after Syzygy days is represented by the negative value of "Δy". The increase of M2 component after Syzygy days' speeds up the rotational speed of Earth gradually in order to maintain the angular momentum of the system.

It is observed that a few months prior to the occurrence of big earthquakes the vertical shift percentage (Δy) has shown sharp variation from positive to negative or vice versa. This sharp variation in "Δy" can provide a useful clue about the possible magnitude of the impending earthquake.

The effect of TAI could be coupled with the anomalous deviation of atmospheric parameters like relative humidity, air temperature, and outgoing long-wave as follows. Anomalous deviations in atmospheric parameters like relative humidity (RH), air temperature (AT) at surface level, and outgoing long-wave radiation (OLR) were observed after the observation of TAI. As the stress due to the tidal irregularity transformed into seismic strain, the tectonic activity gets increased along the potentially vulnerable region. The increased tectonic activity paved way for the release of gases like radon from the voids of the rocks. These radioactive gases migrated through the cracks of the rocks and moved up toward the surface of the Earth, where they ionize the atmospheric air molecules. Each ion is acting as condensation centers for water vapors (Sven mark et al. 2007), hence the anomalous drop in relative humidity (RH) and the latent heat flux get released into the atmosphere. The release of latent heat leads to the increase in air temperature (AT) and due to convection process the hot air moves up. As the ionized air moves up more and more water molecules get condensed over it, more latent heat is released. This process leads to the release of large amount heat energy above the cloud level, and hence, anomalous thermal flux is in the form of outgoing long-wave radiation (OLR).

3.2 Anomalous Outgoing Long-Wave Radiation (OLR)

The electromagnetic radiation from the Earth's surface into the atmosphere in the form of the thermal radiation is called outgoing long-wave radiation (OLR). The energy flux transmitted by outgoing long-wave radiation from the Earth's surface to the atmosphere is measured in W/m^2. OLR is measured at an altitude of 10 to 12 km using satellite technology. Near or over the seismically active regions an anomalous spike in OLR flux was observed and spreads over thousands of square kilometers area. Several scientists have observed transient thermal anomalies prior to the big devastating earthquakes. For the clear sky condition, the raw OLR data is processed for 10 to 13 μm wavelength region. Several studies on OLR flux, have found that the anomalous variations were observed around 1 to 45 days prior to the devastating earthquakes. The anomalous variations of OLR flux at the seismically active regions from the mean OLR flux of previous 10 years can be found as follows,

$$\overline{E_{ij\tau}} = \sum_{=1}^{n} E_{ij\tau} \tag{2}$$

where "n" is the number of predefined previous years for which mean OLR flux is computed for a given location (i, j) and time (τ).

$$\text{Flux index}\left(\Delta E_{ij\tau}\right) = \frac{E_{ij\tau} - \overline{E_{ij\tau}}}{\sigma_{ij\tau}} \tag{3}$$

where

$\Delta E_{ij\tau}$ – flux index for a given latitude(i), longitude(j), and time of data aquisition (τ)

E_{ij} – Current OLR flux computed for a give location(i, j) and time (τ)

$\overline{E_{ij\tau}}$ – Mean OLR flux calculated for a given location(i, j) and time(τ)

The anomalous nature of the energy flux index "$[\Delta E_{ij\tau}]^*$" can be identified by filtering out the energy flux index value below $+ 2\sigma$ level of mean OLR flux, also it helps to ascertain the duration of the anomalous flux observed.

$$\text{If}\, \Delta E_{ij\tau} \geq \overline{E_{ij\tau}} + 2\sigma\, \text{then} \Delta E_{ij\tau} = [\Delta E_{ij\tau}]^{*} \text{ELSE}\, \Delta E_{ij\tau} = 0 \quad (4)$$

where

$$[\Delta E_{ijk}]^{*} = \text{Anomalous energy flux index observed for a given location \& time}$$

From the analysis of "anomalous energy flux index", the potential earthquake-vulnerable zone can be identified. From the time of observation of thermal anomalies, the possibility of occurrence earthquake is imminent from a day to 45 days. Here, the authors have used OLR datasets with the resolution of $1° \times 1°$ provided by the NOAA National Climatic Data Centre [23].

4 Discussion

4.1 Kuril Island 2006–2007 Doublet Earthquake

Earthquakes usually occur as sequences featuring a few (or no) relatively small fore-shocks, a main shock, and a rapidly decaying number of aftershocks, with typically about a unit of magnitude larger than the largest aftershock [24]. An occurrence of comparable size is soon followed by a large earthquake, either on an adjacent portion of the fault that weakened initially or on a different fault, and such incidents are called doublets [25, 26]. The occurrence on November 15, 2006 has a fault inclination of strike 215° and dip 15°, with almost pure thrust movement with a rake of 92° and moment magnitude of = 8.4. On the other hand, for the January 13, 2007 has a normal faulting event of strike 43°, a dip of 59°, and a rake of −115°. The tidal amplitude irregularity (TAI) getting decelerated of Kuril Island 2006–2007 doublet occurred from 40 to 55 days preceding the earthquake, and an event was found on 22/09/2006 which was 55 days before an earthquake occurred with a value of −185.99 and −179.443, respectively. This accelerates further on 06/10/2006 and occurs 40 days before an earthquake with a value of 55.41765 and 54.653756 (Fig. 2). After this, TAI phenomenon revealed in Fig. 1 will have the normal changes of Earth tide shifts. There are significant OLR anomalies before earthquakes in which it shows 10–11 days prior to the earthquake. On October 3, 2006 the current OLR recorded was 250.6262 W/m^2 with anomalous index value (δE) of 2.4548. For the second time the OLR anomaly was observed on November 4, 2006 with the anomalos index value of 2.706048 (Fig. 3). For the second earthquake the OLR anomaly was observed 10 days before. The anomalous index value of 2.2617 was observed on that day (Fig. 3).

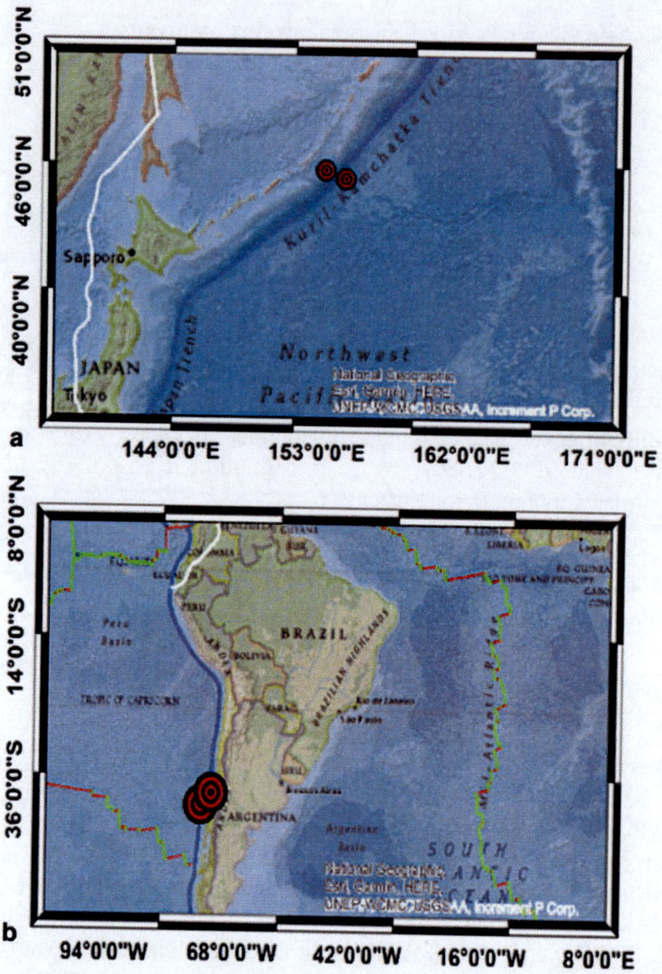

Fig. 1 Map showing locations of the earthquakes of different region (red circle) listed in Table 1. **a** Map showing the epicenters of Kuril doublet earthquakes, **b** Map showing the epicenters of Chile earthquakes

4.2 *Bio-Bio Chile Earthquake*

South-central Chile is a seismically active region with a convergence of almost 70 mm /yr., about twice that of the subduction zone of Cascadia. Large-scale earthquakes occurred along the 1,500 km long coastline in 1835, 1906, 1928, 1960, 1985, and 2010 [27]. On February 27, 2010, the central-south region of Chile was struck by a magnitude of MW 8.8 earthquake with an epicenter latitude of −36.122 and a longitude of −72.898 at the coast in the Maule region. The earthquake was an occurrence of interplate subduction along the boundary between the Nazca Plate and the

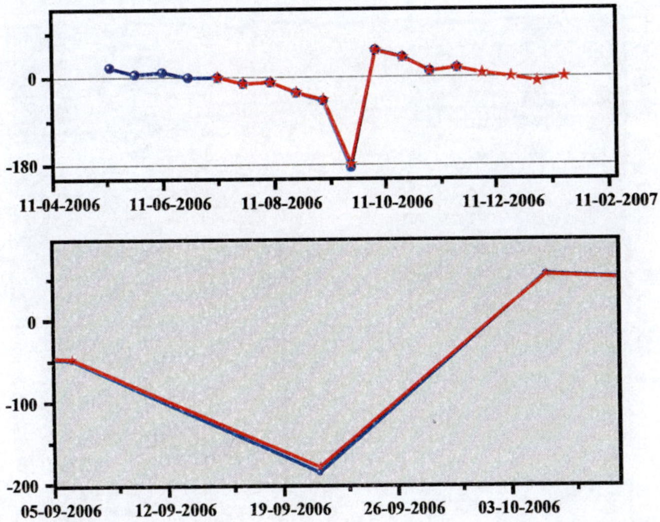

Fig. 2 Tide amplitude irregularities diagram showing the evolution of the 2006–2007 Kuril Islands earthquake sequence along the trench

South American Plate. The magnitude of Mw 7.4 earthquake happened as a result of a shallow normal fault near the Peru–Chile Trench, which represents the location where the Nazca Plate begins its eastward descent below South America. Focal mechanism solutions suggest the rupture emerged on either an approximately north- or south-striking, moderately dipping normal fault. The earthquake is an aftershock of the M 8.8 Maule earthquake that occurred just two hours earlier. The anomaly in TAI has been observed before the Bio-Bio Chile earthquakes, which occurred on February 27, 2010. The first anomaly was observed on 18/9/2009, which is 163 days prior to the occurrence of earthquake. The anomalous value of -50.6101 and -45.684 were observed. The second anomaly was observed on 03/10/2009, which is 148 days prior to the earthuqake. The anomalous values were 37.55534 and 35.71806. The figure shows TAI variations (Fig. 4). The OLR monthly data have been analyzed. We found anomalous OLR variation on September 18, 2009, which holds the current OLR flux of 289.2355 W/m^2 with a ∂E index value found to be 2.1959. Prior to 26 days of the main shock, the current OLR flux of 256.1967 W/m^2 occurred with a ∂E index value of 1.996786 on February 01, 2010 (Fig. 5).

5 Conclusion

Increased tectonic activity reduces the volume voids in the underlying rocks, thus, gases like radon get leaked from the rocks which are carried by the gases like carbon

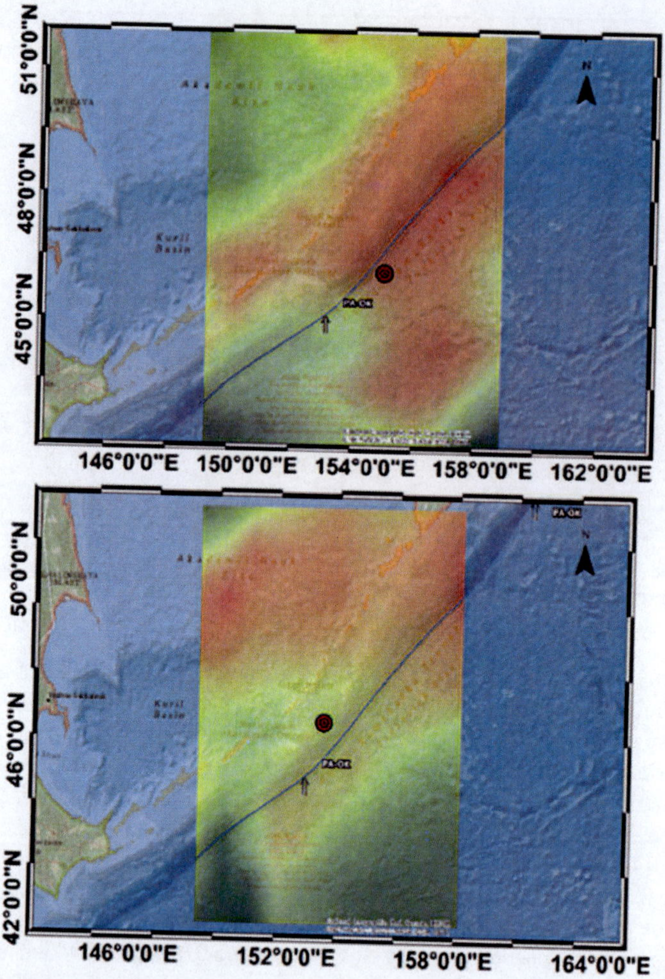

Fig. 3 Outgoing long-wave radiation diagram showing the evolution of the 2006–2007 Kuril Islands earthquake sequence

dioxide and moved up to the surface. At the Earth's surface, ionization of atmospheric air molecules happens due to the decayed radioactive radon. In the present work, earthquake analyzes occurred in various regions and different types of earthquakes. We observed a relationship between SET, OLR, and magnitude. The results indicate that solid Earth tides (SET) proceed to tectonic plate interlocking resulting in the release of huge amounts of heat energy due to the latent heat phenomenon leading to irregular tectonic activity causing the emanation of radon gas and abnormal thermal infrared from the epicenter area in the form of outgoing long-wave radiation. By analyzing the tidal stress of Kuril island doublet earthquake and Bio-Bio, Chile region earthquake, the greatest dip in doublet earthquake indicates the higher

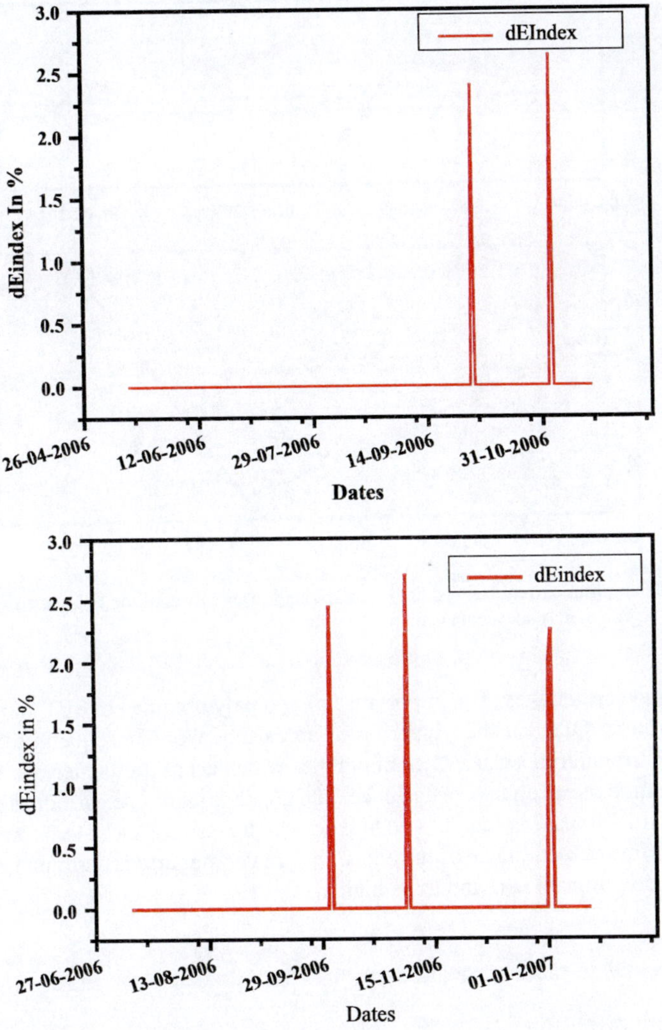

Fig. 3 (continued)

tidal stress thereby higher magnitude earthquakes occur of about $8.4M_{bw}$. The TAI deceleration energy difference of doublet earthquake is 6.54628 (in percent) and the acceleration is 0. 763,895. But the Bio-Bio, Chile region earthquake with the magnitude of Mw 8.8 the deceleration TAI's energy gap is 4.92614 and acceleration is 1.837271 (in percent). The greater difference between the syzygy components indicates the possibility of occurrence of doublet earthquakes with higher magnitude. This can be inferred from the two doublet earthquakes discussed above. The diffrence between the Syzygy component of Bio-Bio, Chile earthquakes is more than

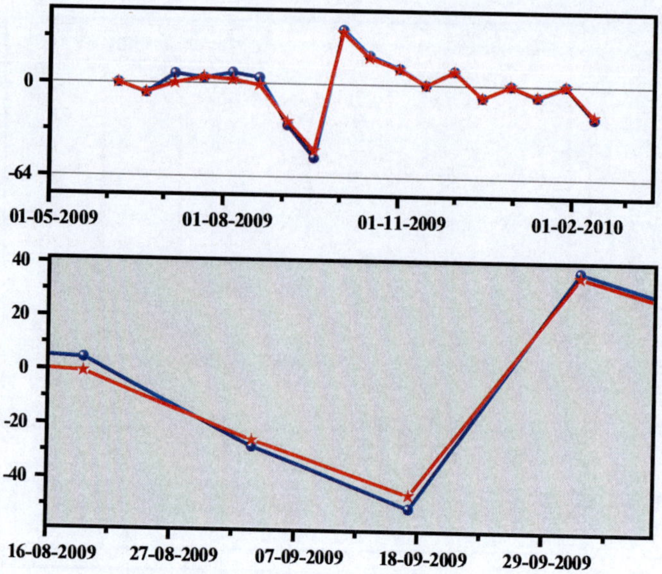

Fig. 4 Tide amplitude irregularities diagram showing the evolution of the February 27, 2010 Bio-Bio, Chile region earthquake sequence

Kuril island earthquakes. The importance of anomaly obtained in SET shows a very high impact of OLR on the triggering of an earthquake. Consequently, SET tidal amplitude irregularities cause OLR anomalies which act as a short-term precursor to detect the time of earthquake occurrence. The greater magnitude earthquake happens when the tidal impact is found to be higher, which acts as a short-term precursor to detect the time of earthquake occurrence. The greater magnitude earthquake happens when the tidal impact is found to be higher.

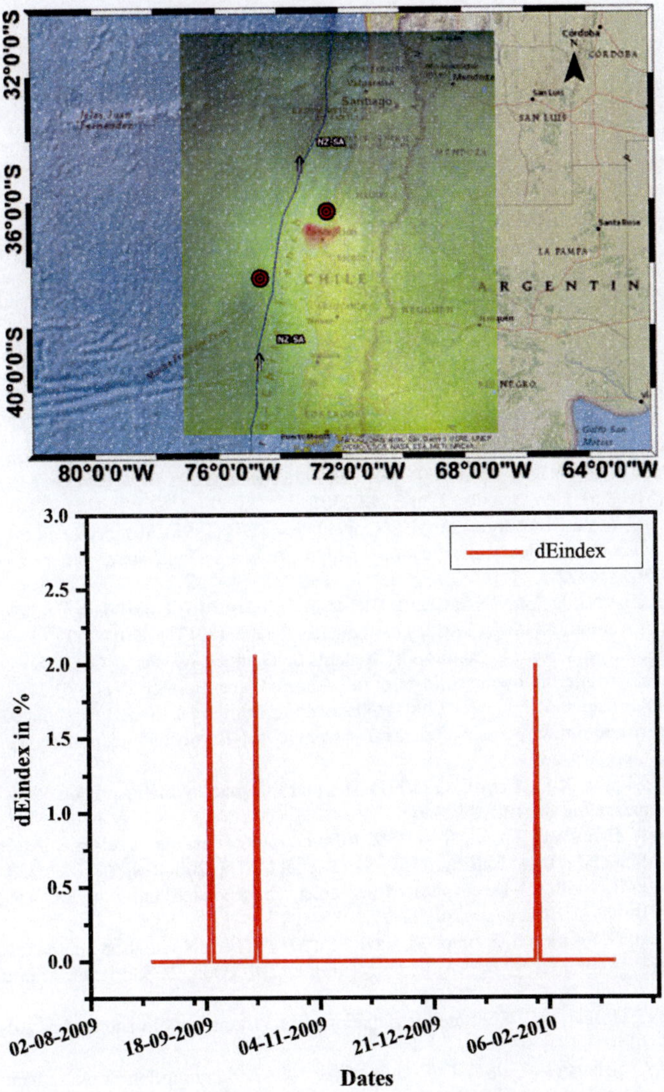

Fig. 5 Outgoing long-wave radiation diagram showing the evolution of the February 27, 2010 Bio-Bio, Chile region earthquake sequence

Acknowledgements We greatly indebted to Ministry of Earth Sciences for financial assistance (Project No: MoES/P.O (seismo)/1(343)/2018). The authors would thank to acknowledge NOAA for making OLR data available to the user community.

References

1. Kagan YY, Jackson DD (1999) Worldwide doublets of large shallow earthquakes. Bull Seismol Soc Am 89(5):1147–1155
2. Nissen E, Elliott JR, Sloan RA, Craig TJ, Funning GJ, Hutko A, Parsons BE, Wright TJ (2016) Limitations of rupture forecasting exposed by instantaneously triggered earthquake doublet, Nature geoscience, Published Online: 8 Feb 2016. https://doi.org/10.1038/Ngeo2653
3. Ruiz S, Tavera H, Poli P, Herrera C, Flores C, Rivera E, Madariaga R (2017) The deep Peru 2015 doublet earthquakes. Earth Planet Sci Lett 478:102–109. https://doi.org/10.1016/j.epsl.2017.08.036
4. William SD, Wilcock (2009) Tidal triggering of earthquakes in the Northeast Pacific Ocean. J Int 179:1055–1070
5. Emter D (1997) Tidal triggering of earthquakes and volcanic events, in Tidal Phenomena, Lecture Notes Earth Sci In: Wilhelm H (eds) vol 66, pp 293–309. Springer, New York
6. Tanaka S, Ohtake M, Sato H (2002) Evidence for tidal triggering of earthquakes as revealed from statistical analysis of global data. J Geophys Res 107(B10):1055–1070
7. Tanaka S (2010) Tidal triggering of earthquakes precursory to the recent Sumatra megathrust earthquakes of 26 December 2004 (Mw 9.0), 28 March 2005 (Mw 8.6), and 12 September 2007 (Mw 8.5). Geophys. Res Lett 37:L02301
8. Vidale JE, Agnew DC, Johnston MJS, Oppenheimer DH (1998) Absence of earthquake correlation with Earth tides: an indication of high preseismic fault stress rate. J Geophys Res 103:24567–24572
9. van der Elst NJ, Delorey AA, Shelly DR, Johnson PA (2016) Fortnightly modulation of San Andrea's tremor and low-frequency earthquakes. PNAS 113(31):8601–8605
10. Métivier L, de Viron O, Conrad CP, Renault S, Diament M, Patau G (2009) Evidence of earthquake triggering by the solid earth tides. Earth Planet Sci Lett 278(3–4):370–375
11. Carreno E, Capote R, Yague A (2001) Observations of thermal anomaly associated to seismic activity from remote sensing. General Assembly of European Seismology Commission, Portugal, pp 265–269
12. Quing Z-J, Xu, X-D, Dian C-G (1991) Thermal infrared anomaly -precursor of impending earthquakes. Chin Sci Bull 36:319
13. Salman A, Egan WG, Tronin AA (1992) Infrared remote sensing of seismic disturbances. In: Polarization and remote sensing. SPIE, San Diego, CA, pp 208–218. ((4):319–323)
14. Liebmann B, Smith CA Description of a complete (Interpolated) outgoing longwave radiation dataset. Bull American Meteorol Soc 77:1275–1277
15. Ouzounov D, Pulinets S, Kafatos M, Taylor P (2017) Thermal radiation anomalies associated with major earthquakes. NASA Technical Report 20170008528 Retrieved from https://ntrs.nasa.gov/search.jsp?R=20170008528
16. Ohring G, Gruber A (1982) Satellite radiation observations and climate theory. Adv Geophys 25:237–304
17. Gruber A, Krueger A (1984) The status of the NOAA outgoing longwave radiation dataset. Bull Am Meteorol Soc 65:958–962
18. Liu D, Kang C (1999) Thermal omens before earthquakes. ACTA Seismological Sinica 12(6):710–715
19. Ouzounov D, Liu D, Kang C, Cervone G, Kafatos M (2007) Taylor, P, Outgoing Long Wave Radiation Variability from IR Satellite Data Prior to Major Earthquakes. Tectonophysics 431:211–220
20. Ouzounov D, Pulinets S, Hattori K, Kafatos M, Taylor P (2011) Atmospheric signals associated with major earthquakes. In: M Hayakawa M (ed) Frontier of earthquake short-term prediction study. Japan, pp 510–531
21. Xiong P, Shen XH, Bi YX, Kang CL, Chen LZ, Jing F, Chen Y (2009) Study of outgoing longwave radiation anomalies associated with Haiti earthquake. Natural Hazards Earth Syst Sci 10:2169–2178

22. Fujii Y, Satake K (2007) Tsunami sources of November 2006 and January 2007 great Kuril earthquakes. Bull Seism Soc Amer. Accepted
23. Lee HT, NOAA CDR Program (2011) NOAA climate data record (CDR) of daily outgoing longwave radiation (OLR), Version 1.2. [Indicate subset used]. NOAA National Climatic Data Center. doi:https://doi.org/10.7289/V5SJ1HH2 (access date)
24. Charles J (2008) Ammon, Hiroo Kanamori & Thorne Lay, A great earthquake doublet and seismic stress transfer cycle in the central Kuril Islands. Nature 451:561–566
25. Lay T, Kanamori H (1980) Earthquake doublets in the Solomon Islands. Phys Earth Planet Inter 21:283–304
26. Xu Z, Schwartz SY (1993) Large earthquake doublets and fault plane heterogeneity in the northern Solomon Islands subduction zone. Pure Appl Geophys 140:365–390
27. Cisternas M, Atwater BF, Torrejón F, Sawai Y, Machuca G, Lagos M, Husni M (2005) Predecessors of the giant 1960 Chile earthquake. Nature 437(7057):404–407

Anomalous Thermal Signatures and Triggering Effect of Solid Earth Tides—A Case Study on Main Central Thrust (MCT) Earthquakes Since 1991 ($M \geq 6.0$)

N. Venkatanathan, Ramya Jeyaraman, and M. Senthil Kumar

Abstract During the latter of the last century, scientists used remote sensing to understand the structural and geomorphology of seismically vulnerable regions. In recent days, remote sensing technology is applied to study the dynamics of atmospheric parameters like outgoing longwave radiation (OLR). Anomalous variations in the OLR index vary between 2.07 and 2.21 were observed 3–30 days prior to the occurrence of MCT (Himalayan region) earthquakes. This could be attributed to the emission of radon gas from the lithosphere due to increased tectonic activity. Similarly, during the lunar phase, the $M2$ component of the solid earth tides varies harmonically, and any deviation in this variation is termed as "tidal amplitude irregularity" (TAI). Irregularities in tidal amplitude influences the rotational speed of Earth, which in turn affects the tectonic plate motion. Thus, along the seismically vulnerable regions, the earthquakes are triggered. The TAI with the irregularity change varies from 15 to 40% was observed 3–6 months prior to the occurrence of earthquakes. From the analysis of earthquakes of the MCT region since 1991 with the magnitude above 6.0, the authors inferred that the OLR anomaly and TAI can be used as an effective tool to forecast devastating earthquakes on a short-term basis.

Keywords Main central thrust (MCT) · Earthquake forecasting · OLR · TAI · GIS

1 Introduction

1.1 Solid Earth Tides and Triggering Effect of Earthquake Occurrence

Most of the devastating earthquakes happen on the edge of tectonic plates. These plates are constantly moving and interact with each other in many ways. They can move apart from each other creating a new land called a divergent plate boundary.

N. Venkatanathan (✉) · R. Jeyaraman · M. Senthil Kumar
SASTRA Deemed to be University, Thanjavur, Tamil Nadu, India
e-mail: venkatanathan@eee.sastra.edu; physics16972@gmail.com

T. G. Sitharam et al. (eds.), *Seismic Hazards and Risk*, Lecture Notes in Civil Engineering 116, https://doi.org/10.1007/978-981-15-9976-7_33

Plates can also push up against each other called convergent plate boundary. Sometimes the plates slowly slide past each other called transform boundary. The relationship between the earthquakes and the syzygy days has been made by previous researchers [1–3] and the triggering of the seismic tremor by means of tidal load [4]. Other studies have reported small positive correlations [5–9] under specific conditions. Wilcock [10] found a weak correlation between the influence of tide wave and occurrences of the earthquake in continental areas. Solid earth tides have a triggering effect on the occurrence of earthquakes [11]. The focus is made on the spatiotemporal variation of solid earth tide, and the occurrence of earthquake correlation is measured in and around the focal regions of the catastrophic earthquakes occurred in the Sumatra region from 2004–2007 (Mw = 8.5–9.0). The result indicates a strong tidal effect appeared for several to ten years before the occurrence of the earthquakes [5].

Ide [9] has confirmed that larger earthquakes are likely to happen during the period of high tidal stress limited to specific regions or circumstances. In any three-dimensional body, stresses generated by the tidal effects are affected by an external inhomogeneous gravity field by rotating planets or Moons. A semidiurnal solid tide triggers earthquakes. The correlation study is made on earthquakes and solid earth tides, and it concluded that earthquake is triggered by semidiurnal tides, and it can be helpful in exploring the possible times of subsequent aftershocks. The correlation between earthquake sequence in 2011 Christchurch, New Zealand, and the solid earth tides shows that the larger aftershocks were mainly triggered by the semidiurnal solid tides, and the most of earthquakes (Ms $\geq$ 7 worldwide since 1900) are analyzed, and it was found that the number of earthquakes that occurred near the 0°, 90°, 180° and 270° phases of the semidiurnal solid earth tide ($M2$ tidal component) exceeds the number that occurred during other phases [12]. The seismic activity is influenced by the stress caused by the body tides within earth. The investigation is made on the triggering of earthquakes by the effects of the Moon and the Sun by performing a statistical comparison of variations in tidal activities and temporal distribution. The magnitude of the triggering effects of earth tides is different spherical harmonics of tides, and magnitude of the triggering effects of earth tides significantly depends on latitude [13].

The advancements in remote sensing instruments of recent years have helped in scientific understanding of atmospheric earthquake signals. Satellite thermal imaging data can be used as an effective tool in the detection of OLR anomaly [14]. OLR data could be considered as an effective precursor to detect anomalies in the seismic

Table 1 List of MCT earthquakes occurred in the Himalayan region since 1991 with magnitude >6 (data provided by USGS http://earthquake.usgs.gov)

Event	Origin time	Latitude	Long	Depth	Mag	Place
19-10-1991	21:23:14.300Z	30.78	78.774	10.3	6.8	Uttarkashi, Uttarakhand
28-03-1999	19:05:11.030Z	30.512	79.403	15	6.6	Chamoli, Uttarakhand
22-07-2007	23:02:14.790Z	30.881	78.239	19	5.1	Kharsali, Uttarakhand

waves. OLR anomalies were processed by using time series, comparative method and Eddy field. The time series, comparative and eddy field is used as a tool for analyzing outgoing longwave radiation data, and the relationship between anomaly and the occurrence of the earthquake has been presented. OLR singularities can be considered as an effective indicator to detect seismic anomalies [15].

1.2 Anomalous Thermal Signatures

Thermal radiation is an important factor in the observation of earthquake precursory phenomena. Rapid variations in the tectonic plate movements lead to the anomalous positive deviations in the atmospheric parameters. For observing OLR anomaly, geostationary satellite data can be used [16]. To forecast the earthquake on a short-term basis with reasonable accuracy, multi-parameters should be taken into consideration which includes total electron content, radon gas emanation and EM emissions; there exists a substantial relation between parameters like OLR and infrasound waves [17]. Jing [18] uses a time series analysis is used as a processing method so as to eliminate the interference of season and terrain.

By means of analyzing the deviation index, the thermal abnormal region is obtained. It has been found that the thermal radiation anomalies appear several months or days prior to this event over the epicenter region [18]. The data include outgoing longwave radiation (OLR), GPS/TEC, lower earth orbit ionosphere tomography and critical frequency of F2. The joint analysis of atmosphere/ionosphere parameters is done with the variations in temporal and spatial characteristics. It has been showed that the occurrence of anomalies over atmosphere/ionosphere is consistent near the earthquake's epicenter [19].

The interdisciplinary approach for pre-earthquake signals such as outgoing longwave radiation (OLR), relative humidity, geo-electrical measurements (GEM), radon gas emission, proton (H+ ion) permeability can be used as a precursor. Anomalous drop in RH and rise in OLR lead to the drift in H+ ions from the core to the surface of the earth triggers the increase in tectonic activity, and thereby, radon gas emission is induced. This results in ionization of air molecules in the atmosphere which leads to falling in relative humidity in the earth's surface. The anomaly in OLR flux happens as a result of latent heat release due to the upward movement of ionized air molecules from the earth's surface [20].

In the present study, we have correlated stress due to the solid earth tides during Syzygy period with the seismic activity. The continuous deceleration of dy/dt leads to an increase in tidal stresses and accelerates the interlocking of the tectonic plate interface. Therefore, such accelerated deformation can lead to a change in the stress state on the whole seismic region; thereby, maximum energy is released and hence increases the probability of earthquakes. Due to this triggering effect of solid earth tides, the plates try to adjust to stay in its original position, it experiences an acceleration, then the force acted upon the plate will be mass times the acceleration, and its direction is opposite to that of acceleration. Hence, it is clearly indicated that if it has more mass thereby causes higher inertial force by which maximum energy is released through which variations in earth's rotation lead to change in tidal stress state which acts as a triggering factor for an occurrence of earthquake.

2 Study Area

On October 19, 1991, an earthquake with the magnitude 6.8 occurred in Garhwal Himalayas in Northern India, which has the 30.78 N latitude and 78.774 E longitude, an earthquake with a magnitude of 6.8 Mwb with strike 330, dip 11, rake 117 (USGS.gov). The region lies in the collision junction of the Indian plate and Asian plate. Uttarkashi lies in the main alpine Himalayan belt, one of the most vulnerable earthquake regions in the world. The strong ground shaking caused in the districts of Uttarkashi, Tehri, Chamoli, in the state of Uttarpradesh. This thrust event occurred along the north of the main central thrust (MCT) fault sequence which defines a boundary between greater and lesser Himalaya. The depth of shallow type earthquake is found in this region which lies within 20 km of the crust. The probabilities of occurrence of landslides are high in this region. The reverse faulting mechanism is identified in the fault plane of this region. The main shock is located on northeast edge of this zone [21].

On March 28, 1999, Chamoli earthquake is yet another important earthquake in Northern India from the viewpoint of the Himalayan region, occurs in the Chamoli district in the Indian state of Uttarkhand, which has the epicenter in 30.512 N latitude and 79.403 E longitude with strike 290, dip 14, rake 90 (USGS.gov). The magnitude of an earthquake is 6.6 Mwb. It struck the Central Himalaya, the region which is prone to earthquakes. It has been placed in the highest seismic zone (Zone V) of the Himalaya. The event was located in active MCT zone and may have caused a low angle fault [22].

In the early morning hours of July 22, 2007, an earthquake occurred in Garwal Himalaya near the Kharsali region (30.881 N latitude and 78.239 E longitude), an earthquake with a magnitude of Mwb5.1 (USGS.gov) which has strike 290 dip 7, rake 101. The deformation occurred by reverse faulting with a significant strike-slip component. It occurred close to the MCT where seismic gap exists [23]. It has been inferred that the Kharsali earthquake occurred due to a northerly dipping low angle thrust fault at a depth of 19 km (Fig. 1).

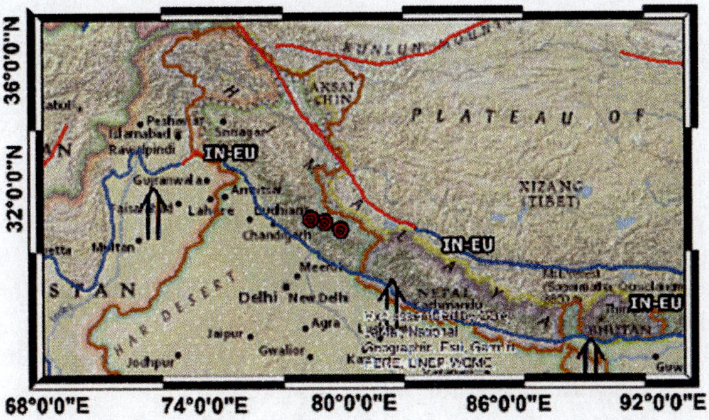

Fig. 1 Map showing locations of the earthquakes of Uttarakhand region (red dotted) listed in Table 1

3 Methodology

3.1 Tidal Amplitude Irregularities (TAI)

Scientists have been attempting to link the solid earth tides and earthquakes for more than a century. Results of earlier studies indicate that the "Strong Earth Tides" are the origin of the major earthquakes; in this paper, the authors found a strong association between the irregularity of the amplitude of solid earth tides, the presence of anomalous atmospheric precursors and the frequency of earthquake. The normal range of solid earth tide is ±20 cm and can sometimes reach 30 cm or more. Solid earth tides were measured using a model developed by Dennis Milbert, NASA, and based on an edited version of the source code "dehanttideinelMJD.f," the guidelines of the 2003 IERS conventions have been followed.

As the Earth rotates under the tidal force owing to the Sun and Moon, each point on Earth will undergo harmonic fluctuations and regularly displace the Earth's surface vertically up and down (two highs and two lows per day). The $M2$ component of solid earth tides has maximum amplitude on syzygy days and has minimum amplitude during the Quadrature phase of the Moon. The $M2$ portion slowly decreases as the Moon orbits from syzygy to Quadrature position, and it increases gradually as the Moon moves from Quadrature to syzygy again. Any deviation from this harmonic variation is termed "irregularity of tidal amplitude" (TAI). The TAI was probably caused by the tidal friction created in the system "Earth–Moon–Sun." The tidal forces synchronize the rotation and orbital motion of the tidally locked celestial bodies. As the celestial bodies orbit is eccentric, the velocity of the orbit varies with the distance. Therefore, the tidal bulge produced on the Earth's surface would not point toward the line joining Moon and Sun centers, the tidal bulge would either lead or lag in space and time. To preserve the system's angular momentum, the Earth's rotational velocity is modified. A backward torque slows down the earth's rotation, so before the syzygy days, the maximum $M2$ component of one half of the lunar cycle was observed. Likewise, if the maximum $M2$ component of the one half of the lunar cycle was measured after syzygy days, then the rotation of the Earth slowly accelerates, i.e., the changes in the $M2$ component of the solid earth tide lead to a change in the Earth's rotational velocity. A change in the earth's rotational speed determines the probability of collision with the tectonic plate due to their differential inertia, just as the passengers collide with each other when the bus driver unexpectedly applies the brake. Besides, the friction of the tidal bulge being pulled back and forth through the Earth's solid body heats the Earth's innermost. Therefore, these stresses transformed slowly into the seismic strain, and when this strain goes above the critical level along the seismically sensitive area, earthquake is triggered.

Instead of observing the peak amplitude of the $M2$ component of the tide on syzygy days, it is observed either prior to the syzygy days or after it. The percentage of error in the vertical shift of $M2$ component (dE) can be calculated as follows:

$$\mathrm{dE} = \left(\frac{A_{M2}^{\uparrow} - A_{M2}^{\leftrightarrow}}{A_{M2}^{\leftrightarrow}} \right) \times 100\% \tag{1}$$

where

dE = error percentage in the vertical displacement of $M2$ component of solid earth tide
$A_{M2}^{\uparrow}$ = highest amplitude of the $M2$ component either prior or after syzygy days
$A_{M2}^{\leftrightarrow}$ = actual amplitude of the $M2$ component on syzygy days.

The positive "dE" suggests that, on syzygy days, the maximum tidal amplitude was observed prior to the syzygy days and thus a lesser $M2$ portion, the Earth should slow its rotation. After syzygy days, measuring the full tidal amplitude is expressed by the negative value of "dE." The increase in the $M2$ portion after syzygy days slowly speeds up Earth's rotational speed in order to maintain the system's angular momentum.

It is found that the vertical shift percentage (dE) showed strong variance from positive to negative or vice versa a few months prior to the occurrence of major earthquakes. This sharp "dE" variance can provide a useful hint as to the potential severity of the imminent earthquake.

3.2 *Anomalous Outgoing Longwave Radiation (OLR)*

Outgoing longwave radiation (OLR) is considered the electromagnetic radiation from the earth's surface into the atmosphere in the form of thermal radiation. The energy flux that is transferred from the earth's surface to the atmosphere by outgoing long-wave radiation is expressed in W/m^2. Using satellite technology, OLR is measured at an altitude of 10–12 km.

This anomalous increase in OLR flux has been found near or above the seismically active regions and extends over thousands of square kilometers. Many scientists before the major destructive earthquakes reported transient thermal anomalies. The raw OLR data were processed for a wavelength region of 10–13 μm for clear sky conditions. Numerous OLR flux studies found the anomalous variations were detected between 1 to 45 days before the devastating earthquakes.

The anomalous variations of OLR flux at the seismically active regions from the mean OLR flux of previous 10 years can be found as follows:

$$\overline{E_{xy\tau}} = \sum_{\tau=1}^{n} E_{xy\tau} \tag{2}$$

where "*n*" is the number of previous years for which mean OLR flux is computed for a given location (*x*,) and time (*t*).

$$\text{Flux index } \left(\text{dE}_{xy\tau}\right) = \frac{E_{xy\tau} - \overline{E_{xy\tau}}}{\sigma_{xy\tau}} \tag{3}$$

where

$\text{dE}_{xy\tau}$—flux index for a given latitude (*i*), longitude (*j*) and time of data acquisition (τ)
$E_{xy\tau}$—current OLR flux computed for a give location (*i*, *j*) and time (τ)
$\overline{E_{ij\tau}}$—mean OLR flux calculated for a given location (*i*, *j*) and time (τ).

The anomalous nature of the energy flux index "$[\text{dE}_{xy\tau}]^*$" can be identified by filtering out the energy flux index value below $+2\sigma$ level of mean OLR flux, and also, it helps to ascertain the duration of the anomalous flux observed.

$$\text{If } \text{dE}_{ij\tau} \geq \overline{E_{ij\tau}} + 2\sigma \text{ then } \text{dE}_{ij\tau} = \left[\text{dE}_{ij\tau}\right]^* \text{ ELSE } \text{dE}_{ij\tau} = 0 \tag{4}$$

where

$\left[\text{dE}_{ij\tau}\right]^*$ = Anomalous energy flux index observed for a given location and time.

The possible earthquake-vulnerable zone can be identified from the study of the "anomalous energy flux index." Earthquake occurrence is imminent from one day to 45 days from the time of detection of thermal anomalies. The authors used OLR data sets with a resolution of 1° × 1° given by the NOAA National Climatic Data Center [24].

4 Results and Analysis

4.1 *October 19, 1991: Uttarkashi Earthquake, India Region*

The study is made on the tidal amplitude irregularities and OLR before the devastating earthquake at Uttaranchal, India.

The deceleration has happened from 10 to 170 days prior to the occurrence of an earthquake; the anomaly has been observed on August 10, 1991, which is 70 days before the occurrence of an earthquake with the value of −13.95770587. It further decelerates on August 25, 1991, that happens 55 days before the earthquake with the value of −10.76748292. It further decelerates on September 8, 1991 (41 days) with the value of −34.82335615. The next acceleration happens with a value of 31.20271801 on September 23, 1991 (26 days) prior to the occurrence of devastating earthquakes. There was the greatest dip in the anomaly from 31.20271801 to 10.23772 on October 7, 1991, which happens 12 days prior to the occurrence of the main shock (Fig. 2).

Anomalous variations in the thermal energy were observed above the cloud level were observed in the form of OLR wave emissions. The increased tectonic activity due to SET triggers the anomaly in outgoing longwave radiation. The OLR signals were analyzed for the past six months from the main shock, and the anomalous signals were observed a few days before the main shock. On 09-10-1991, the anomalous energy index of 2.0901 was observed 10 days prior to the main shock (Fig. 3).

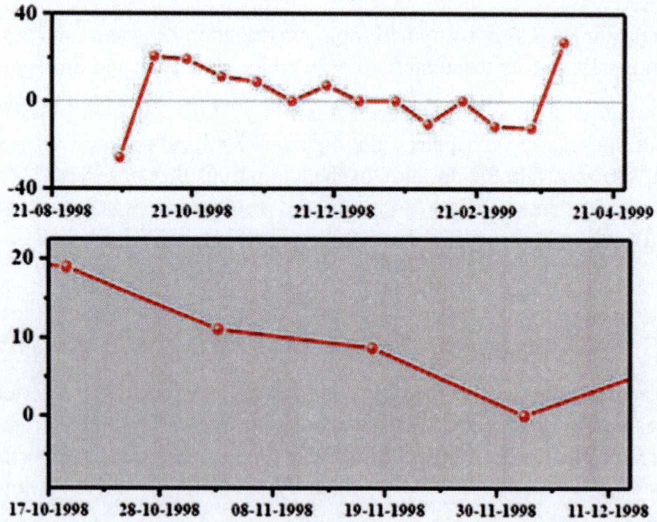

Fig. 2 Time series for the tidal amplitude irregularities during syzygy before the Uttarkashi M6.8 earthquake. Here, *x*-axis represents the syzygy dates, and *y*-axis represents velocity [d*y*/d*t* (1991) in %]

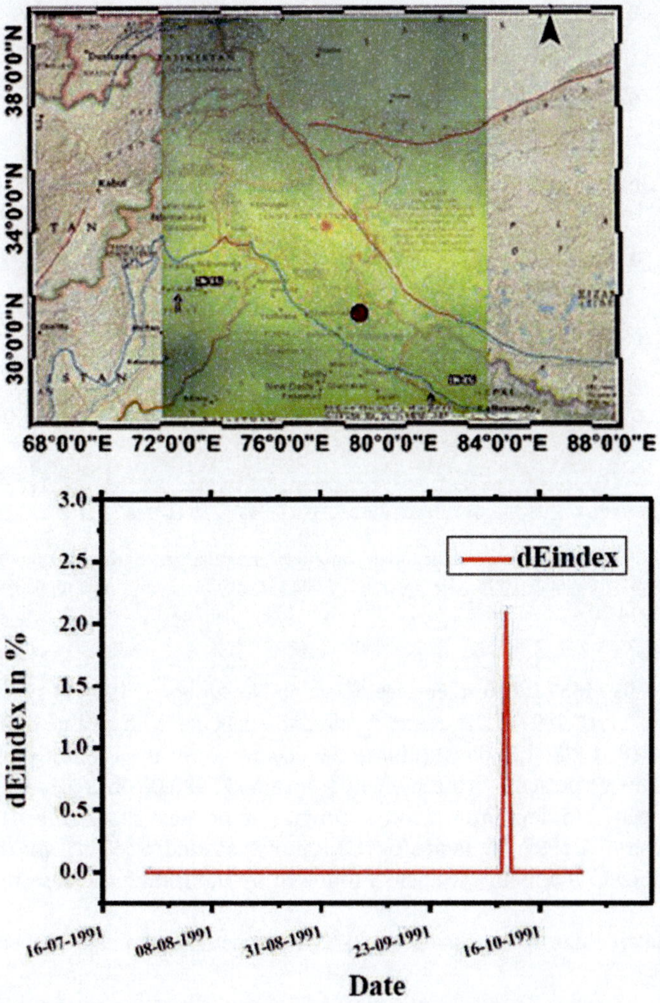

Fig. 3 Anomalous OLR flux graph variation before the Uttarkashi M6.8 earthquake (epicenter is marked with red in map)

4.2 March 28, 1999: Chamoli Earthquake, India Region

The study is made on the tidal amplitude irregularities before the devastating earthquake which occurs at the Chamoli, Uttaranchal, India, on March 28, 1999.

The deceleration has happened from 10 to 170 days prior to the occurrence of earthquake; an anomaly has been observed on October 5, 1998, which is 174 days before the occurrence of earthquake with the value of −20.48001617. It further decelerates on October 19, 1998, that happens 160 days before earthquake with the

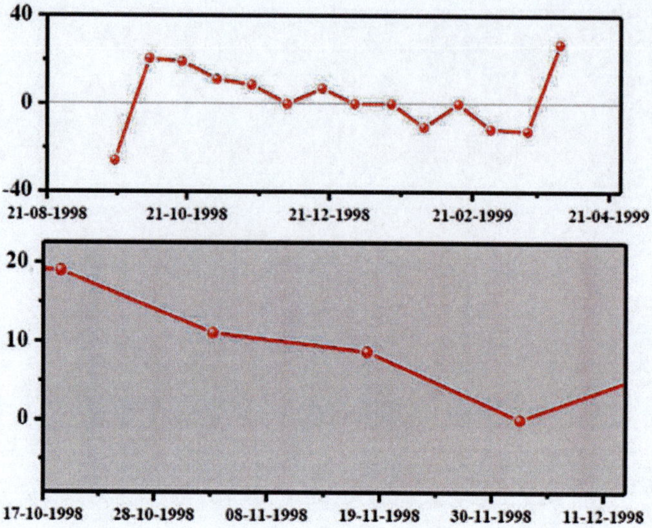

Fig. 4 Time series analysis for the tidal amplitude irregularities during syzygy before the Chamoli M6.6 earthquake. Here, *x*-axis represents the syzygy dates, and *y*-axis represents velocity [dy/d*t* (1999) in %]

value of 18.99716374. It further decelerates on November 3, 1998 (145 days) with the value of 11. 04889741. The next deceleration happens with a value of 8.6437 on November 18, 1998 (130 days) prior to the occurrence of devastating earthquakes. There was the greatest dip in the anomaly from 8.6437 to 0.00 on December 3, 1998, which happens 115 days prior to the occurrence of the main shock (Fig. 4).

On January 14, 1999, the anomalous OLR flux was found to be 1.98 which happens 73 days prior to the main earthquake followed by the further increase in tectonic activity.

Has shown abnormal emission on 30-01-1999 with an anomalous energy index of 2.23 (Fig. 5).

4.3 *July 22, 2007: Kharsali Earthquake, India Region*

The analysis is made on the Kharsali, India region (30.881 N latitude and 78.239 E longitude), an earthquake with a magnitude of Mwb5.1. Prior to this earthquake, both SET and OLR anomalies were observed.

The deceleration has happened from 50 to 120 days prior to the occurrence of earthquake, an anomaly has been observed on 2/4/2007 which is 111 days before the occurrence of an earthquake with a value of 27.17. It further decelerates on 17/4/2007 that happens 96 days before earthquake with the value of 11.21. It further decelerates on 2/5/2007(81 days) with the value of 8. 559. The next deceleration happens

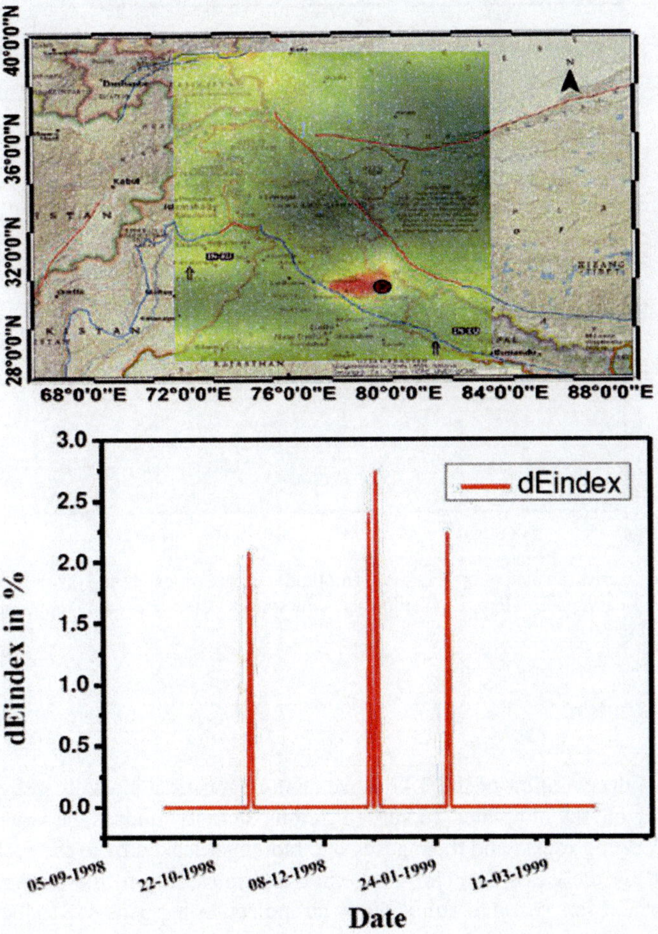

Fig. 5 Anomalous OLR flux graph variation before the Chamoli M6.6 earthquake (epicenter is marked with red in map)

with a value of 7.18 on 16/5/2007(67 days) prior to the occurrence of devastating earthquakes (Fig. 6).

The anomalous OLR flux was observed twice prior to the main shock. From the analysis of outgoing longwave, the anomalous OLR flux was observed twice prior to the main shock. We have found the abnormal OLR scenario on May 23, 2007, due to increased tectonic activity. The anomalous energy index of 2.59869 was observed 60 days prior to the earthquake. The second anomaly was observed on July 12, 2007 with the anomalous index of 2.1908, which is 10 days prior to the main shock. The spatial distribution of the anomalous flux over and near the epicenter region gives clues on impending earthquakes of a particular region (Fig. 7).

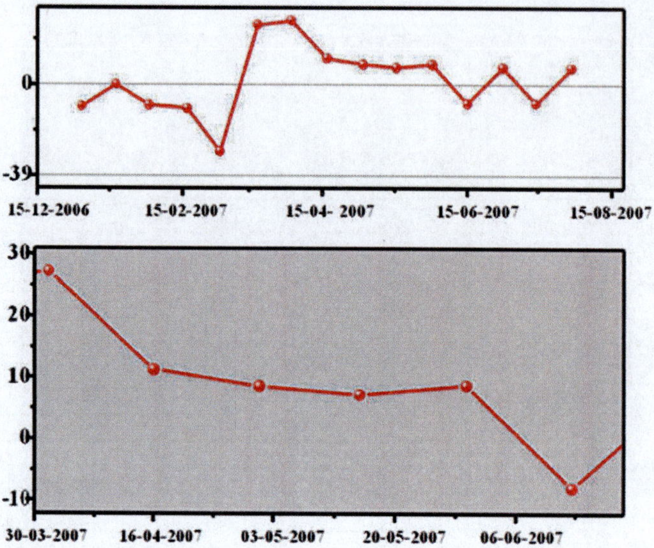

Fig. 6 Time series analysis for the tidal amplitude irregularities during syzygy before the Kharsali M5.1 earthquake. Here, *x*-axis represents the syzygy dates, and *y*-axis represents velocity [d*y*/d*t* (2007) in %]

5 Conclusion

Continuous deceleration of the TAI movement creates tidal stress thereby leads to interlocking of tectonic plates. Increased tectonic activity reduces the volume voids in the underlying rocks, and thus, gases like radon get leaked from the rocks which are carried by the gases like carbon dioxide and moved up to the surface. At the Earth's surface, ionization of atmospheric air molecules happens due to the decayed radioactive radon.

To study the effects of seismic activities on OLR, the OLR analysis was made 6 months before the occurrence of earthquakes in that area so as to seek short-term seismic anomalies. Anomalous thermal variations were observed at an above cloud level in the form of longwave IR emissions. The initiation of the rupture process in the rocks gives rise to the abnormal infrasound variation; thereby, critical level of stress occurs along the fault interface [25].

The analysis of OLR anomaly discovered which could be related to the time of occurrence of earthquakes. Our analysis of the continuous outgoing longwave earth radiation (OLR) indicates anomalous variations prior to a number of medium to large earthquakes. It will take some time to transform the built-up stress into elastic strain energy. The large amount of inertia of huge rocks involved in the collision process; hence, the elastic strain energy requires time to rupture the rocks in that region [26].

In the present work, by analyzing the earthquake occurred in 2007 and 1991, we identified a relationship between SET, OLR AND time of occurrence of the

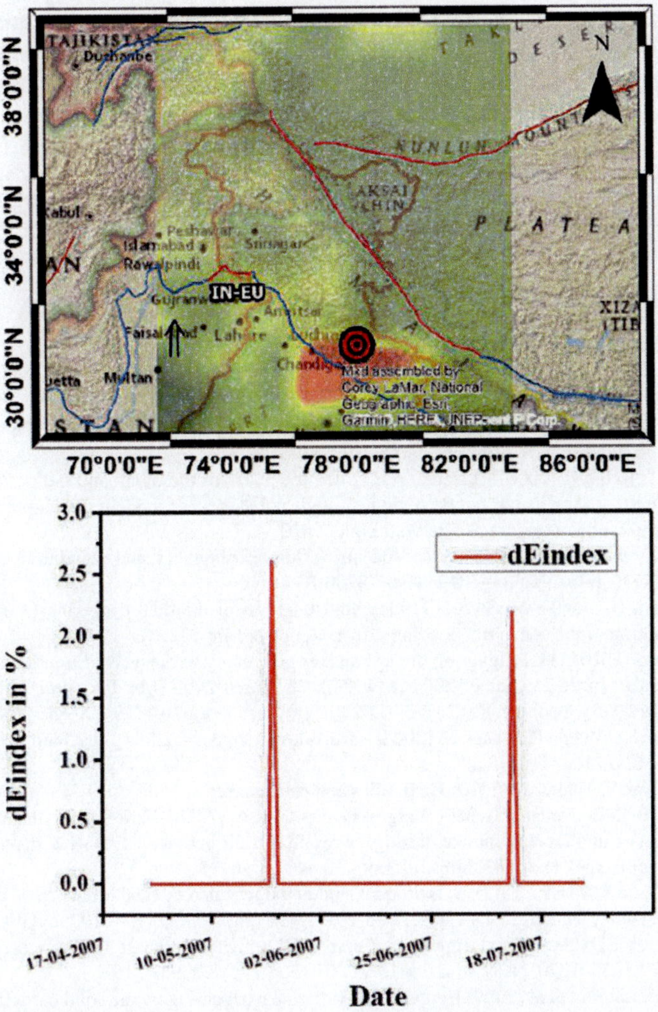

Fig. 7 Anomalous OLR flux graph variation before Kharsali M5.1 earthquake (epicenter is marked with red in map)

earthquake. The results indicate that solid earth tides (SET) lead to the interlocking of tectonic plates; thereby, enormous amount of heat energy is released due to the latent heat phenomenon which leads to an abnormal tectonic activity which triggers the emanation of radon gas and abnormal thermal infrared from the epicenter region in the form of outgoing longwave radiation.

By analyzing the tidal stress of 1991, 1999 and 2007, the greatest dip in 1991 indicates the higher tidal stress; thereby, higher magnitude earthquakes occur of about 6.8 Mbw. The significance of anomaly obtained in SET evidences a very high impact of OLR on earthquake triggering. Hence, tidal amplitude irregularities of SET trigger OLR anomalies which act as a short-term precursor for detecting the time of occurrence of earthquakes. When the tidal triggering is found to be stronger, the larger magnitude earthquake will occur.

Acknowledgements We greatly indebted to the Ministry of Earth Sciences for financial assistance (Project No: MoES/P. O(seismo)/1(343)/2018). The authors would thank to acknowledge NOAA for making OLR data available to the user community.

References

1. Lin CH, Yeh YH, Chen YI, Liu JY, Chen KJ (2003) Earthquake clustering relative to lunar phases in Taiwan. Terr Atmos Ocean Sci 14:289–298
2. Iwata T, Katao H (2006) Correlation between the phase of the moon and the occurrences of micro earthquakes in the tamba region through point-process modeling. Geophys Res Lett 33:L07302. https://doi.org/10.1029/2005gl025510
3. Gackstatter F (2007) Lunisolar effect on spring tides, earthquakes, and tsunamis. J Coastal Res 23:528–530. https://doi.org/10.2112/06-0700.1
4. Tsuruoka H, Ohtake M, Sato H (1995) Statistical test of the tidal triggering of earthquakes contribution of the ocean tide loading effect. Geophys J Int 122:183–194
5. Tanaka S (2010) Tidal triggering of earthquakes precursory to the recent Sumatra mega thrust earthquakes of 26 December 2004 (Mw 9.0), 28 March 2005 (Mw 8.6), and 12 September 2007 (Mw 8.5). Geophys Res Lett 37:L02301. https://doi.org/10.1029/2009gl041581
6. Cochran ES, Vidale JE, Tanaka S (2004) Earth tide can trigger shallow thrust fault earth-quakes. Science 306:1164–1166
7. Kolvankar V, Thakur N (2010) Earth tides and earthquakes
8. Arabelos DN, Contadakis ME, Vergos G, Spatalas S (2016) Variation of the Earth tide-seismicity compliance parameter during the recent seismic activity in Fthiotida, central Greece. Ann Geophys 59(1):S0102. https://doi.org/10.4401/ag-6795
9. Ide S, Yabe S, Tanaka Y (2016) Earthquake potential revealed by tidal influence on earthquake size frequency statistics. Nat Geosci Lett 2016:2796. https://doi.org/10.1038/NGEO
10. Wilcock WSD (2009) Tidal triggering of earthquakes in the Northeast Pacific Ocean. Geophys J Int 179:1055–1070. https://doi.org/10.1111/j.1365-246x.2009.04319.x
11. Métivier Laurent et al (2009) Evidence of earthquake triggering by the solid earth tides. Earth Planet Sci Lett 278:370–375
12. Chen L, Chen JG, Xu QH (2012) Correlations between solid tides and worldwide earthquakes MS $\geq$ 7.0 since 1900, Nat Hazards Earth Syst Sci 12:587–590. www.nat-hazards-earth-syst-sci.net/12/587/2012/. https://doi.org/10.5194/nhess-12-587-2012 © Author(s) 2012. CC Attribution 3.0 License
13. Varga P, Grafarend E (2019) Influence of tidal forces on the triggering of seismic events. https://doi.org/10.1007/978-3-319-96277-1_6
14. Carreno E, Capote R, Yague A (2001) Observations of thermal anomaly associated to seismic activity from remote sensing. General Assembly of European Seismology Commission, Portugal, 10–15 September, pp 265–269
15. Chen Y, Shen X, Jing F, Xiong P (2010) Application of outgoing longwave radiation data for earthquake research. In: IEEE international conference on intelligent computing and intelligent systems, Xiamen, 2010, pp 46–48. https://doi.org/10.1109/icicisys.2010.5658836
16. Venkatanathan N, Natyaganov V (2014) Outgoing longwave radiations as pre-earthquake signals: preliminary results of 24 September 2013 (M 7.7) earthquake. Curr Sci 106

17. Venkatanathan N, Yichun Y, Lv J (2017) Observation of abnormal thermal and infrasound signals prior to the earthquakes: a study on Bonin Island earthquake M7.8 (May 30, 2015). Environ Earth Sci 76. https://doi.org/10.1007/s12665-017-6532-x
18. Jing F, Shen X, Kang C (2012) Outgoing long wave radiation variability feature prior to the Japan M9.0 earthquake on March 11, 2011. In: IEEE international geoscience and remote sensing symposium, Munich, 2012, pp 1162–1165. https://doi.org/10.1109/IGARSS.2012.6351341
19. Ouzounov Dimitar et al (2011) Atmosphere-ionosphere response to the M9 Tohoku earthquake revealed by multi instrument space-borne and ground observations: Preliminary results. EarthqSci 24:557–564. https://doi.org/10.1007/s11589-011-0817-z
20. Natarajan V, Bobrovskiy V, Shopin S (2019) Satellite and ground-based observation of pre-earthquake signals—a case study on the Central Italy region earthquakes. Indian J Phys
21. Rajendran K, Rajendran CP, Jain SK, Murty CVR, Arlekar JN (2000) The Chamoli earthquake, Garhwal Himalaya: field observations and implications for seismic hazard. Curr Sci 78(1):45–51
22. Shrikhande M, Rai D, Narayan JP, Das J (2000) The March 29, 1999 Earthquake at Chamoli, India
23. Paul A, Kumar N (2010) Estimates of source parameters of M 4.9 Kharsali earthquake using waveform modelling. J Earth Syst Sci 119:731–743. https://doi.org/10.1007/s12040-010-0050-5
24. Lee H-T, NOAA CDR Program (2011) NOAA Climate Data Record (CDR) of Daily Outgoing Longwave Radiation (OLR), Version 1.2. NOAA National Climatic Data Center. https://doi.org/10.7289/v5sj1hh2
25. Ouzounov D, Liu D, Kang C, Cervone G, Kafatos M, Taylor P (2007) Outgoing long wave radiation variability from IR satellite data prior to major earthquakes. Tectonophysics 431:211–220
26. Venkatanathan N, Philipoff P (2018) Observation of surface and atmospheric parameters using "NOAA 18" satellite: a study on earthquakes of Sumatra and Nicobar Is regions for the year 2014 ($M \geq 6.0$). Nat Hazards 92. https://doi.org/10.1007/s11069-018-3242-y (2018)

Correction to: Ultrasonic P Wave Velocity Measurement of Sedimentary Rocks in Laboratory and Its Dependency on Different Parameters by Using Multivariate Regression Analysis

Siddharth Garia, Arnab Kumar Pal, Archana M. Nair, and K. Ravi

Correction to:
Chapter "Ultrasonic P Wave Velocity Measurement of Sedimentary Rocks in Laboratory and Its Dependency on Different Parameters by Using Multivariate Regression Analysis" in: T. G. Sitharam et al. (eds.), *Seismic Hazards and Risk*, Lecture Notes in Civil Engineering 116, https://doi.org/10.1007/978-981-15-9976-7_30

The original version of the book was published without the reference to the source of Figure 3 in Chapter 30, which has now been updated as follows: Figure 3 is adapted from a journal publication by Garia et al., 2021 in the Journal of Petroleum Science and Engineering-ELSEVIER. The book and the chapter have been updated with the change.

The updated version of this chapter can be found at
https://doi.org/10.1007/978-981-15-9976-7_30

T. G. Sitharam et al. (eds.), *Seismic Hazards and Risk*, Lecture Notes in Civil Engineering 116, https://doi.org/10.1007/978-981-15-9976-7_34